Handbook on Integrating Smart Technologies for Sustainable Development

The text discusses the latest research on essential issues and prospects surrounding the deployment of intelligent technologies, such as infrastructural needs, data privacy issues, and legislative considerations. It includes case studies of businesses using innovative technology for sustainable development.

This book:

- Investigates the relationship between intelligent technology and sustainable development, offering a thorough summary of how cutting-edge technological advancements might support an environmentally friendly future.
- Discusses the role of using smart technologies in revolutionizing waste collection and recycling systems.
- Explores ethical considerations in the implementation of smart technology and sustainability.
- Highlights the latest innovations in smart transportation and digital twin technology.
- Showcases the use of smart technologies in areas including energy management, transportation, waste management, water conservation, and healthcare.

It is primarily written for senior undergraduates, graduate students, and academic researchers in fields including electrical engineering, electronics, communication engineering, computer science and engineering, information technology, civil engineering, and environmental engineering.

Smart Technologies for Engineers and Scientists

Series Editor: Mangey Ram

Smart Technologies including artificial intelligence, the internet of things (IoT), machine learning, and big data computing have played an important role in the advancement of processes and the functioning of machines. Smart systems are being used in a wide range of industries such as transportation, energy, healthcare, safety and security, logistics, and manufacturing. This book series discusses the use of smart technologies in diverse areas of research including electric vehicles, advanced manufacturing techniques, industry 4.0, and nanoelectronics. Aimed at senior undergraduate, graduate students, and academic researchers, the proposed series will cover the most efficient and innovative technological solutions for industrial applications. It will further cover key concepts such as 3D printing, 5G technology, augmented reality, virtual reality, battery management systems, robotics, bioinformatics, and quantum computing.

Blockchain Technology
Transforming Businesses and Shaping the Future
Edited by Jay Daniel, Ashutosh Samadhiya, and Jose Arturo Garza-Reyes

Internet of Things in Modern Computing
Theory and Applications
Edited by Vinay Chowdary, Abhinav Sharma, Naveen Kumar, and Vivek Kaundal

Sustainable Energy Solutions with Artificial Intelligence, Blockchain Technology, and Internet of Things
Edited By Arpit Jain, Abhinav Sharma, Vibhu Jately, and Brian Azzopardi

Technology Developments in Computer Intelligence and their Applications in the era of Industry 5.0
Edited by Mukul Kumar Gupta, Abhinav Sharma, Vinay Rishiwal, and Meera C S

Radio Frequency and Microwave Design for Next Generation Wireless Applications
Edited by Rajkishor Kumar, Avinash Chandra, Indrasen Singh, Vijay Kumar, and Yu-Chen Hu

Handbook on Integrating Smart Technologies for Sustainable Development
Edited by Radhika Thapar, A. V Senthil Kumar, Fahad Alturise and Omar S. Saleh

For more information about this series, please visit: https://www.routledge.com/Smart-Technologies-for-Engineers-and-Scientists/book-series/CRCSTFES

Handbook on Integrating Smart Technologies for Sustainable Development

Edited by
Radhika Thapar, A. V. Senthil Kumar,
Fahad Alturise and Omar S. Saleh

CRC Press
Taylor & Francis Group
Boca Raton London New York

CRC Press is an imprint of the
Taylor & Francis Group, an **informa** business

Designed cover image: Radhika Thapar

First edition published 2026
by CRC Press
2385 NW Executive Center Drive, Suite 320, Boca Raton FL 33431

and by CRC Press
4 Park Square, Milton Park, Abingdon, Oxon, OX14 4RN

CRC Press is an imprint of Taylor & Francis Group, LLC

ISBN: 9781032958804 (hbk)
ISBN: 9781032958811 (pbk)
ISBN: 9781003586951 (ebk)

DOI: 10.1201/9781003586951

Typeset in Sabon
by KnowledgeWorks Global Ltd.

Contents

About the Editors

Radhika Thapar is currently an Associate Professor in the Department of Management, Rukmini Devi Institute of Advanced Studies, India. She has a doctorate in computer science, MPhil in computer science, master's in computer applications, and master's in business administration. She has 19+ years' experience in teaching undergraduate and graduate courses in computer applications, information technology, and management studies. She has actively participated in continued learning through conferences, networking events, and professional research. Dr Thapar has contributed immensely to quality research through paper publications in national and international conferences and journals, having indexing in SCOPUS, ABDC, and UGC Care. Her areas of research include human resource development, software engineering, and business intelligence. She has been reviewer and session chair for various conferences.

Dr. A. V. Senthil Kumar is Principal, Nehru Institute of Information Technology and Management, Coimbatore, India. He has been a Professor and Director, PG and Research Department of Computer Applications, Hindusthan College of Arts & Science, Coimbatore, Tamilnadu for more than 15 years and a Senior Grade Lecturer in CMS College of Science and Commerce for 14 years. He has five years of industrial experience and 29 years of teaching experience. He earned a Doctor of Science (D.Sc in Computer Science). He has to his credit 88 Book Chapters, 234 papers in International and National Journals, 88 papers in International Conferences in International and National Conferences, and edited 17 books and 3 Text books. He is an associate editor of IEEE Access and an editor-in-chief for many journals and Key Member for India, Machine Intelligence Research Lab (MIR Labs). He is an editorial board member and reviewer for various international journals. He is also a committee member for various international conferences. He has guided 16 Ph.D scholars and is guiding 3 Ph.D scholars now.

Fahad Alturise is a Professor in the Department of Cybersecurity, College of Computer, Qassim University. He has 16 years of teaching and research experience. He holds a PhD in information technology from Flinders

University, where he was also a member of the Golden Key International Honour Society, recognizing his outstanding academic achievements. His primary research interests include e-learning, e-services, e-government, IoT, ICT adaptation, IT security, AI, ML, and software engineering. He has published more than 100 papers in international journals/conference proceedings. He was also a member of the Australian Computer Society (ACS) for 4 years.

Omar S. Saleh is Associate Professor at the Ministry of Higher Education and Scientific Research, Iraq, where he contributes to the Studies, Planning, and Follow-Up Directorate. He holds a BSc in computer engineering (University of Technology, Baghdad, 2006), an MSc in software engineering (Staffordshire University, APU Malaysia, 2012), and a PhD in Information Technology (Universiti Utara Malaysia, UUM). An active researcher in blockchain, cloud computing, software architecture, and IoT, Dr. Saleh has published over 30 peer-reviewed articles, including journal papers, conference proceedings, and book chapters. He serves as a reviewer for top-tier indexed journals and participates in organizing IEEE and other international conferences, including his role on the Technical Program Committee of the IEEE International Conference on Blockchain.

Contributors

Leela Glory A.
Department of Education
Bharathiar University
Coimbatore, Tamil Nadu, India

Wajahat Ali
Department of Statistics & Operations Research
Aligarh Muslim University
Aligarh, Uttar Pradesh, India

Namratha B.
Department of Chemistry
The Yenepoya Institute of Arts, Science, Commerce and Management, Yenepoya (Deemed to be University)
Mangaluru, Karnataka, India

M. Balasubramaniam
Department of Education
Bharathiar University
Coimbatore, Tamil Nadu, India

Divanshu Bansal
Department of Pharmacology
Chitkara College of Pharmacy
Chitkara University
Rajpura, India

Bhawna
Department of Pharmacology
Chitkara College of Pharmacy
Chitkara University
Rajpura, India

Divyedeep Bishnoi
Department of Pharmacology
Chitkara College of Pharmacy
Chitkara University
Rajpura, India

Anand Joseph Daniel
Department of Computer Science and Engg
Hindustan Institute of Technology and Science
Chennai, Tamil Nadu, India

Dimple Dhiman
Department of Pharmacology
Chitkara College of Pharmacy,
Chitkara University
Rajpura, India

A Leela Glory
Department of Education,
Bharathiar University
Coimbatore, Tamil Nadu, India

Anuj Kumar Goel
Trinity Western University
Langley, British Columbia, Canada

Anushka Goel
Apeejay School of Management
Delhi, India

Vinodkumar Jadhav
Associate Professor
MIT World Peace University
Pune, India

Rekha K.
Department of Educational Technology
Bharthiyar University
Coimbatore, Tamil Nadu, India

Vishwanath Karad
MIT World Peace University
Pune, Maharashtra, India

Ujjwal Kaushik
Department of Pharmacognosy, Chitkara College of Pharmacy, Chitkara University
Rajpura, India

Karan Khajuria
Department of Computer Science & Engineering
Model Institute of Engineering & Technology
Jammu, India

MaheshVasantrao Kulkarni
Department of Mechanical Engineering
Pune, India.

Neha Kukreti
Amity Institute of Pharmacy, Amity University
Noida, Uttar Pradesh

Biplab Sambit Kumar
Department of Mechanical Engineering,
Synergy Institute of Technology
Bhubaneswar, Odisha, India

NamrataVasant Kumar Lotia
Department of Mechanical Engineering
Anjuman College of Engineering and Technology
Nagpur, Maharashtra, India

Saniya Marwah
Department of Management Studies, RDIAS, 2A & 2B
Rohini, Delhi, India

Akshata Musale
Department of Civil Engineering
Jain College of Engineering and Technology
Hubbali, Karnataka, India

Jayanthi M. G.
Department of CSE
Cambridge Institute of Technology
Bengaluru, Karnataka, India

Ramesh Chandra Nayak
Department of Mechanical Engineering
Synergy Institute of Technology
Bhubaneswar, Odisha, India

Sankar Padmanabhan
Department of Electronics and Communication Engineering
Hindustan Institute of Technology and Science
Chennai, Tamil Nadu, India

Pushpitha Reddy Patur
Data Analytics
Houston, TX

Shreekanth M Prabhu
Department of CSE
Cambridge Institute of Technology
Bengaluru, Karnataka, India

Manju Priya R.
School of Commerce
JAIN (Deemed-to-be University)
Bangalore, Karnataka, India

Raji Rajan
School of Commerce
JAIN (Deemed-to-be University)
Bengaluru, Karnataka, India

Latha Ramamoorthy
Leading Banking Organization
USA

Janardhana Kumar Reddy P
Department of Education
Bharathiar University
Coimbatore, Tamil Nadu, India

Yashaswini S.
Department of CSE
Cambridge Institute of Technology
Bangalore, Karnataka

N. Sathyanarayana
School of Commerce
Jain (Deemed-to-be University)
Bengaluru, Karnataka, India

Sajid Hameed Siddiqui
Department of Mechanical Engineering
Anjuman College of Engineering and Technology
Nagpur, Maharashtra, India

Maddireddi S. S. V. Srikumar
Global Business School
KLEF(Deemed to be University)
Hyderabad, India

Jabeen Sultana
Computer Science Department
College of Computer and Information Sciences
Imam Mohammad Ibn Saud Islamic University (IMSIU)
Saudi Arabia

Udita Taneja
University School of Management Studies
Guru Gobind Singh Indraprastha University
Delhi, India

S. Thangarajathi
Department of Educational Technology
Bharathiar University
Coimbatore, Tamil Nadu, India

Radhika Thapar
Department of Management Studies
RDIAS
Delhi

Aashmi V.
Department of Education
Bharathiar University
Coimbatore, Tamil Nadu, India

Suryavanshi Bhagyeshkumar Vijaybhai
Department of Mechanical Engineering
Birla Vishwakarma Mahavidhyalya (BVM) Engineering College
Anand, Gujarat, India

Chapter 1

Impact of fintech adoption drivers on sustainable financial inclusion

Saniya Marwah and Radhika Thapar

1.1 INTRODUCTION

The growth of sustainability in contemporary banking is largely driven by technology-based solutions (Li et al. 2021). In addition to internet and mobile banking, traditional banks have gained popularity. This shift has resulted in reduced service costs in rural and semi-urban areas, enhancing the sustainability of financial services (Chouhan et al. 2020; Schuetz and Venkatesh 2020). The industry is making two complementary contributions to create a more inclusive financial sector (Anagnostopoulos 2018; Cecchetti and Schoenholtz 2020). First, financial inclusion and sustainability can serve as bridges to connect individuals. Second, economic growth attracts new participants to both the sustainable financial system and the broader economy (Mader 2018; Schuetz and Venkatesh 2020). The push for greater access to sustainable financial services has led to the emergence of new types of banks (Banna et al. 2021). Digital solutions that enhance business relationships, improve service efficiency, and increase accessibility to meet evolving needs—while prioritizing sustainability—are becoming essential. In this landscape, fintech companies are positioned to disrupt traditional banking, offering the impetus needed for modernization. By promoting innovation, fintech firms can lower operational costs, enhance financial inclusion, and create a more sustainable, customer-focused banking environment. Traditional banks and financial institutions can capitalize on their existing customer base to develop digital solutions that strengthen business relationships. For example, Barclays, the largest bank in the United Kingdom, has launched its sixth global fintech innovation center in India, while Goldman Sachs' Principal Strategic Investments Group (GSPSI) is interested in investing in Bengaluru's burgeoning fintech startup scene. According to Dang and Nguyen (2021), traditional financial institutions and Indian fintech startups can collaborate and coexist, both benefiting from the opportunities offered by technological innovation.

The Reserve Bank of India (RBI) has established a multidisciplinary committee to explore the country's fintech sector, aiming to identify potential risks and create new frameworks that will help the financial system adapt

DOI: 10.1201/9781003586951-1

effectively. As part of its initiatives, the RBI has supported the implementation of the Unified Payments Interface (UPI) within the financial advisory ecosystem. Clients often hesitate to adopt new technologies because they have confidence and trust in the existing systems. Users need to feel comfortable with their privacy before they can effectively use new technologies. Even if these new options are easier than traditional methods, building customer trust will take time. Achieving this vision requires the introduction of new business models. To understand the impact of technology-oriented services in the banking sector, it's essential to explore the various factors that constitute the primary objectives of this study.

1.2 REVIEW OF LITERATURE

Advancements in financial technology can be achieved through sustainable and cost-effective solutions in the financial sector. It can reduce transaction fees (Black and Babin 2019). Sales are boosted for small businesses as they lower the costs of traditional payment methods (Frost et al. 2019). Aron (2018) examined the role of mobile money and presented evidence supporting its operational impact on financial systems.

According to Mbiti and Weil (2013) and Wieser et al. (2019), the rise in the adoption of sustainable fintech services has led to a decline in traditional savings practices and a rise in remittance transactions. In the Indian context, fintech startups and established financial institutions are often collaborating to provide sustainable and cost-effective solutions (Dang and Nguyen 2021). Nowadays, access to the technology-driven financial services is critical for the purpose of enhancing the role of financial inclusion.

Ensuring fintech accessibility for older populations requires the development of more intuitive and user-friendly platforms. Digital safety norms should be prioritized in the regions with low financial literacy rates (Nguyen 2022). Research by Banna et al. (2021) indicates that financial technology innovation (FTI) influences banks' risk-taking behavior—an effect that has grown stronger in the context of Industry 4.0.

Breach of trust has been raised due to the misuse of data, and threat to financial services can lead to breach of trust on fintech services (Gogia and Chakraborty 2022). In order to ensure the preservation of confidential information, the cybersecurity threats need to be addressed (Rauniyar et al. 2021). The rights of fintech users can be protected through various stringent rules and regulations (Mosteanu and Faccia 2021). The trust and confidence in fintech services can be built through data protection and privacy legislation (Mosteanu and Faccia 2021). The regulatory frameworks can be created through innovation while protecting consumer rights and financial stability (Arnone et al. 2024). Integration can be successfully achieved through

Figure 1.1 Word cloud analysis in "NVivo" software.

efficient financial systems (Haddad and Hornuf 2019). The issues of financial inclusion are connectivity and digital divide in rural areas (Arslanian and Fischer 2019). Limited access to smartphones, poor internet, and online applications can hinder the adoption of fintech solutions (Zetzsche et al. 2017). The monetary benefits can be provided through restriction from conventional financial frameworks through versatile banking (Zetzsche et al. 2017). The engagement of business people and private ventures through fintech is creating economies (Gogia and Chakraborty 2022) (Figure 1.1).

1.3 RESEARCH GAP AND OBJECTIVES

This study aimed to bridge a slit in the current corpus of research on financial inclusion and sustainability. The current study used to assess the impact of financial technology with regard to financial inclusion and sustainability on banking services in developing economies as mentioned in below table (Table 1.1).

Table 1.1 Scales adopted

Constructs	*Codes*	*Recent literature using the dimensions*
Behavioral intention	BI	Asif et al. 2023
Social influence	SI	
Service trust	ST	
Usability	UB	
Use of fintech for sustainable financial inclusion	FTFI	

1.4 HYPOTHESIS AND CONCEPTUAL FRAMEWORK

The quantitative technique involves confirmatory factor analysis and structural equation modeling (SEM), which were utilized in quantification of the effect of technology-oriented sustainable financial inclusion. The diagram depicts the four dormant variables ("behavioral intention", "service trust", "usability", and "social impact") as circles and the 12 manifest variables as rectangles.

H1: Behavioral intention has a significant impact on sustainable financial inclusion.
H2: Service trust has a significant impact on sustainable financial inclusion.
H3: Usability has a significant impact on sustainable financial inclusion.
H4: Social influence has a significant impact on sustainable financial inclusion.

Hence, Figure 1.2 was proposed.

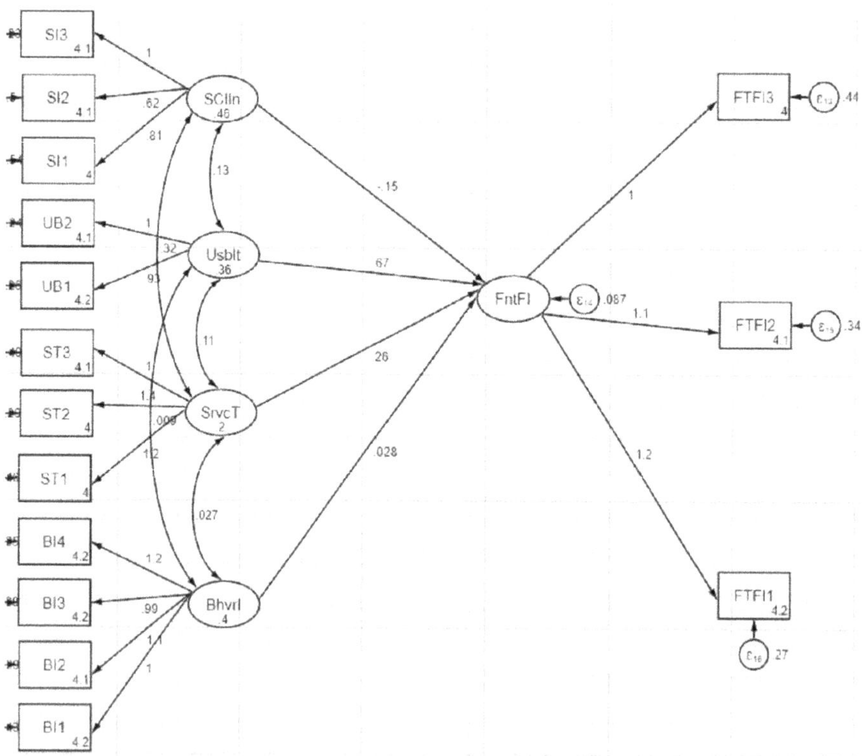

Figure 1.2 Proposed model.

1.5 RESEARCH METHODOLOGY

A descriptive approach has been adopted in the research. The representative of the study is selected based on people who use sustainable banking services. The research method is "qualitative" and "quantitative" in nature. Respondents from Delhi were selected at random using a process known as random sampling. The total number of respondents was 300. All the questions were close-ended and one choice was selected based on 5-point Likert scale. The data analysis has been carried out with factor analysis and SEM using Software Stata/IC 15.0.

1.6 RESULTS

As per Table 1.2, 300 usability questionnaires were collected, including 148 (50.0%) for females and 152 (50.0%) for males. The respondents' occupation includes business/self-employed, government service, homemakers, private service, and student. The respondents' monthly income ranges are 25,000–50,000, 50,000–1,00,000, 1,00,000–1,50,000, 1,50,000–2,00,000, and above 2,00,000.

The index of reliability of 0.60 or higher is considered to be "acceptable" (Figure 1.3).

The Kaiser-Meyer-Olkin (KMO) value of 0.633 is considered acceptable, which surpasses the 0.6 benchmark; the sampling adequacy is deemed sufficient. Also, the significance level of Bartlett's test of sphericity is below 0.05, which indicates the presence of meaningful relationships between variables, making factor analysis appropriate (Figure 1.4).

Table 1.2 Sample characteristics

Sample categories	*Frequency*	*Percentage*
Gender		
Male	148	51.0
Female	152	49.0
Age (years)		
25–35	126	42.7
36–45	122	40.0
46–55	30	10.0
Above 56	22	7.3
Occupation		
Business/self-employed	118	40
Government service	62	20
Homemaker	40	13.33
Private service	80	26.67

```
. alpha( BehaviouralIntention SocialInfluence ServiceTrust Usability UseofFintechforFinancialSer)

Test scale = mean(unstandardized items)

Average interitem covariance:     .1741038
Number of items in the scale:            5
Scale reliability coefficient:      0.6113

.
```

Figure 1.3 Cronbach's alpha.

```
. factortest BehaviouralIntention SocialInfluence ServiceTrust Usability UseofFintechforFinancialSer

Determinant of the correlation matrix
Det                  =      0.519


Bartlett test of sphericity

Chi-square           =           193.728
Degrees of freedom   =                10
p-value              =             0.000
H0: variables are not intercorrelated


Kaiser-Meyer-Olkin Measure of Sampling Adequacy
KMO                  =      0.633
```

Figure 1.4 KMO and Bartlett's test of sphericity.

A positive correlation, i.e. 0.0945, has been recorded between behavioral intention and fintech inclusion and sustainability. A positive correlation, i.e. 0.1459, has been recorded between social influence and fintech inclusion. A positive correlation, i.e. 0.2367, has been recorded between service trust and fintech inclusion. A positive correlation, i.e. 0.4673, has been recorded between usability and fintech inclusion (Figure 1.5).

R-squared, i.e. 20%. The adjusted R-squared is recorded as 25.5. For significance testing, the p-value is less than the significance level, and the regression test is significant (Figure 1.6). The regression equation can be interpreted as:

$$y = a + bx$$

```
. pwcorr BehaviouralIntention SocialInfluence ServiceTrust Usability UseofFintechforFinancialSer, list
```

	Behavi~n	Social~e	Servic~t	Usabil~y	UseofF~r
Behavioura~n	1.0000				
SocialInfl~e	0.3855	1.0000			
ServiceTrust	0.2785	0.3752	1.0000		
Usability	0.1043	0.1210	0.1842	1.0000	
UseofFinte~r	0.0945	0.1459	0.2367	0.4673	1.0000

Figure 1.5 Correlational analysis.

```
Contains data
  obs:           336
 vars:            22
 size:       439,152
--------------------------------------------------------------------------------
              storage   display    value
variable name   type    format     label      variable label
--------------------------------------------------------------------------------
BI1             str108  %108s                 I intended to contribute to the expansion of access to financial services throug
BI2             str108  %108s                 I will always give first priority to using mobile services based on financial te
BI3             str62   %62s                  I intend to keep implementing fintech for financial inclusion
BI4             str90   %90s                  It is my intention to contribute to financial inclusion through the application
SI1             str94   %94s                  Financial technology and services for the financially excluded are things I am s
SI2             str116  %116s                 Peers who have an impact on my decisions recommended that I try out financial in
SI3             str135  %135s                 It is more likely that I will use financial inclusion services based on fintech
ST1             str95   %95s                  Services for the financially excluded that are based on fintech have been proven
ST2             str90   %90s                  Financial technology (fintech)-based services for the undeserved must be handled
ST3             str110  %110s                 Due to my prior positive experience with such services, I have faith in services
UB1             str97   %97s                  When it comes to financial inclusion, I am likely to use services powered by fin
UB2             str118  %118s                 I regularly make use of services that promote financial inclusion that are enabl
UB3             str75   %75s                  Several of the services that are based on fintech are quite important to me
N               byte    %10.0g
O               byte    %10.0g
P               byte    %10.0g
Q               byte    %10.0g
R               byte    %10.0g
S               byte    %10.0g
T               byte    %10.0g
U               byte    %10.0g
V               byte    %10.0g
--------------------------------------------------------------------------------
```

Figure 1.6 Variable name and Variable label.

Ten questions about independent factors were asked in the recent study from Asif et al. (2023).

The likelihood that it is more than just measurement error increases with uniqueness. All the variables in this problem have values less than 0.30, which is generally regarded as high, or values greater than 0.6. If the uniqueness is low, the components can adequately describe the variable.

Independence implies sphericity, which is necessary for conducting factor analysis on a correlation matrix. Successfully passing this test ensures factor analysis results meaningful. When using the maximum likelihood (ml) option in Stata, it provides a likelihood-ratio test that evaluates the fit of the proposed factor model against the saturated model, along with the "standard" output.

The relationship between social impact and usability of fintech products was weak (0.11). The relationship between "usability" and "service trust" (0.12) was similarly connected, but the relationship was less than the one between "social influence" and "usability" factors. The association between "service trust" and "behavioral intention" (0.025) was determined to be the weakest of the dormant variable correlations. The relationship between "social influence" and "service trust" was not as strong, i.e. 0.22. The relationship between the "usability" of fintech products and "behavioral intention" is 0.007. The relationship between "social influence" and the "use of fintech for sustainable financial inclusion" is recorded as 0.11. The linkage between "usability" and the "use of fintech for sustainable financial inclusion" is recorded as 0.57. The linkage between "service trust" and the "use of fintech for sustainable financial inclusion" is recorded as 0.36. The association between "behavioral intention" and the "use of fintech for sustainable financial inclusion" is recorded as 0.026 (Figure 1.7).

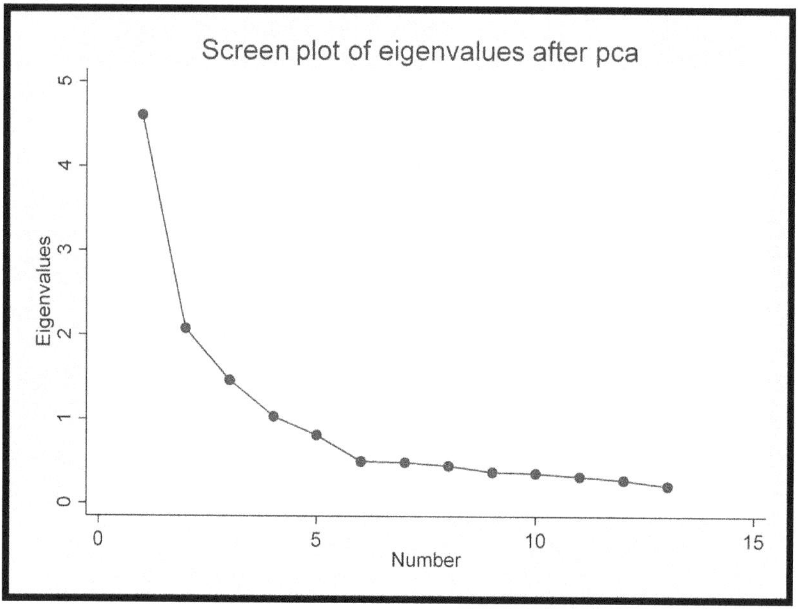

Figure 1.7 Screen plot.

1.7 CONCLUSION

In India, fintech is increasingly recognized as both accessible and user-friendly, transforming the way individuals approach financial services. Influenced by their peers, many people feel encouraged to engage with fintech-driven financial inclusion solutions, fostering a culture of adoption and familiarity. A significant number of users report that utilizing these services has seamlessly integrated into their daily lives, becoming second nature. However, despite this growing acceptance, there remains a widespread agreement on the importance of exercising caution when using fintech-based financial inclusion services. The rapid expansion of these tools necessitates a thoughtful approach to ensure users are aware of potential risks and make informed decisions. The findings of the study underscore that fintech is playing a pivotal role in advancing financial inclusion across the Indian economy. It not only empowers individuals with better access to financial services but also champions sustainable financial practices, fostering an environment where responsible financial behavior is encouraged and supported (Thomas and Hedrick-Wong 2019).

REFERENCES

Anagnostopoulos, Ioannis. 2018. "Fintech and Regtech: Impact on Regulators and Banks." *Journal of Economics and Business* 100: 7–25. doi:10.1016/j.jeconbus.2018.07.003.

Arnone, Massimo, Alberto Costantiello, Angelo Leogrande, Syed Kafait Hussain Naqvi, and Cosimo Magazzino. 2024. "Financial Stability and Innovation: The Role of Non-Performing Loans." *FinTech* 3 (4): 496–536. doi:10.3390/fintech3040027.

Aron, Janine. 2018. "Mobile Money and the Economy: A Review of the Evidence." *The World Bank Research Observer* 33 (2): 135–88. doi:10.1093/wbro/lky001.

Arslanian, Henri, and Fabrice Fischer. 2019. "Applications of Artificial Intelligence in Financial Services." In *The Future of Finance*, edited by Henri Arslanian and Fabrice Fischer, 179–97. Cham: Springer International Publishing. doi:10.1007/978-3-030-14533-0_15.

Asif, Mohammad, Mohd Naved Khan, Sadhana Tiwari, Showkat K. Wani, and Firoz Alam. 2023. "The Impact of Fintech and Digital Financial Services on Financial Inclusion in India." *Journal of Risk and Financial Management* 16 (2): 122. doi:10.3390/jrfm16020122.

Banna, Hasanul, M. Kabir Hassan, and Mamunur Rashid. 2021. "Fintech-Based Financial Inclusion and Bank Risk-Taking: Evidence from OIC Countries." *Journal of International Financial Markets, Institutions and Money* 75: 101447. doi:10.1016/j.intfin.2021.101447.

Black, William, and Barry J. Babin. 2019. "Multivariate Data Analysis: Its Approach, Evolution, and Impact." In *The Great Facilitator: Reflections on the Contributions of Joseph F. Hair, Jr. to Marketing and Business Research*, edited by Barry J. Babin and Marko Sarstedt, 121–30. Cham: Springer International Publishing. doi:10.1007/978-3-030-06031-2_16.

Cecchetti, Stephen G., and Kermit L. Schoenholtz. 2020. "Finance and Technology: What Is Changing and What Is Not." *CEPR Discussion Paper No. DP15352*, October. https://ssrn.com/abstract=3723541.

Chouhan, Vineet, Bibhas Chandra, Pranav Saraswat, and Shubham Goswami. 2020. "Developing Sustainable Accounting Framework for Cement Industry: Evidence from India." *Finance India* 34 (4).

Dang, Van Cuong, and Quang Khai Nguyen. 2021. "Internal Corporate Governance and Stock Price Crash Risk: Evidence from Vietnam." *Journal of Sustainable Finance & Investment* 14 (1): 24–41. doi:10.1080/20430795.2021.2006128.

Frost, Jon, Leonardo Gambacorta, Yi Huang, Hyun Song Shin, and Pablo Zbinden. 2019. "BigTech and the Changing Structure of Financial Intermediation." *Economic Policy* 34 (100): 761–99. doi:10.1093/epolic/eiaa003.

Gogia, Jyoti, and Debasis Chakraborty. 2022. "Open Banking: A Revolution in the Tech-Fin Industry." *International Journal of Electronic Banking* 3 (2): 100. doi:10.1504/IJEBANK.2022.122224.

Haddad, Christian, and Lars Hornuf. 2019. "The Emergence of the Global Fintech Market: Economic and Technological Determinants." *Small Business Economics* 53 (1): 81–105. doi:10.1007/s11187-018-9991-x.

Mader, Philip. 2018. "Contesting Financial Inclusion." *Development and Change* 49 (2): 461–83. doi:10.1111/dech.12368.

Mbiti, Isaac, and David N. Weil. 2013. "The Home Economics of E-Money: Velocity, Cash Management, and Discount Rates of M-Pesa Users." *American Economic Review* 103 (3): 369–74. doi:10.1257/aer.103.3.369.

Mosteanu, Narcisa Roxana, and Alessio Faccia. 2021. "Fintech Frontiers in Quantum Computing, Fractals, and Blockchain Distributed Ledger: Paradigm Shifts and Open Innovation." *Journal of Open Innovation: Technology, Market, and Complexity* 7 (1): 19. doi:10.3390/joitmc7010019.

Nguyen, Quang Khai. 2022. "The Effect of FinTech Development on Financial Stability in an Emerging Market: The Role of Market Discipline." *Research in Globalization* 5: 100105. doi:10.1016/j.resglo.2022.100105.

Rauniyar, Kanchan, Komal Rauniyar, and Deependra Kumar Sah. 2021. "Role of FinTech and Innovations for Improvising Digital Financial Inclusion." *International Journal of Innovative Science and Research Technology* 6 (5): 1419–24.

Schuetz, Sebastian, and Viswanath Venkatesh. 2020. "Blockchain, Adoption, and Financial Inclusion in India: Research Opportunities." *International Journal of Information Management* 52: 101936. doi:10.1016/j.ijinfomgt.2019.04.009.

Wieser, Christina, Miriam Bruhn, Johannes Philipp Kinzinger, Christian Simon Ruckteschler, and Soren Heitmann. 2019. "The Impact of Mobile Money on Poor Rural Households: Experimental Evidence from Uganda." *World Bank Policy Research Working Paper No. 8913*. https://openknowledge.worldbank.org/handle/10986/31978.

Zetzsche, Dirk A., Ross P. Buckley, Douglas W. Arner, and Janos Nathan Barberis. 2017. "From FinTech to TechFin: The Regulatory Challenges of Data-Driven Finance." *SSRN Electronic Journal*. doi:10.2139/ssrn.2959925.

Chapter 2

E-learning platforms and digital resources

Transforming education

Rekha K. and S. Thangarajathi

2.1 INTRODUCTION

The educational landscape is experiencing a significant metamorphosis, propelled by the swift advancement of digital technologies. E-learning platforms and digital resources have emerged as crucial instruments, fundamentally altering the modalities through which knowledge is disseminated, accessed, and assimilated (Law et al. 2008; Selwyn 2016). These technologies have transitioned from being mere adjuncts to becoming essential elements of modern educational frameworks, accommodating a variety of learning preferences and responding to the requirements of a globally interconnected society.

The transition toward technology-mediated learning signifies a fundamental shift, transcending conventional classroom environments to adopt flexible, accessible, and interactive educational experiences (Moore & Kearsley 2011; Simonson et al. 2011). This transformation carries profound implications for pedagogical practices, necessitating that educators modify their instructional strategies to utilize the distinctive advantages afforded by digital tools. Concurrently, learners are evolving into proactive participants in their educational trajectories, empowered to investigate, collaborate, and innovate within vibrant digital contexts.

This chapter seeks to furnish a thorough examination of the multifaceted realm of e-learning platforms and digital resources. It will explore the pedagogical benefits afforded by these technologies, including individualized learning experiences, heightened engagement, and augmented accessibility. Moreover, it will undertake a critical evaluation of the obstacles encountered in their deployment, encompassing concerns related to digital equity, technological infrastructure, and the imperative for effective educator training. Ultimately, the chapter addresses the essential elements that facilitate the successful assimilation of e-learning platforms and digital resources, presenting insights into best practices and strategies aimed at optimizing their influence on student learning.

DOI: 10.1201/9781003586951-2

2.2 TYPES OF LEARNING PLATFORMS AND DIGITAL RESOURCES

2.2.1 Massive open online courses

Characteristics: Examine the fundamental characteristics of massive open online courses (MOOCs), which encompass attributes such as unrestricted access, extensive participant engagement, digital delivery mechanisms, and frequently, the incorporation of video lectures, interactive discussion forums, and automated evaluation methods. Analyze the various classifications of MOOCs along with their distinctive educational methodologies (Daniel 2012; Yuan & Powell 2013).

Advantages: Examine the advantages inherent in MOOCs, including the provision of equitable access to high-quality educational resources for a worldwide demographic, the adaptability in the rate of learning, and the prospects for continuous education throughout one's lifetime. Emphasize their significance in the democratization of education and in facilitating avenues for professional advancement (Daniel 2012; Yuan & Powell 2013).

Limitations: Thoroughly analyze the obstacles related to MOOCs, such as low completion rates, insufficient personalized engagement, the risk of digital disparities, and the challenges in maintaining quality and accreditation (Daniel 2012).

2.2.2 Learning management systems

Functionalities: Outline the essential features of learning management systems (LMS), including course content organization, communication features (such as forums and chat), assignment submission and evaluation, student monitoring, and analytics. Provide examples of well-known LMS platforms like Moodle, Canvas, and Blackboard (Coates et al. 2005; Watson & Watson 2007).

Role in Course Delivery: Investigate the ways in which LMS platforms enhance online and blended learning by organizing course materials, delivering multimedia content, and managing online activities. Analyze the support these platforms provide for various teaching methodologies (Coates et al. 2005; Watson & Watson 2007).

Assessment: Examine the evaluation features of LMS, focusing on the implementation of quizzes, tests, rubrics, and digital grading tools. Explore how LMS can facilitate prompt feedback and monitor student advancement (Coates et al. 2005).

2.2.3 Interactive simulations and games

Potential for Engaging Learners: Analyze the application of simulations and games in developing immersive and engaging educational experiences. Explore how these tools can promote active learning, enhance problem-solving abilities, and cultivate critical thinking (Plass et al. 2015).

Fostering Deeper Understanding: Examine the ways in which simulations and games enable learners to delve into intricate concepts, test various scenarios, and visualize abstract notions. Additionally, consider the application of gamification strategies to boost motivation and engagement (Plass et al. 2015).

2.2.4 Virtual and augmented reality

Applications in Education: Investigate the various uses of virtual reality (VR) and augmented reality (AR) in educational settings, such as virtual excursions, immersive simulations, interactive training programs, and the representation of intricate data. Examine how these technologies can improve spatial reasoning, boost engagement, and enhance knowledge retention (Dede 2009; Bailenson 2018; Radianti et al. 2020).

Potential to Enhance Learning Experiences: Examine the capabilities of VR and AR in enhancing immersive, interactive, and tailored educational experiences. Explore the obstacles and prospects linked to the adoption of these technologies within learning environments (Dede 2009; Bailenson 2018; Radianti et al. 2020).

2.3 OTHER RELEVANT PLATFORMS AND RESOURCES

2.3.1 Educational apps: The mobile classroom

Mobile applications have transformed education by shifting it from conventional classrooms to a more dynamic, accessible, and personalized learning experience. The rise of smartphones and tablets has significantly contributed to the expansion of educational apps, which address a wide variety of subjects and accommodate different learning styles.

Language Learning: Applications such as Duolingo, Babbel, and Memrise leverage gamification and interactive activities to enhance the process of language learning, making it both enjoyable and efficient. These platforms provide organized lessons, vocabulary exercises, and pronunciation training, enabling users to progress at their own speed. Additionally, they frequently include functionalities like speech recognition and personalized learning to customize the experience according to each user's requirements.

Mathematics: Applications like Photomath and Khan Academy offer detailed solutions to mathematical problems, complemented by video tutorials and practice exercises. These tools assist students in grasping intricate mathematical concepts, enhancing their problem-solving abilities, and boosting their self-confidence. Similarly, platforms such as brilliant.org emphasize the development of problem-solving skills and the cultivation of mathematical intuition.

Science: Applications such as **Star Walk 2** and **Periodic Table 2023** provide interactive visualizations and simulations that facilitate the exploration of scientific concepts. These applications enhance the learning experience in subjects like astronomy, chemistry, biology, and physics, making them more engaging and accessible. Additionally, the rise of apps featuring virtual laboratories enables safe experimentation from home.

The availability of these applications on mobile devices facilitates learning anytime and anywhere, thereby making education more adaptable and user-friendly. This emerging trend highlights the rising incorporation of technology into contemporary educational methods, improving learning results and nurturing a continuous passion for education.

2.3.2 Online libraries

The emergence of online libraries and digital repositories has significantly transformed our approach to accessing and using educational resources. Their significance includes the following:

Democratizing Access to Knowledge: Digital repositories are essential for the preservation and distribution of important historical and cultural materials. They enable the swift sharing of research results and educational resources, thereby enhancing the speed of knowledge exchange. These libraries help to expand their reach. Digital repositories can house vast collections of books, journals, research papers, and multimedia content, far exceeding the capacity of physical libraries. This expanded reach facilitates lifelong learning and research for a global audience.

Enhancing Educational Opportunities: Online libraries play a crucial role in enhancing research by offering robust search capabilities that allow users to efficiently find pertinent information. These digital repositories grant access to academic articles and research data, thereby aiding scholarly activities across all educational levels. Additionally, they accommodate a variety of learning preferences, presenting resources in multiple formats such as text, audio, and video to suit different learning styles. Accessibility features, including text-to-speech options and customizable font sizes, further support individuals with disabilities. Digital libraries help dismantle obstacles by removing geographical constraints, enabling individuals in remote locations or with mobility challenges to access a vast array of information. They frequently offer resources that might be financially out of reach through conventional methods.

Key Benefits:

Accessibility and Convenience: Round-the-clock access from any location with internet connectivity.

Cost-Effectiveness: Lower expenses related to physical storage and upkeep.

Searchability and Organization: Effective information retrieval using sophisticated search tools.
Up-to-Date Information: Availability of the most recent studies and publications.
Preservation: Preserving valuable and delicate digital materials.

Essentially, online libraries and digital repositories serve as crucial resources for contemporary education, promoting a more inclusive and dynamic learning atmosphere.

2.3.3 Open educational resources

Open educational resources (OERs) signify a transformative change in education, providing a vast array of openly licensed materials for teaching, learning, and research. These resources include textbooks, lesson plans, multimedia content, and software, all available for free use, adaptation, and sharing. The advantages of OERs are diverse, greatly enhancing accessibility, affordability, customization, and supporting the overarching objectives of open access and equitable education.

Accessibility:

Breaking Down Barriers: OERs remove the financial obstacles linked to conventional educational materials, thereby making high-quality education available to students from various socioeconomic backgrounds. This is especially important for learners in developing nations and those experiencing financial difficulties.
Universal Access: The digital format of numerous OERs facilitates seamless distribution and accessibility across multiple platforms and devices. This guarantees that learners who have restricted physical access to conventional resources can still interact with high-quality educational materials.
Accessibility for Diverse Learners: OERs can be modified to accommodate the requirements of learners with disabilities. For instance, text-based OERs can be transformed into audio formats or enlarged for students with visual impairments.

Affordability:

Cost Reduction: One of the primary advantages of OERs is the considerable decrease in educational expenses. By using freely available and openly licensed materials, both students and institutions can achieve significant savings compared to the costs associated with traditional textbooks and proprietary resources.
Institutional Savings: Educational institutions can achieve cost savings by implementing OERs, enabling them to redirect funds to essential areas like research, infrastructure, and student support services.

Reduced Financial Burden: OERs reduce the financial strain on students and their families, allowing them to concentrate on their studies without the anxiety of high textbook expenses.

Customization:

Adaptability: OERs can be customized, adjusted, and localized to fit particular educational settings and cultural contexts. Instructors have the ability to modify content to address the distinct requirements of their learners, thereby enhancing relevance and fostering engagement.

Flexibility: The open licensing of OERs facilitates the development of tailored learning experiences and the incorporation of various viewpoints and narratives into educational content.

Continuous Improvement: OERs promote a collaborative method for creating content, enabling both educators and learners to enhance and refine existing materials. This ongoing process guarantees that educational resources stay up-to-date and pertinent.

Role in Promoting Open Access and Equitable Education:

Open Access: OERs play a fundamental role in the open access movement, which champions unrestricted access to knowledge and information. By encouraging the development and utilization of openly licensed materials, OERs foster a more transparent and accessible educational environment.

Equitable Education: OERs enhance equitable access to information by enabling learners to interact with educational content freely. This cultivates a lifelong learning environment and encourages active involvement in the generation and sharing of knowledge.

Global Collaboration: OERs promote worldwide collaboration between educators and students, allowing for the exchange of effective teaching strategies and the creation of culturally appropriate educational content. This international sharing of knowledge fosters mutual understanding and collaboration across cultures.

Supporting Innovation: OERs promote innovation in teaching and learning by offering educators the freedom to explore new teaching methods and technologies, thereby creating a vibrant and continuously developing educational environment.

Empowering Educators: OERs enhance the autonomy of educators, allowing them to tailor their curriculum and instructional materials. Educators have the ability to choose, modify, and develop resources that reflect their teaching philosophies and cater to the unique requirements of their students.

In summary, OERs provide a revolutionary method for education, enhancing accessibility, cost-effectiveness, personalization, and fair access to information. By adopting OERs, both educational institutions and individuals can play a role in fostering a more open, inclusive, and sustainable learning environment.

2.3.4 Educational video platforms

Educational video platforms such as YouTube and Vimeo have emerged as significant resources in contemporary education, providing various methods to enrich learning experiences. Following is an overview of their applications.

2.3.4.1 *YouTube*

YouTube offers a vast array of educational resources, featuring an extensive collection of videos on nearly every subject. From lectures and tutorials to documentaries and practical demonstrations, learners can access a rich source of information. It serves as an invaluable tool for enhancing education, enabling students to solidify their understanding of classroom material or delve deeper into various topics. YouTube's no-cost access enhances its availability, eliminating financial obstacles to education. This feature is especially advantageous for students from underprivileged backgrounds or those in search of cost-effective learning solutions. The video format accommodates various learning preferences, offering visual and auditory learners captivating educational opportunities. Additionally, the platform's interactive elements, including comments and discussions, promote collaborative learning.

2.3.4.2 *Vimeo*

Vimeo delivers premium educational resources for learners, recognized for its emphasis on high-quality video, which is essential for educational materials that demand clear visuals and sound. It is a preferred choice among educators and institutions looking for a professional platform to host and distribute educational videos. Vimeo provides enhanced privacy settings, enabling educators to limit access to their videos and ensure secure sharing, a crucial aspect for institutions that must safeguard sensitive educational information. Additionally, Vimeo includes features that support professional educational applications, such as fewer ad interruptions and superior embedding tools, making it a popular option for many educational organizations to host their course content.

2.3.5 Collaborative tools

Collaborative platforms such as Google Docs have transformed our work and learning processes, enhancing teamwork and facilitating smooth information exchange.

2.3.5.1 *Google Docs*

Google Docs allows multiple users to collaborate on the same document in real time, enabling them to view each other's edits instantly. This feature resolves version control challenges and enhances the collaborative writing process. Other platforms, such as Microsoft 365's Word Online and Zoho Writer, provide similar real-time collaboration capabilities. These applications are particularly beneficial for group assignments, research projects, and joint report writing, as they support collective brainstorming, editing, and feedback, ensuring that all contributions are included. Tools like commenting, suggesting changes, and maintaining version history promote effective feedback and revision workflows, fostering constructive criticism and collaborative document enhancement. With cloud-based storage, documents can be accessed from any location with internet access, removing the necessity for physical file transfers and guaranteeing that all users are working with the most current version. Google Docs also integrates smoothly with other Google Workspace applications, including Google Drive, Google Sheets, and Google Slides, thereby boosting overall productivity. Numerous other collaborative platforms similarly offer integration with additional useful tools.

Other Collaborative Platforms:

Project Management Tools (e.g., Trello, Asana, and Jira): These platforms enable effective task management, project oversight, and team collaboration. They empower teams to allocate tasks, establish deadlines, and track progress, ensuring efficient project completion.

Communication Platforms (e.g., Slack and Microsoft Teams): These platforms offer instant messaging, file sharing, and video conferencing features. They enhance team communication and collaboration, facilitating smooth information exchange.

Whiteboarding Tools (e.g., Miro and Mural): These tools offer digital whiteboards that facilitate visual collaboration, brainstorming, and planning, making them particularly beneficial for visual thinkers and teams working on intricate concepts.

Code Collaboration Platforms (e.g., GitHub and GitLab): These platforms enable developers to work together on coding projects, monitor modifications, and oversee version control. They are crucial for software development and open-source initiatives.

Online Design Tools (e.g., Canva and Figma): These platforms enable numerous individuals to work together on design projects simultaneously, proving to be highly beneficial for marketing teams and creative professionals.

General Benefits of Collaborative Tools:

Enhanced Communication: Collaboration tools enhance effective and transparent communication among team members.

Increased Productivity: These tools improve overall productivity by optimizing workflows and removing unnecessary steps.

Improved Teamwork: They cultivate a spirit of collaboration and collective responsibility, resulting in improved teamwork outcomes.

Remote Work Enablement: These tools are crucial for remote work, allowing teams to collaborate efficiently from any location.

Democratization of Input: Collaborative tools enable contributions from every team member rather than just a limited group.

Collaborative tools are essential in contemporary educational and professional settings, fostering effective teamwork, improved communication, and heightened productivity.

2.4 PEDAGOGICAL ADVANTAGES AND CHALLENGES OF MODERN EDUCATIONAL PRACTICES

Contemporary educational methods, influenced by technological advancements and changing teaching theories, provide numerous benefits but also pose considerable challenges. Grasping both aspects is essential for successful application and enhancement.

2.4.1 Advantages

Increased Access to Education: Digital education platforms, OERs, and video materials eliminate geographical and financial obstacles, broadening access to education for marginalized groups. Adaptable learning timelines and self-directed modules cater to a variety of learner requirements, including individuals with disabilities or limited time availability (Means et al. 2010; Sangrà et al. 2012).

Personalized Learning Experiences: Adaptive learning technologies and data analytics facilitate personalized instruction that accommodates individual learning preferences and speeds. Customized learning pathways and tailored content empower students to concentrate on the areas where they require the greatest assistance (Johnson et al. 2014).

Enhanced Student Engagement and Motivation: Engaging and enjoyable learning experiences are enhanced through interactive multimedia content, gamification, and virtual simulations. Additionally, collaborative online projects and peer-to-peer learning cultivate a sense of community and collective purpose (Dede 2009; Plass et al. 2015).

Improved Collaboration and Communication: Collaborative tools such as Google Docs, shared whiteboards, and online discussion forums enhance teamwork and communication. These online platforms link students and educators from various backgrounds, fostering cross-cultural understanding and broadening global perspectives ((International Society for Technology in Education 2016).

Development of 21st-Century Skills: Learning environments that incorporate technology enhance critical thinking, problem-solving abilities, digital literacy, and skills in information management. Tools for online collaboration and communication promote teamwork, adaptability, and effective communication skills, which are vital for achieving success in today's workforce (International Society for Technology in Education 2016).

2.4.2 Challenges

Equity and Accessibility Issues: The gap in digital access leads to inequalities in technology availability, internet access, and digital skills, putting students from low-income or rural backgrounds at a disadvantage. To provide equal opportunities for students with disabilities, it is essential to thoughtfully incorporate assistive technologies and adhere to inclusive design standards (Zhu et al. 2009; Selwyn 2016).

Concerns about Quality and Reliability of Resources: The vast array of online resources demands strong critical evaluation skills to differentiate between trustworthy information and misleading or subpar content. Ensuring the quality and reliability of online educational materials involves thorough curation and review procedures (Selwyn 2016).

Potential for Isolation and Lack of Social Interaction: Overdependence on online education may result in social isolation and diminished in-person interactions, which can adversely affect students' social and emotional growth. It is essential to strike a balance between online and traditional learning methods to cultivate a sense of community and encourage social engagement (Zhu et al. 2009).

Need for Effective Pedagogical Design and Teacher Training: Effectively incorporating technology into education necessitates thoughtful pedagogical planning and alignment with educational goals. Teachers must engage in continuous professional development to gain the skills and expertise required to create and implement engaging and impactful online learning experiences (Law et al. 2008; International Society for Technology in Education 2016).

Ensuring Data Privacy and Security: The gathering and utilization of student data in digital learning settings pose significant privacy and security issues. It is crucial to establish strong data protection protocols and follow ethical standards to protect student information.

2.5 CRITICAL FACTORS FOR SUCCESSFUL IMPLEMENTATION

Digital Literacy: At the core of successfully integrating e-learning platforms and digital resources is the cultivation of strong digital literacy skills among both educators and learners. This includes not only the technical proficiency to operate digital tools but also the critical thinking abilities required to assess online information, effectively navigate digital spaces, and comprehend the ethical considerations surrounding technology use. Educators should be adept at leveraging various digital platforms and resources, while learners must acquire the skills to responsibly search, analyze, and synthesize digital information. Fostering a culture of digital literacy guarantees that technology serves as a means for meaningful learning rather than just a passing trend (International Society for Technology in Education 2016).

Equitable Access: It is essential to guarantee fair access to technology and digital resources to prevent worsening current educational inequalities. This requires tackling the digital divide by ensuring all students, irrespective of their socioeconomic status or location, have reliable internet access and the necessary devices. Additionally, e-learning platforms and digital content must be designed with accessibility in mind, making them usable for students with disabilities. To accomplish this, strategies may include enabling offline resource access, providing technical assistance, and creating platforms that work with assistive technologies (Means et al. 2010; Selwyn 2016).

Teacher Training and Professional Development: The effective incorporation of technology in education relies on educators' capacity to utilize digital tools proficiently in their teaching methods. Continuous professional development is vital to provide teachers with the necessary pedagogical insights and technical expertise to create engaging and impactful online learning experiences. This encompasses training on specific e-learning platforms, online facilitation techniques, and assessment strategies for student learning in digital settings. Ongoing support and collaborative opportunities among educators are critical to nurturing a culture of innovation and ensuring that teachers feel confident and skilled in employing technology to improve student learning (Law et al. 2008; International Society for Technology in Education 2016).

Pedagogical Design: Effective pedagogical design is crucial for utilizing digital resources to improve learning outcomes instead of merely substituting conventional teaching methods. Learning activities should harness the distinctive advantages of digital tools, including interactivity, multimedia, and collaborative capabilities. This requires a shift from passive information consumption to the development of dynamic and engaging learning experiences that foster critical thinking, problem-solving, and teamwork. Educators need to thoughtfully evaluate how technology can accommodate various learning styles and establish inclusive and accessible educational settings (International Society for Technology in Education 2016).

Assessment and Evaluation: Robust assessment and evaluation techniques are essential for gauging student learning in online and blended learning settings. This requires a shift from conventional tests and quizzes to a diverse array of assessment methods that effectively capture the comprehensive nature of student learning. Digital tools can enhance the implementation of authentic assessments, including e-portfolios, projects, and presentations, which enable students to showcase their knowledge and skills in significant ways. Learning analytics can offer critical insights into student progress and engagement, allowing educators to deliver timely feedback and modify their teaching approaches as necessary. Additionally, assessing e-learning platforms and digital resources is vital to ensure their effectiveness and alignment with educational goals (Russell & Cunningham 2014).

2.6 FUTURE TRENDS AND DIRECTIONS

Artificial Intelligence (AI) in Education: The incorporation of AI in education presents significant opportunities for enhancing learning experiences. AI-driven tools can enable personalized learning by adjusting content and pacing to meet the unique needs of each student, delivering tailored feedback, and pinpointing areas where students may struggle. Intelligent tutoring systems can provide individualized instruction and assistance, serving as virtual mentors to help students navigate challenging concepts. Additionally, AI can streamline assessment procedures, allowing educators to dedicate more time to meaningful student interactions. Analyzing student data with AI can yield valuable insights into learning behaviors and guide teaching strategies, resulting in more effective and efficient educational methods (UNESCO 2019; Hwang et al. 2020).

The Metaverse and Immersive Learning: The rise of the Metaverse and immersive technologies, including VR and AR, presents new opportunities for developing captivating and authentic educational experiences. By replicating real-life situations, the Metaverse enables learners to navigate intricate settings, perform virtual experiments, and participate in immersive simulations. This approach can boost student involvement, enhance information retention, and promote a more profound comprehension. Furthermore, the capacity to design interactive and collaborative virtual environments facilitates more vibrant and engaging educational settings, dismantling conventional learning obstacles and nurturing a sense of presence and connection among learners (Bailenson 2018; Radianti et al. 2020).

The Role of Blockchain in Education: Blockchain technology could transform education by establishing secure and transparent credentialing systems. Digital credentials based on blockchain can guarantee the authenticity and portability of academic records, making it easier to verify qualifications and support lifelong learning. Additionally, blockchain can improve the safeguarding of learner data by offering a decentralized and secure method for storing and managing sensitive information. This technology also has the potential to create transparent and fair systems for funding education and allocating educational resources, thereby enhancing accountability and efficiency in the education sector (Buterin 2014).

Emerging Technologies: In addition to AI, the Metaverse, and blockchain, various emerging technologies are set to influence the future of education. For instance, the Internet of Things (IoT) can foster interconnected learning environments, allowing for the collection and analysis of data to enhance learning experiences. IoT devices can also support personalized learning by adjusting to the unique needs and preferences of each student. Moreover, progress in fields such as natural language processing, data analytics, and cloud computing will further drive the creation of innovative educational tools and platforms. As these technologies evolve, it is essential to investigate their potential uses in education and formulate strategies for their ethical and effective implementation.

2.7 CONCLUSION

The ongoing transformation of the educational landscape, driven by the relentless advancement of digital technologies, marks a pivotal shift from traditional learning paradigms to more dynamic, accessible, and personalized experiences. This chapter has explored the multifaceted realm of e-learning platforms and digital resources, highlighting their diverse forms—from the global reach of MOOCs and the structured environments of LMS to the immersive potential of interactive simulations, games, VR, and AR. We also delved into the ubiquitous presence of educational apps, the expansive knowledge offered by online libraries, the democratizing power of OERs, and the collaborative synergy fostered by various digital tools.

The pedagogical advantages of these modern educational practices are clear: they significantly increase access to education, enable truly personalized learning experiences, enhance student engagement and motivation, improve collaboration and communication, and crucially, foster the development of essential 21st-century skills. However, alongside these benefits, significant challenges persist. Issues of digital equity and accessibility, concerns regarding the quality and reliability of online resources, the potential for social isolation, and the vital need for effective pedagogical design and continuous teacher training coupled with the imperative of data privacy and security, all demand careful consideration.

Ultimately, the successful integration of e-learning platforms and digital resources hinges on several critical factors: cultivating strong digital literacy among all stakeholders, ensuring equitable access to technology, providing robust teacher training and professional development, implementing thoughtful pedagogical design, and establishing effective assessment and evaluation methods. Looking ahead, the horizon of education is shaped by exciting trends, including the increasing integration of AI for hyper-personalization, the emergence of the Metaverse and immersive learning for unprecedented engagement, and the transformative potential of blockchain technology for secure credentialing and resource allocation. As these and other emerging technologies continue to evolve, their ethical and effective application will be paramount in shaping a truly inclusive, engaging, and impactful future for learning worldwide.

REFERENCES

Bailenson, Jeremy. 2018. *Experience on Demand: What Virtual Reality Is, How It Works, and What It Can Do*. New York: W. W. Norton.

Buterin, Vitalik. 2014. "A Next-Generation Smart Contract and Decentralized Application Platform." *Ethereum White Paper*. https://ethereum.org/en/whitepaper/.

Coates, Hamish, Richard James, and Gabrielle Baldwin. 2005. "A Critical Examination of the Effects of Learning Management Systems on University Teaching and Learning." *Tertiary Education and Management* 11 (1): 19–36. https://link.springer.com/article/10.1007/s11233-004-3567-9

Daniel, John. 2012. "Making Sense of MOOCs: Musings in a Maze of Myth, Paradox and Possibility." *Journal of Interactive Media in Education* 2012 (3): 18. https://doi.org/10.5334/2012-18.

Dede, Chris. 2009. "Immersive Interfaces for Engagement and Learning." *Science* 323 (5910): 66–69. https://doi.org/10.1126/science.1167311.

Hwang, Gwo-Jen, Haoran Xie, and Mark Warschauer. 2020. "A Critical Review of Emerging Learning Technologies Based on a Meta-Analysis of the Literature." *Computers & Education* 157: 103944. https://doi.org/10.1016/j.compedu.2020.103944.

ISTE (International Society for Technology in Education). 2016. *ISTE Standards for Students*. Washington, DC: ISTE.

Johnson, Larry, Samantha Adams Becker, Victoria Estrada, and Alex Freeman. 2014. *The NMC Horizon Report: 2014 Higher Education Edition*. Austin, TX: The New Media Consortium.

Law, Nancy, Willem J. Pelgrum, and Tjeerd Plomp. 2008. *Pedagogy and ICT Use in Schools around the World: Findings from the IEA SITES 2006 Study*. Hong Kong: CERC-Springer.

Means, Barbara, Yukie Toyama, Robert Murphy, Marianne Bakia, and Karla Jones. 2010. *Evaluation of Evidence-Based Practices in Online Learning: A Meta-Analysis and Review of Online Learning Studies*. Washington, DC: U.S. Department of Education.

Moore, Michael G., and Greg Kearsley. 2011. *Distance Education: A Systems View of Online Learning*, 3rd ed. Belmont, CA: Wadsworth Cengage Learning.

Plass, Jan L., Bruce D. Homer, and Charles J. Kinzer. 2015. "Foundations of Game-Based Learning." *Educational Psychologist* 50 (4): 258–83. https://doi.org/10.1080/00461520.2015.1122533.

Radianti, Jaziar, Tim A. Majchrzak, Jennifer Fromm, and Isabell Wohlgenannt. 2020. "A Systematic Review of Immersive Virtual Reality Applications for Higher Education: Design Elements, Lessons Learned, and Research Agenda." *Computers & Education* 147: 103778. https://doi.org/10.1016/j.compedu.2019.103778.

Russell, Glenn, and Ian Cunningham. 2014. "Online Assessment in Higher Education: Beyond the Technology." *International Journal of E-Learning & Distance Education* 29 (1): 1–11.

Sangrà, Albert, Dimitrios Vlachopoulos, and Nati Cabrera. 2012. "Building an Inclusive Definition of E-Learning: An Approach to the Conceptual Framework." *International Review of Research in Open and Distributed Learning* 13 (2): 145–59. https://doi.org/10.19173/irrodl.v13i2.1161.

Selwyn, Neil. 2016. *Education and Technology: Key Issues and Debates*, 2nd ed. London: Bloomsbury Academic.

Simonson, Michael, Sharon Smaldino, Michael Albright, and Susan Zvacek. 2011. *Teaching and Learning at a Distance: Foundations of Distance Education*, 5th ed. Boston: Pearson.

UNESCO. 2019. *Artificial Intelligence in Education: Challenges and Opportunities for Sustainable Development*. Paris: UNESCO.

Watson, William R., and Sunnie Lee Watson. 2007. "An Argument for Clarity: What Are Learning Management Systems, What Are They Not, and What Should They Become?" *TechTrends* 51 (2): 28–34. https://doi.org/10.1007/s11528-007-0023-y.

Yuan, Li, and Stephen Powell. 2013. *MOOCs and Open Education: Implications for Higher Education*. White Paper. Bolton, UK: University of Bolton.

Chapter 3

Security and reliability of 6G and IoT networks

Vinod Kumar Jadhav

3.1 OVERVIEW

Today, we are at a stage of 5G that aids businesses and daily life in several ways, but the buzzy era of sixth generation (6G) is approaching a reality. Following the path of 5G, the most talked about and anticipated part of 6G will be a previously unseen performance, with data speeds of tortoise-like slowness, extremely low latency (less than 1 ms), and trillions of devices connected to the network. These innovations will not only facilitate continuity of communication but also set new benchmarks for network dependability and security. Minimizing latency in processing and transferring huge amounts of data will revolutionize everything from healthcare to manufacturing, improving many societal outcomes in an unprecedented manner. The Internet of Things (IoT) ecosystem is expanding rapidly with billions of devices exchanging sensitive information between each other, which adds to the need for secure and reliable networks. That being said, when we are entering the era of 6G, we have to solve many problems related to trust, data protection, and system integrity. With technologies like quantum encryption, artificial intelligence (AI)-enriched threat detection, and blockchain-centered authentication, 6G offers vigorous resistance to cyberthreats (Li et al 2021). It will also help ensure that data communicated between devices is secure from interception or alteration, ushering in a new phase of confidence in interconnected systems. Besides, the reliability of 6G technology will also help in which it will be a major bottleneck for many businesses in the present time, as business must continue under extreme circumstance (like peak hour) or in a widely wired environment (like natural environment). For mission critical applications, in particular, where some outages can impact lives and livelihoods, this kind of resilience is even more critical. Even IoT ecosystems where devices need to be always connected to make reliable will also have the much needed stability to work even in the most adverse of scenarios. In addition, 6G networks are anticipated to play a pivotal role in ushering to the futuristic era featuring safe, durable, and smart communication structures within the evolutionary technological paradigm (Khan et al 2020). They will drive the apps we are familiar with but they will also power innovations that have seemed distant until now, such as self-driving

 DOI: 10.1201/9781003586951-3

cars and personalized medicine. With a lot of promise and potential, things will set the tone for a future of faster, reliable, and secure (yes, the most secure) connectivity as we shift gears from 5G to 6G.

The SDN-enabled enhancement with standard cloud computing improves the performance of the IoT system to become more agile and evolving toward an autonomous system is being implemented through mobile Internet technology (Snehi et al 2022). This has unlocked a range of new services like mobile shopping, smart home automation, mobile payment system, and interactive gaming systems (Katz, Matinmikko-Blue and Latva-Aho 2018). The 6G of wireless networks approached, and the priority has changed from classic problems through speed and data transfer speed part to a much bigger problem of performing IoT security and reliability. The growing volume of connected devices is accompanied by broader entry points for potential attack, and the balance that IoT ecosystems have to strike between communicating at all times and enforcing stringent security protocols is one of the most pertinent challenges. Sixth-generation networks are about to change the game again. Boasting ultra-low latency, massive device connectivity, and advanced AI-oriented capabilities, 6G will unlock the potential of the IoT across all layers of the application and business scenarios. But like any technology, these innovations have their weaknesses, as well. The growth of heads and complexities systems can also threaten the integrity of IoT that provides a new way for malicious actors (hackers) to attack IoT globally. Therefore, the transition toward 6G as well as IoT has shifted the focus to the deployment of high-performance, secure, and reliable next-generation wireless communication systems to maximize the advanced multifunctional potential that these elements provide (Zhao et al 2020). This includes a multi-faceted approach to protecting sensitive user data, communication between devices, and mitigation of service outages, among others. Such promise of 6G to transform industries and fuel smart applications across various sectors is significant yet the promise also brings a plethora of risk that leaves the mobility networks vulnerable to safeguard against rising number of cyberattacks on both data and infrastructure. Only then can we guarantee that the next wave of IoT innovation will be safe, secure, and sustainable (Figure 3.1).

3.2 ARCHITECTURE

The 6G network architecture in Figure 3.2 shows three primary layers: the service layer, the function layer, and the resource layer. Collectively, they all provide the efficiency, security, and reliability of the network for advanced technologies and business models, all while delivering high performance. Now let's discuss each layer in detail:

Service layer: Provides interfaces that integrate the services and subsystems that form the basis of any network communications, which may include and not limited to networks defined by the 3GPP (3rd Generation

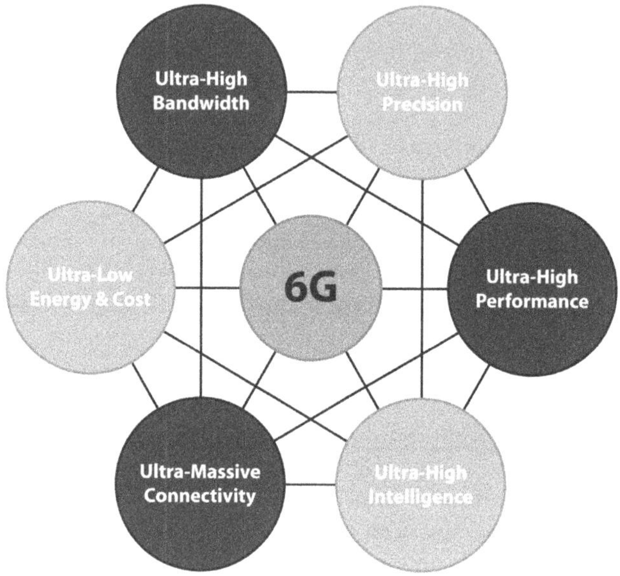

Figure 3.1 Overview of 6G network.

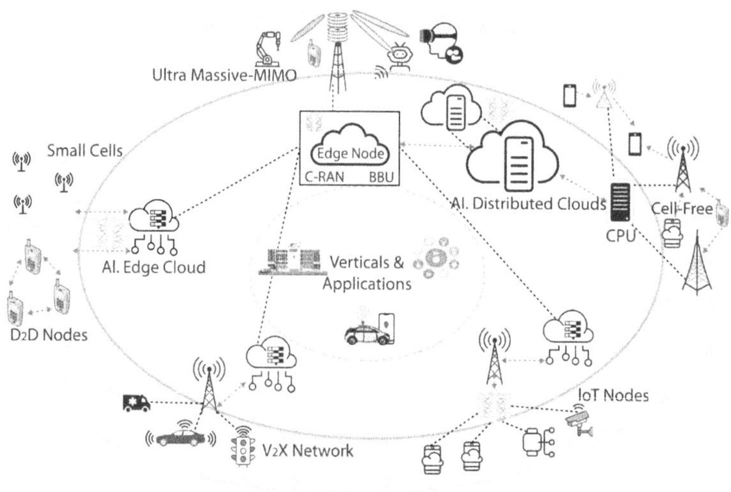

Figure 3.2 6G architecture.

Partnership Project) such as those services forming the IP Multimedia Subsystem (IMS) within 5G or services external to the networks (e.g., OTT (over the top) networks and services) adding additional functionalities and layers of complexity. As we move toward 6G, we anticipate significant evolution in this service layer, meaning the components enabling emerging technologies and business models, such as highly immersive XR, digital humans, and multi-sensory interconnections (Corici et al 2021). The technological advances needed for these new services impose more demanding roles to the network: higher levels of transmitted capacity and more advanced sensing and computing capabilities to manage their complexity. With the advent of 6G networks, it is more important than ever to secure and maintain the reliability of the service layer. This is because the communication services it underpins will cover applications that are both highly sensitive and resource intensive. Such as with real-time immersive experiences or feeding and operating digital avatars, which will require stable and secure connections for smooth operation at scale. I hope a technology stack supporting these capabilities at the service layer that is resilient to cyberattacks, data breaches, and other security threats, all while maintaining high levels of service availability. The two-pronged approach to security and reliability will prove crucial as 6G networks support increasingly sophisticated use cases and the continued rise in demand for smooth and high-quality communication experiences.

Function layer: aRGIS single-cell and multi-cell network technical solutions are applied to the function layer of core network (CN), RAN, and UE. With the deployment of 6G networks, it will be further enhanced to enable new services and more complex functionalities. As such, conventional communication functions, like access management, mobility management, session management, policy control, and user plane data transmission, will still be relevant but enhanced to meet the high performance requirements of 6G, while new functions must be developed to support next-generation technologies, including sensing, computing, and data management, as well as advanced security methods and devices for almost zero power IoT (AZP IoT). The new features will lead to Integrated Sensing and Communication (ISAC), which combines sensing and communication for real-time data exchange, and will drive the convergence of mobile networks with advanced computing and data processing technologies (Uusitalo et al 2021). Also, it combines the data in cross domain and applies intelligent network functions. Reliability and performance of 6G network will rely on handling of each component right from security to computing to information management functions in a seamless manner, with designed configurations capable of being performed satisfactorily with divided functions and specifications. This is an important capability for ensuring, at scale for the many use cases that 6G is expected to accommodate, reliability, security, and performance of the network.

Resource layer: The resource layer next provides basic support; it manages the spectrum resources under the pressure of massive bandwidth, and will also manage the critical computing and storage resources in 6G networks (Qamar et al 2020). This expansion is vital to support the growing demand for data-intensive and processing-intensive services facilitation by 6G, such as complex AI applications and automated systems. The efficient and secure operation of the network relies on the proper management of these resources. That is, immunity from anyone reading, writing, or deleting the data, nothing without a certificate it can create and forward to itself, mounted to its storage node. Moreover, adaptable resource allocating will be required when system performance drops or risks network failures that could affect service reliability. Thus, the protocols of that resource layer, which will be the backbone of both the above-mentioned service and function layer, need to be robust enough to take care of not only the data security and reliability but also real-time processing enabled by several emerging technologies. So, mesh all regions in such a strong glare to the 6G network's successful superstructure, as resource multiple utilization directly transfers the service quality and ushers in data protection and stability of the network.

Consideration of different functional components to support operations can aid in visualizing the overall architecture of a 6G network. Although 5G standards and their lifecycle serve as the basis for the communication functionalities within 6G, the concept is likely to integrate multiple new elements, even if many are still up for debate. It includes the main components from the service-based architecture (SBA) including data functions and sensing functions to AZP IoT functions, computing functions, and security and intelligence functions.

1. **Data functions:** The CN data plane plays a key role in important functions of the 6G network, such as controlling and sending data, as well as storing data. They enable the handling and transfer of data throughout the network. The data plane interacts closely with various functions of the peer-to-peer protocol in the RAN and UE, enabling seamless communication. Since 6G networks are projected to support a staggering amount of data traffic, certain new data plane function needs to be adopted which can accommodate this surge. However, as infrastructure will be required for more advanced applications and a wider array of connected devices, this expansion is crucial to buttress the WAN's reliability and performance. With the development of these data plane functions, upcoming networks will be scalable and adaptive, streamlining data over the complete system.
2. **Functions of senses:** Provides awareness of its surroundings to the network. Since collection functions effectively, consider the processing of environment data for efficient performance of smart networks. The incorporation of sensing capabilities in 6G can enable a reactive approach to adapting the network to changing environmental conditions, enhancing both its security through instant threat detection as well as its reliability through adjustment of network performance in function of the domain conditions.

3. **AZP IoT functions:** AZP IoT functions are intended to control devices with a low energy usage level. They are responsible to manage AZP IoT devices, decompose tasks, and AZP IoT-related service control. They also take care of small data forwarding based on control planes. These low-power operations are vital for extending the reach of the network, especially in remote or energy poor regions. Moreover, these capabilities enhance security by keeping energy consumption low, thus minimizing attack surfaces for possible vulnerabilities.
4. **Computing functions:** Computing functions are an important part of 6G, not only in providing computational resources, like with network management, but also for external AI services and other external functions. This makes the 6G system versatile, delivering much better performance and reliability thanks to the initial computation resources collected through this network, coupled with vast quantities of data. It now allows for an additional security layer, as the many AI algorithms can be used to identify device anomalies and predict security breaches.
5. **Security functions:** 6G considers the future risk environment where cyberthreats are complex; hence, the architecture offers significant, stronger, and more sophisticated security functions. These functions are intended to give improved encryption, authentication, and threat detection mechanisms. Since it is predicted that 6G networks will enable many more technologies such as very high automated control systems or implemented IoT devices, there should be a lot of robust security features to stay harmless from illegal access as well as mitigate threat risk. Security of data and services in a 6G network will serve as one of its cornerstone services.
6. **Implementation of intelligence functions (for network):** AI-led intelligence functions are core/major operations of 6G networks. They aid in training and inference for AI models, as well as for network control and optimization recommendations. Using AI, 6G networks are capable of intelligently adapting to an ever-changing environment, recognizing potential security threats and allocating resources accordingly. They will aid in monitoring and responding to real-time security threats, increasing resilience within the network.
7. **New messaging framework:** One of the design features of 6G networks is a new messaging framework that increases the overall efficiency of running data. Whereas the SBA of the 5G era is mostly employed in the control plane, with the new messaging framework in the 6G era necessitating large-scale data transfers in CN. It will improve the performance and reliability of the network, allowing it to handle the exponentially growing data traffic projected for the future. Moreover, the messaging framework is essential for leveraging the unique features of 6G, including ultra-low-latency communication and massive IoT connectivity.
8. **Peer-to-peer functions in RAN and UE:** In order to evolve communication functions, RAN and UE will adopt peer-to-peer communication in

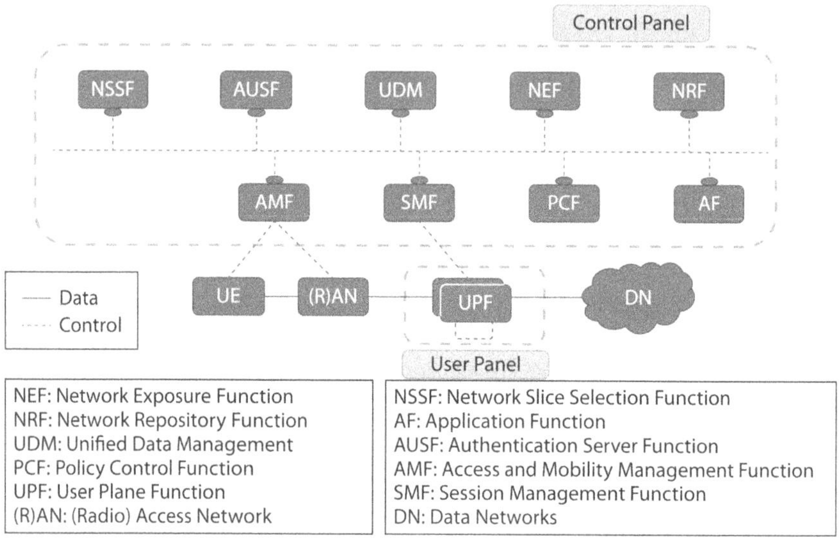

Figure 3.3 Service-based 6G architecture.

the 2030 era to allow direct communication between devices. As such, it's anticipated that these functions will enable better communication procedures resulting in efficiency and robustness for the 6G network. Additionally, peer-to-peer communication supports more secure and resilient network structures via distributed operations with fewer central points of failure (Figure 3.3).

3.3 USE CASES

Although 5G networks can accommodate a range of trade-offs including latency, energy consumption, cost, hardware complexity, throughput, and end-to-end reliability, they typically cater to individual needs via different configurations. Some functionalities in 5G such as mobile broadband and ultra-reliable low-latency communications (URLLC) are configured through isolated setups. Sixth generation is bringing a more uniform integrated solution to cater to diverse and stringent network requirements. Sixth generation will integrate ultra-high reliability, capacity, efficiency, and ultra-low latency in a seamless and unified manner, contrasting with 5G. In the 2030 perspective, economic, social, technical, and environmental factors will largely determine the mission for 6G, including performance, security, and reliability. Security will be key most importantly; security will be a primary focus in protecting sensitive data and maintaining the resilience of the communication infrastructure against cyberattacks. Furthermore, 6G will be required to be able to reliability to support critical applications such as autonomous systems,

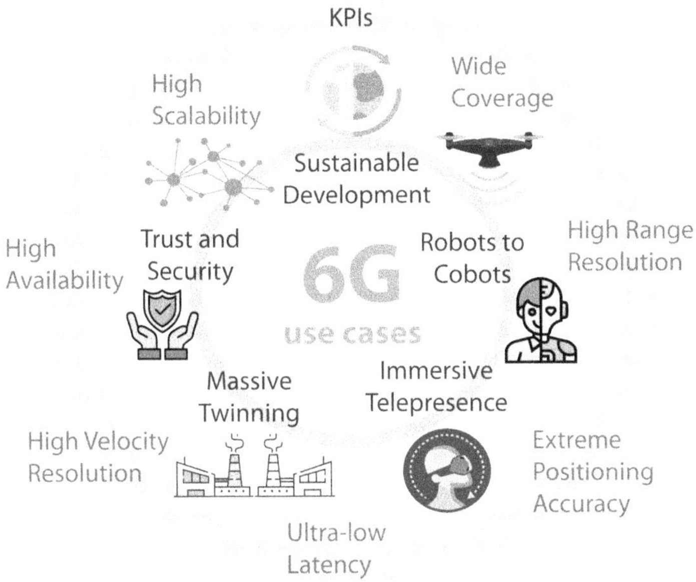

Figure 3.4 Use cases.

Internet of vehicles, smart grid, smart city, healthcare, etc. In such context, 6G will evolve to address them through the delivery of ultra-high reliability and performance, across a broad spectrum of use cases. Expected future 6G services would include these use cases due to their broad applicability and complementary nature. They will often need high security, reliable networks, or efficient resource management. This will require the network to be able to support a huge volume of data, provide low-latency communication, and be resilient against routs of communication traffic. Various key performance indicators (KPIs) just evaluating the performance of those use cases provide an overview of how good the network is in terms of security, reliability, efficiency, speed, etc. The 6G network must be able to adapt and scale so that it can provide high security and reliability, catering to the new variety of demands from next-generation technologies and applications (Figure 3.4).

3.3.1 Augmented reality and virtual reality

The shift from 4G to 5G networks was designed to be a step change in wireless telecommunications, enabling for one of the most demanding applications by then video over wireless. With the growth of multimedia and streaming services, the demand for higher bandwidth drove the deployment of new spectrum technologies, like millimeter waves (mmWaves) to provide more network capacity. The move for 5G was this evolution as it needed to theoretically support multi-gigabit-per-second data rates to sustain the rising demand for high-quality multimedia service and content. On the other hand, the rise of augmented reality (AR) and virtual reality (VR) applications, even though they are significantly more data-hungry compared to conventional multimedia content, is expected to be the next wave of 5G networks' data consumption. Due to their immersive nature and the demand for vast amounts of data, AR and VR are anticipated to be among the first applications of the 5G era. As these technologies become more common, they will put a heavy load on the 5G network, eventually overwhelming it, just as video-over-wireless stressed 4G networks. These complexities, including a network of and beyond 1 terabit per second (Tb/s), far above and beyond the 20 gigabit per second (Gb/s) target for 5G, and real-time interaction (the need for low latency) that exerts an influence on AR/VR users, will require an even more sophisticated and beyond a 5G now form of network in the future. Unlike traditional multimedia content that could be compressed for a more effective transmission, AR/VR is unable to audit such a compression, as the data are time-sensitive. When we move into 6G networks, data rate per user must be beyond gigabit-per-second figures, a stark contrast from the relatively achievable 5G goal of 100 megabit-per-second precision. Sixth-generation security has to ensure that the integrity of real-time AR/VR interaction is kept pristine but also that this is done in the face of a vast data throughput and associated cyberthreat surface area. In addition, the demands for consistently engaged, real-time service posed by immersive technologies will test the resiliency of the network that would need to deliver them. For example, within the context of AR/VR, 6G will need to provide not only higher capacity and speed but also improved security protocols and reliability mechanisms to accommodate the complex requirements of new applications.

3.3.2 Holographic telepresence (teleportation)

While holographic telepresence (teleportation) takes a huge leap in interactions across space and time through realistic three-dimensional (3D) holographic representations of individuals. Nonetheless, with such a dimension of remote reliance in 6G networks, engineers will face significant challenges in security and reliability. The data throughput requirements for a 3D holographic display are enormous without compression, a raw hologram, including colors, full parallax, and a refresh rate of 30 fps, would require a staggering 4.32 terabits per second (Tb/s) of bandwidth. More data alone is not sufficient; the latency requirements for such systems are expected to go down to sub-millisecond, which is equally important to facilitate real-time interaction. Moreover, with few angles required in current VR/AR technologies, that display would need thousands of synchronized view angles spanning to make an immersive experience. In order to enable a truly immersive experience remotely, the envisioned 6G networks will also need to support visual and auditory data in addition to digitized transmission of all five human senses. This would exponentially increase the aggregate data rate. Given these holographic telepresence breakthroughs, data transfer security is a top priority. Biometric data is particularly sensitive (audio, video, and in some cases, sensory data), and it stands to reason that it should be submitted through secure channels to prevent eavesdropping, hacking, and data tampering. Sixth-generation mobile networks will also need to ensure that such large amounts of data can be transmitted without loss or delay. A unique challenge of the product is to make sure the network backbone can be stable and steady under high traffic. Therefore, these problems of data security and network reliability must be solved in order for holographic telepresence to be successfully introduced into 6G networks.

3.3.3 eHealth

The future of eHealth services BA S eliminating time and space constraints in providing eHealth services. The physical limitations imposed on global healthcare delivery will all but be eliminated and remote surgery will move from a dream to reality. The healthcare service itself will be significantly augmented by workflow optimization. But the main limitation of such a future development is the absence of real-time haptic feedback during remote surgery, which challenges the quality of remote interaction. The advanced eHealth services will bring further challenges in securing the integral network infrastructure and its proper functioning along with these technological hurdles (Katz, Matinmikko-Blue and Latva-Aho 2018). Chief among these concerns is the need to satisfy very tight quality of service (QoS) requirements, including consistent, seamless connectivity with ultra-high reliability (99.99999% uptime), near-zero latency (sub-millisecond response time), and the seamless support of mobility for users on the go. Such use cases necessitate the need for secure, reliable, and resilient networks that can deliver performance under even the most rigorous conditions. Sixth-generation networks are expected to play a key role in tackling those challenges with more spectrum and more intelligence available. With the advent of 6G, advanced features including URLLC and enhanced error correction techniques will enable a high level of security and reliability, allowing for the possible QoS needs for eHealth applications to be met consistently. The spectral efficiency of 6G networks is also expected to improve significantly, with five to ten times higher rates achievable than is possible now. This, in turn, enables a more efficient use of the scarce spectrum resource, which is necessary in an effort to deploy the capabilities required for eHealth services and provide the requisite data security and reliability. Sixth-generation networks will ensure that eHealth services are secure, reliable, and as best possible to meet the strict standards inherent in this critical application (Figure 3.5).

3.3.4 Pervasive connectivity

Mobile will keep growing rapidly and mobile traffic will triple between 2016 and 2021. This deluge will take the number of mobile devices to record

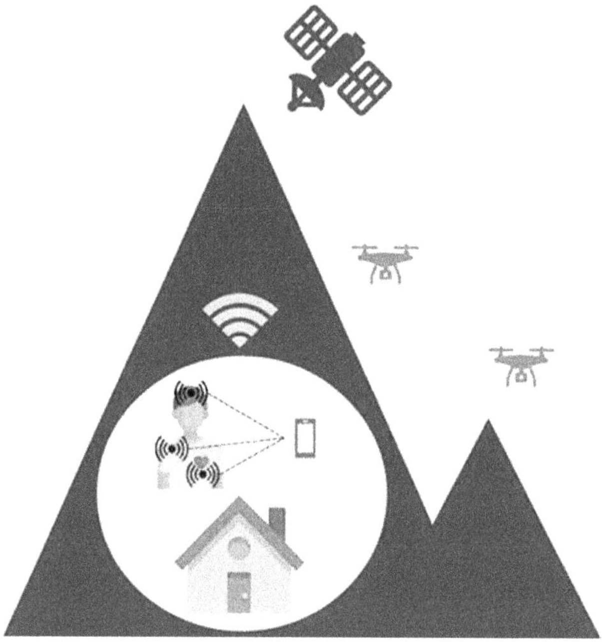

Figure 3.5 6G technologies in health.

highs: up to ten devices per square kilometers against even just ten in certain metropolis such as Singapore. It is predicted that 125+ billion connected devices will exist globally by the year 2030. New 6G networks will have to connect not just personal devices but sensors, vehicles, and components of an entire smart city infrastructure. This connectivity surge will exert tremendous stress on existing networks, some of which will find it difficult to ensure reliable connectivity to all devices while complying with strict application requirements for speed, latency, and QoS. Sixth-generation networks should be significantly more energy efficient (10–100 times over 5G networks), to enable scalable and cost-effective deployments with a reduced carbon footprint, in light of the growing global demand for urban and rural connectivity. Although 5G networks are growing, mostly outdoors, and often use mmWave spectrum frequencies, they can have limited indoor coverage because high-frequency radio waves do not penetrate easily (or at all) through most materials, such as concrete walls or other dielectric materials. While current 5G networks tend to have certain expansive coverage areas, 6G networks will target ubiquitous and seamless connectivity in diverse settings with limitations of neither indoor or outdoor constraints. This means that in every use case, 6G systems should be able to deliver reliable, affordable, high-quality connectivity, and be resilient to failures and external threats. That means 6G

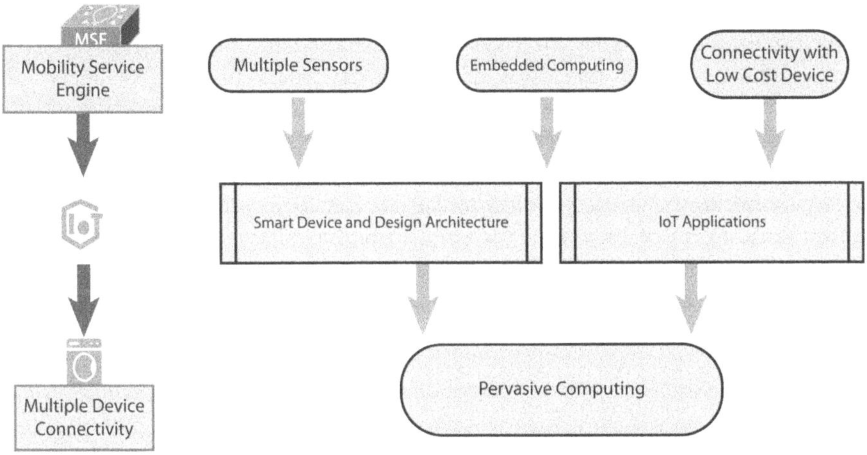

Figure 3.6 Pervasive connectivity.

networks must not only be able to handle skyrocketing demand for mobile traffic but also be secure and robust enough to support vital infrastructure from healthcare to transportation and, at the same time, operate with the highest levels of energy efficiency. Such security and reliability efforts will be critical in meeting the smart cities requirements while unlocking other future ADAS, IoT, and AI-specific applications (Figure 3.6).

3.3.5 Industry 4.0 and robotics

Thus, the connection that 6G will provide can fulfill the mandate of the Industry 4.0 revolution that was set off by 5G; the industrial revolution put in place by the introduction of Industry 4.0 relates to the digitization of manufacturing, driven by applications based on cyber-physical and IoT systems (Vaigandla et al 2021). With 6G, the gap between the physical factory floor and the virtual computational space will be easily bridged. Such integration will allow for cost-effective, flexible, high-efficiency, Internet-based systems of diagnosis, maintenance, operations, and direct inter-machine communication. In this regard, however, the introduction of automation brings its own challenges, notably for reliable and synchronous communication. Some of these challenges correspond to the requirements of ultra-low latency, time synchronization, and communication standards that enable cyber-physical system operations. Therefore, deploying 6G network there will serve as a key point to fulfill these demands. In particular, it is anticipated to support industrial control systems that demand microsecond-level delay jitter for real-time operations. Moreover, 6G will provide the bandwidth necessary for high-bandwidth applications such as AR and VR, both of which are becoming "must-haves" for industrial training, inspection, and operational processes.

Such applications require gigabit-per-second peak data rates, and 6G will offer the underlying fabric to make certain that such high demands will reliably, and securely, be met. Similar to security and reliability, 6G needs to ensure that these operations take place with no disruption or cyberattack. For mission-critical applications, the network not only can be highly secure but also has to be resilient, protecting sensitive industrial data while providing continuous, faultless communication that allows fast decision-making and coordination of machines on a real-time basis. Sixth generation will be an enabler of Industry 4.0, allowing advanced manufacturing ecosystems to provide secure and efficient services with unprecedented responsiveness and flexibility.

3.3.6 Unmanned mobility

When realized, the shift to an entirely robotic transport system will transform everyone's travel experience, resulting in safer, more efficiently managed traffic flows, and vulnerable to endless infotainment choices with global markets expected to exceed US$7 trillion. But the level of connectivity required for autonomous vehicles (AVs) is quite challenging, especially instant availability and stable security. AVs operate much more safely and reliably when ultra-high reliability and extremely low latency are guaranteed, with an availability of >99.999% and latency <1 ms, even ultra-high mobility scenarios; for example, vehicles moving up to 1000 km/h upon connection, however, achieving these essential requirements based on the current techniques is a significant challenge. The amount of data generated by each vehicle is expected to increase dramatically to terabytes of data per hour of driving as sensor-based data approaches the goal for self-driving technologies. This will far surpass the capabilities of today's networks, requiring a far higher data transfer rate. In addition to the already existing use cases for flying vehicles, such as emergency response, construction, and many more scenarios, these activities have much more potential that needs to be explored. This operational use case for swarms of drones operating in concert will add pressure to connectivity, with networks needing to deliver capacity orders of magnitude greater than those necessary to support 5G's tech rollout. As such, 6G networks can be an essential player in enabling a secure and reliable vehicular communication network. This will seamlessly underpin the communication required for AVs and drones and will be delivered through advances in hardware, software, and connectivity solutions that the technology of 6G will enable. The far-reaching use cases made possible by the 6G such as autonomous transport and swarms of drones requires revolutionary innovations in technology and new designs for networks (Vaigandla et al 2021). These innovations will be essential to help unlock the full promise of 6G, in terms of supporting the security and reliability requirements of future transportation systems and providing support for next-generation connectivity.

3.4 KEY TECHNOLOGIES (AI, BLOCKCHAIN, etc.)

Here are a few projected 6G applications that will be supported and enabled by the 6G communications:

- **eMBB-Plus:** Enhancing mobile broadband needs has been a major tooling in 5G which has evolved to eMBB Outside 1.0 in 6G networks to provide further development on the quality of experience (QoE) in the utilization of data and the compliance with more enhanced advanced norms. Essential in bringing this next-generation wireless broadband frontier to reality constitutes essential functions of the next-generation area such as network optimization, handover, and interference management which will merge as a fundamental part of the subsystems. Through big data analytics, these operations will be made smarter, more efficient, and predictive. An interesting feature of eMBB-Plus is that it provides accurate indoor positioning that can have various applications, including AR, smart infrastructure, and location-based services. In addition, the goal is to enable a universally compatible connection on different mobile operating networks on the globe. It is also anticipated that these capabilities would be delivered at an affordable rate, making them available to a greater diversity of network subscribers. These enhancements must, however, be engineered and applied in a way that doesn't degrade the security, secrecy, and privacy of people. With the growing complexity and interconnectedness of 6G networks, it becomes essential to formulate and implement robust strategies to ensure user data security against such threats. This includes the deployment of cutting-edge encryption protocols, secure data-sharing frameworks, and privacy-preserving technologies that ensure absolute confidentiality of subscriber data, protecting them from potential cyber incursions. Moreover, reliability is one of the pillars of the 6G vision and eMBB-Plus must be able to function under various conditions. Due to its redundancy mechanisms, error correction techniques, and AI-assisted diagnostic functions, eMBB-Plus can ensure stable connections and high reliability critical in demanding network circumstances.
- **Big communications (BigCom):** The 6G unified data communication protocol features optimized BigCom for scalable capacity, efficient performance in both the city and the wild, and huge resources allocation. This ensures that anyone, anywhere, at ultra-high data rates will be able to communicate between their subscribers right away. Integral to its performance, 6G aims to capitalize on unconventional technologies such as extremely large bandwidth via Terahertz (THz) waves, which can deliver data at much greater speeds than today's systems (Wang et al 2022). When it comes to security, 6G carries highly sophisticated AI systems which improve not only network functioning but also its security. This AI is able to handle operational and environmental factors so

the network stays resilient, even when facing unpredictable conditions. Fully harnessing the potential of AI and quantum technology, 6G will allow dynamic adjustment to changing demands, implement real-time responses to security threats, and ensure reliable connectivity to users across the globe (Porambage et al. 2021). Aprai Bigning Infrastructure Design BigCom's infrastructure design focuses on creating a secure and reliable communication platform that not only enables a wide variety of services but also prioritizes user privacy and data integrity in light of emergent threat.

- **Number of ultra-reliable low-latency communications (URLLC):** URLLC is one of the architectures that will directly contribute to the realization of extreme reliability and ultra-low-latency communication in 6G networks. Sixth-generation wireless networks enhance the URLLC, the massive machine-type communications (mMTC) capability of 5G, by meeting more stringent reliability and latency requirements. URLLC aims to ever-improving extreme reliability and ultra-low-latency communication for safety-critical aspects in communication systems is a required feature for applications such as driverless or AVs that will be seen in 6G. The above also holds true for URLLC in 6G as it carries a more stringent reliability requirement compared to that of 5G with a goal that stretches the reliability even further (99.999%) to enable different application scenarios/distributed communication relations (IoT), enabling ultra-low latency across all levels with target 99.999%, i.e., the communication can even undergo damage for 1 9, but slurries can still be transmitted directly. This degree of reliability is essential in applications where failure could have dire consequences, including vehicular safety systems, traffic management, and real-time navigation. To the same extent, the latency in 6G URLLC is predicted to fall below 0.1 ms. Such ultra-low latency guarantees that data transmission happens virtually instantaneously, allowing decision-making in real time for AVs and other time-critical solutions. In this chapter, we focus on safety applications of vehicular communication, where ensuring low latency is critical to avoid collision and achieve V2V (Vehicle-to-Vehicle communication) or V2I (Vehicle-to-Infrastructure communication) coordination. In addition, it is essential to address a robust security framework in URLLC in 6G. According to the extensive interconnectedness and information-rich characteristics of vehicular networks, it has become indispensable to preserve confidentiality, integrity, and authenticity of transmissions to avert cyberattacks and threats. This security framework protects against potential vulnerabilities resulting from malicious attacks or hacks, facilitating safe and reliable communication between vehicles, infrastructure, and other entities. In 6G, URLLC is one of the relevant features and its integration with a robust security framework will allow establishing secure yet efficient vehicular communication capabilities in the vehicular scenario. This enhanced localization is not only expected to enhance the autonomous driving technology but also going to facilitate the modern era of urbanization by making it safer, smarter, and more intelligent.

- **Three-dimensional integrated communications (3D-InteCom):** As the advent of 6G networks is approaching, the heights of the communication devices do not play much role in the communication models because most of the empirical propagation models are two-dimensional (2D) models. But the forthcoming 6G era brings the concept of 3D-InteCom, which breaks the conventional 2D paradigm through adding the height dimension. With this novel design paradigm, the vertical deployment of communication nodes (base stations, devices, and relays) will have a significant effect on signal propagation and network performance. In particular, the migration to 3D-InteCom will require adaptations of analysis and modeling efforts that have so far been focused on 2D wireless communication models, primarily through the use of stochastic geometry and graph theory. When 6G networks enter the vertical dimensions, it would be required to extend existing models to include complexity due to the height of communication nodes. Various technologies such as satellite communications, unmanned aerial vehicles (UAVs), and underwater communications have already started to bring 3D thinking into their communication plans (Vaigandla et al 2021). The benefits of considering altitude and positioning in communications networks are highlighted by these technologies, which accelerate the development of al-l oxide via-elevations in 6G like elevation beamforming and full-dimensional multiple-input multiple-output (MIMO) architectures. By providing an optimal way to utilize spatial dimensions of communication and to manage the signal propagation in the 6G plan, the advanced technologies will boost their reliability and capacity. In addition, using 3D space allows the 6G network to achieve better coverage, less interference, and reliable connections even in difficult environments. From a security perspective, the inclusion of 3D-InteCom paves the way toward further increased 6G network reliability and security. By controlling and conforming to signal propagation in 3D space, access points can more precisely lock in devices as well as create enforced source zones to minimize or eliminate eavesdropping or area access to insecure devices. To explore the applications and opportunities of 3D-InteCom, further research is needed in various aspects, including comparatives between 2D and 3D, complexities of 3D network-based solutions, encryption and security protocols, and efficient resource management. The four dimensions will enable dynamic reconfiguration of the network topology and additional spectrum utilization, thereby significantly improving the network performance in 6G. At this stage, analytical models that can be adapted and new technologies will be developed to harness the full potential of 6G networks.
- **Unconventional data communications (UCDC):** Since there is no known precise and complete UCDC definition, we seek for important aspects

like holographic, tactile, and human-bond communications in forthcoming 6G networks as well as their security and availability. Holographic communication, a step ahead, could change the way users interact with each other through more immersive experiences. But these holographic data streams need to be secure if we're going to enable unwanted access to them or manipulation (Kumari, Gupta and Tanwar 2021). While tactile communication can provide sensations of touch, it is also promising for applications in telemedicine and remote control; given the sensitivity of such applications, a fault in tactile communication should not render it unreliable, especially since it forms the fundamentals of medical or industrial operations that may put lives at risk. Communication attuned to human bond transmission of emotional signals and corporeal movement, however, might deliver deeper, more intimate exchanges but relies upon strong privacy protections to safeguard against emotional manipulation or data breaches. For 6G, which targets URLLC and mMTC, UCDC brings a trade-off of prospects and challenges. In order to enable this new form of communication and mold the full potential of this network, 6G networks need to understand advanced security architectures that deliver information transfer over such unconventional communication modes with high confidentiality and reliability with the lowest possible risk from cyberattacks.

- **Holographic communications**: Sixth generation is expected to bring glamor in terms of holographic communications. The hologram is a 3D technology that bends light rays emitted to an object and then records the interference pattern created on a recording device (Porambage et al 2021). The truth is that a mere 3D picture transmission, devoid of a stereo voice, fails to convey the personal presence qualities. In 6G, reconfigurable stereo audio will inspire the establishment of a framework for a platform to be utilized for the capture of multiple in-person appearances based on each arrangement. That is, it gives plenty of leeway for the entities to interact with and alter received holographic data and video, if deemed necessary. Holographic data will require a very high bandwidth and will need to be transferred through reliable network links.
- **Tactile communications**: What is termed holographic communication allows you to communicate a virtual vision of near to real sights of people, events, and environments. The cinematic experience is incomplete without this kind of tactile usage of the Internet which would make it possible to instantaneously convey an image using the second screen. The mentioned technology is anticipated to benefit the applications in teleoperation, cooperative automated driving, and interpersonal communication. Haptic touch could be easily integrated using the same communication networks for these technologies. If realized, this technology would bring about the disappearance of the open systems interconnection network model and the widespread use of a cross-layer

communication system design. This means the cross-layer architecture must cater to the rigorous demands of these technologies. This scenario could inspire novel design of physical layer (PHY) schemes to improve the ability to implement the signaling system designs and support waveform multiplexing. Another point of focus is the how to formulate mechanisms [IR, (de)queue, (de)scheduling, handover, and protocols] that will satisfy the needs of 6G networks. Clearly, no existing wireless communication systems can meet these demands, so an analysis of over-the-air fiber communication systems is necessary.

- **Human bond communications**: Namely, human-centric communication is likely to be a key enabler of 6G communication. Humans will be able to access and/or share physical features or express physical phenomenon as it is, with this technology. The five human senses will certainly be a part of this project. For example, the "communication through breath" project is an example of such technology that enables reading a human bio-profile according to exhaled breath, even the interaction with the human body is described according to inhalation using volatile organic compounds. Such technology, therefore, aids in diagnosing diseases, detecting emotions, collecting biological features, and remotely interacting with the human body. Developing a communication system that can simulate the five human senses requires interdisciplinary research partnerships (Dang et al 2020). This would sooner or later lead to developing hybrid communication technologies that could harvest different physical quantities and deliver them on demand to the selected receiver through secured communication lines.

3.5 ROLE OF ARTIFICIAL INTELLIGENCE IN 6G

Here, for example, machine learning (ML), AI, deep neural network (DNN) technology help drive revolutionary technology which in turn creates avenues for new research opportunities for both 6G communications and IoT to name a few. A number of features such as self-configuration, aggregation, opportunistic set-up, and context awareness are anticipated to be extended in AI-empowered 6G. Simply put, many of the benefits of wide-band radio working in a cognitive to intelligent radio range can be brought to life through AI-empowered 6G with a little help from ML algorithms (Sheth et al 2020). Moreover, the intelligent and reconfigurable materials are expected to establish the hardware basis for the wireless technologies in an advanced setting; they are anticipated as the MIMO 2.0 and investigated comprehensively in. The reviews of ML techniques for device-to-device (D2D) communication, massive MIMP (Multimedia Internet Mail Protocol), and heterogeneous networks are found in, and also introduced a new network architecture in mobile communications to be used for the analytics of big data, where analysis can lead to optimizations for the PHY. In, the authors emphasized the importance

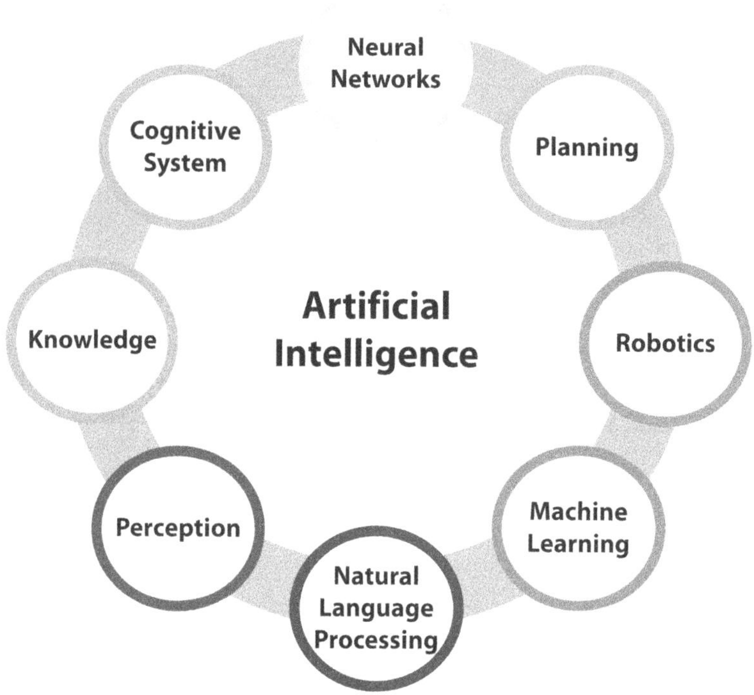

Figure 3.7 Artificial intelligence.

of index modulation for enabling efficient 6G network support. Indeed, great intelligence and ML techniques could potentially not only enhance performance but also affect 6G networks configuration and design (Figure 3.7). The advantages of intelligent 6G functions for wireless and non-wireless communication services are described in the following three categories:

- **Operational intelligence (OI)**: OI is enabled by technologies that help allocate resources more effectively, such as bandwidth or power, to optimize network operations beyond what traditional capabilities could do. This is especially important in the domain of 6G with its heterogeneous, dense, and scalable nature. Due to the complexity of managing these networks, multi-objective optimization strategies have been used, which emphasize the selection of multiple performance metrics. However, these optimizations are usually NP-hard problems, which makes them difficult to measure and apply in real time. New algorithms based on AI and ML methods, in particular on deep reinforcement learning (DRL), can provide hope (Sun et al 2020). Such capability is critical to the set of resource allocation strategies, thus through DRL, we can let decision-makers improve their decisions iteratively through a feedback loop. This use of AI is seen as an enabler of 6G and allows for more dynamic and efficient resource management. Indeed, there is another category of solutions, for example, recent research conducted by Luong et al has pointed out issues, such as data offloading, caching, and adaptive modulation, that can be approached using these technologies, which are necessary to improve network performance in 6G.

- **Environmental intelligence (EI)**: EI is a concept where pervasive and distributed intelligence becomes integrative part of various technological systems. The progress of smart radio spaces, materials, develops the wireless communication area. Rising capabilities of such environments, EI can power endless applications, e.g., cloud computing, data centers, and IoT, including land and aerial unmanned vehicles and robotics. While inheriting self-organizing and self-healing features, these systems can further improve their performance, especially in terms of D2D communication in 6G networks (Rizvi et al 2020). The use of smart frameworks can be extremely helpful in increasing the reliability of these communication systems. Generative sensing techniques present a recent approach in this direction, using the adaptive transformation of radio waves based on this concept. This optimization is enabled through the use of reconfigurable intelligent surfaces (RIS) that configure the radio environment to help meet exact requirements on communication. Such advancements lead to the development of specialized hardware that is tailored to accommodate the principles of EI, ultimately fostering a more intellectual and effective communication ecosystem (Strinati et al 2021). But an essential problem occurs when processing data in DNNs. Edge or cloud devices must perform actual computation of the extracted initial features from data; however, this is a complicated approach because different devices have different capacity for computation and communication. Moreover, these devices are highly heterogeneous and require advanced solutions to process data and convey messages across the entire network, making this task challenging.
- **Service intelligence (SI)**: SI is of great importance for 6G networks, especially for human-centric communication services. These services cover a wide spectrum such as eHealth, localization service (indoor and outdoor), service for managing multiple devices, information extraction, security, etc. Incorporating AI features into these services can translate 6G networks to become more intelligent and user-centric. In the case of positioning systems, deep learning approaches can be applied to optimize performance, particularly in indoor or obstructed environments where normal GPS systems perform poorly. Additionally, intelligent IoT systems built on multi-modal infrastructures for personalized eHealth applications, including personalized health measurement, allow for disease prevention tailored to each individual's needs and preferences. Robust core 6G will require high-performance CNs

that will offer the much-needed bandwidth, speed, and reliability to accommodate complex data processing and communications required for such services. In summary, 6G networks will bring about the advanced SI, which will in large part improve user satisfaction through optimization, the identification of the needs of the users of these networks, their dynamic adaptation, enhanced efficiency, and responsiveness of communication services while improving connectivity and security in various applications.

3.6 HIGH SECURITY, SECRECY, AND PRIVACY

Up to now, most of the works in 4G and 5G communications have been focused to the classical CN parameters such as throughput, reliability, latency, and number of served users with little emphasis put on security, secrecy, and privacy issues. The two most powerful strategies to improve these metric are widely recognized and adopted: increasing the density of the network and deploying a higher frequency to transmit modulated symbols. It's still used for providing transmission security and secrecy in 5G networks that use traditional encryption algorithms based on Rivest-Shamir-Adleman (RSA) public-key cryptosystems. Under the influence of big data and AI technologies, RSA cryptosystems have already lost their security; nevertheless, new privacy protection mechanisms of its type are not fully developed during the 5G era. Therefore, advancements in security, secrecy, and privacy in 6G networks are anticipated to be achieved with these novel technologies (Gui et al 2020). PHYS security mechanisms and quantum key distribution using VLC (Visible Light Communication) must resolve the data security challenges in 6G networks. Advanced frontiers of the QCC (Quantum Key Distribution) models may excel in supplying a full-fledged coverage against various cyberassaults. However, communication data service providers have in fact legally built-up a colossal trove of user information, and instances of leak of private data took place from time to time. As anticipated, this will serve as a soft spot of the people-centric 6G technology and may lead to destructive situations in the absence of efficient precautionary action. To overcome this problem, 6G systems are ensured to achieve absolute anonymization, non-distributed, and untraceability through blockchain technology. On the other hand, recent research efforts have been driving toward actual real-life systems, with multiple access schemes, physical air interfaces, and 6G data center handovers. The network structure of 6G is introduced in 6G, where cellless topology, distributed resource allocation, and 3D super-connectivity are widely predicted to be present in 6G networks. Impact of future generation wireless networks on vertical industries; the chapter postulates that 6G may set off the evolution of the first-ever, barrier-breaking standard capable of entirely replacing the existing industry-specific communication standards, providing a holistic solution enabling uncomplex association of all requirements in vertical industries.

3.7 CONCLUSION

This chapter explores the essential properties of 6G communication with an overview of its key attributes in terms of security, ultra-reliable communication, and network resilience. It highlights the potentially matured techniques that will evolve around the 6G world with focus on security techniques and dependability methods. It sees trade-offs with 6G networks and analysis of issues such as secure data transport, reliability under dynamic scenarios, platform complexity as key challenges across the board. This section also discusses potential solutions, such as sophisticated encryption techniques, fault-tolerant network topologies, and novel risk-management strategies. The chapter consolidates key research findings, trends, and breakthroughs in the domain to assist in the roadmap of future 6G networks, particularly the protection of data integrity and email continuity. Finally, it concludes by summarizing the main insights and stressing the need for security and reliability in the design and deployment of 6G networks. It also requires future research directions to address remaining challenges and unleash the vision of 6G technologies.

REFERENCES

Brito, J. M. C., Mendes, L. L., and Gontijo, J. G. S. "Brazil 6G Project—An Approach to Build a National-Wise Framework for 6G Networks." Proceedings of the 2020 2nd 6G Wireless Summit (6G SUMMIT), Levi, Finland, 17–20 Mar. 2020, pp. 1–5.

Catak, F. O., et al. 2021. "Security Concerns on Machine Learning Solutions for 6G Networks in mmWave Beam Prediction." arXiv:2105.03905.

Corici, M., et al. 2021. "An Ultra-Flexible Software Architecture Concept for 6G Core Networks." Proceedings of the 2021 IEEE 4th 5G World Forum (5GWF), Montreal, QC, Canada, pp. 400–405.

Dang, S., et al. 2020. "What Should 6G Be?" Nature Electronics, vol. 3, pp. 20–29.

Gui, G., et al. 2020. "6G: Opening New Horizons for Integration of Comfort, Security, and Intelligence." IEEE Wireless Communications, vol. 27, pp. 126–132.

Katz, M., Matinmikko-Blue, M. and Latva-Aho, M. 2018. "6Genesis Flagship Program: Building the Bridges Towards 6G-Enabled Wireless Smart Society and Ecosystem." Proceedings of the 2018 IEEE 10th Latin-American Conference on Communications (LATINCOM), Guadalajara, Mexico, pp. 1–9.

Khan, L. U., et al. 2020. "6G Wireless Systems: A Vision, Architectural Elements, and Future Directions." IEEE Access, vol. 8, pp. 147029–147044.

Kumari, A., Gupta, R. and Tanwar, S. 2021. "Amalgamation of Blockchain and IoT for Smart Cities Underlying 6G Communication: A Comprehensive Review." Computer Communications, vol. 172, pp. 102–118.

Li, G., et al. "SecCDV: A Security Reference Architecture for Cybertwin-Driven 6G V2X." IEEE Transactions on Vehicular Technology, vol. 71, 2021, pp. 4535–4550.

Li, W., et al. 2021. "Edge Caching in Blockchain Empowered 6G." China Communications, vol. 18, pp. 1–17.

Liu, J., et al. "Post-Quantum Secure Ring Signatures for Security and Privacy in the Cybertwin-Driven 6G." IEEE Internet of Things Journal, vol. 8, 2021, pp. 16290–16300.

Mousa, A., et al. "Utilizing the eSIM for Public Key Cryptography: A Network Security Solution for 6G." Proceedings ofthe 2020 2nd International Conference on Computer and Information Sciences (ICCIS), Sakaka, Saudi Arabia, 13–15 Oct. 2020, pp. 1–6.
Partala, J. 2021. "Post-Quantum Cryptography in 6G." 6G Mobile Wireless Networks, Springer, Berlin/Heidelberg, Germany, pp. 431–448.
Porambage, P., et al. 2021. "The Roadmap to 6G Security and Privacy." IEEE Open Journal of the Communications Society, vol. 2, pp. 1094–1122.
Porambage, P., et al. 2021. "6G Security Challenges and Potential Solutions." Proceedings of the European Conference on Networks and Communications (EuCNC), Porto, Portugal, pp. 1–6.
Qamar, F., et al. 2020. "Issues, Challenges, and Research Trends in Spectrum Management: A Comprehensive Overview and New Vision for Designing 6G Networks." Electronics, vol. 9, p. 1416.
Rizvi, S., et al. 2020. "Threat Model for Securing Internet of Things (IoT) Network at Device-Level." Internet of Things, vol. 11, p. 100240.
Sheth, K., et al. 2020. "A Taxonomy of AI Techniques for 6G Communication Networks." Computer Communications, vol. 161, pp. 279–303.
Strinati, E. C., et al. 2021. "Wireless Environment as a Service Enabled by Reconfigurable Intelligent Surfaces: The RISE-6G Perspective." arXiv:2104.0665.
Sun, Y., et al. 2020. "When Machine Learning Meets Privacy in 6G: A Survey." IEEE Communications Surveys & Tutorials, vol. 22, pp. 2694–2724.
Uusitalo, M. A., et al. 2021. "Hexa-X the European 6G Flagship Project." Proceedings of the 2021 Joint European Conference on Networks and Communications & 6G Summit (EuCNC/6G Summit), Porto, Portugal, pp. 580–585.
Vaigandla, K. K., et al. 2021. "A Survey on Future Generation Wireless Communications-6G: Requirements, Technologies, Challenges and Applications." International Journal of Advanced Trends in Computer Science and Engineering, vol. 10, pp. 3067–3076.
Wang, J., et al. 2022. "6G THz Propagation Channel Characteristics and Modeling: Recent Developments and Future Challenges." IEEE Communications Magazine.
Yu, Q., et al. 2020. "A Cybertwin-Based Network Architecture for 6G." Proceedings of the 2020 2nd 6G Wireless Summit (6G SUMMIT), Levi, Finland, pp. 1–5.
Zhao, Y., et al. 2020. "A Comprehensive Survey of 6G Wireless Communications." arXiv:2101.03889.

Chapter 4

Harnessing digital twin technology in smart transportation system

Wajahat Ali

4.1 INTRODUCTION

In recent years, rapid urbanization, increasing population density, and the intensification of environmental challenges have put immense pressure on global transportation systems (Duan et al. 2025; Ali et al. 2025a). Conventional transport infrastructure, often rigid and reactive, cannot address modern cities' dynamic needs (Gilrein et al. 2021). As nations strive to create more livable, efficient, and environmentally responsible urban environments, the role of smart transportation technologies becomes increasingly vital (Elassy et al. 2024). Among the cutting-edge solutions emerging in this domain, digital twin (DT) technology stands out as a transformative tool capable of reshaping the way of planning, monitoring, and optimizing transportation systems (Ali et al. 2025b). Figure 4.1 illustrates the interconnected components of a modern urban mobility network. It highlights the integration of DT technology, Internet of Things (IoT) sensors, real-time data analytics, and smart infrastructure.

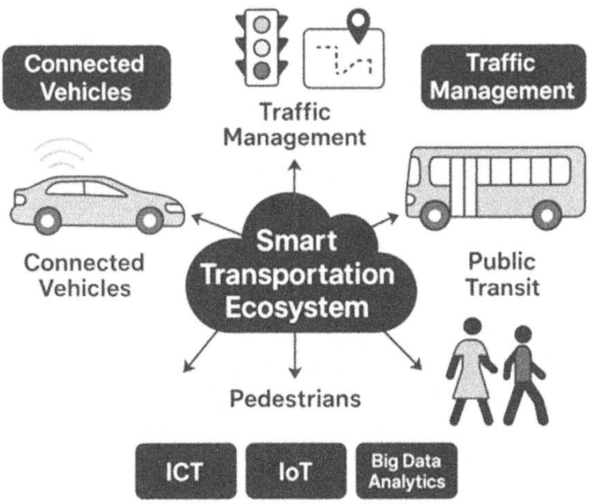

Figure 4.1 Overview of the smart transportation ecosystem.

DOI: 10.1201/9781003586951-4

A DT is a dynamic, digital representation of a physical entity or system that combines sensor data, computation models, and analytics to facilitate real-time monitoring, prediction, and optimization (Javaid et al. 2023). Ongoing synchronization between the physical and virtual components of DTs is crucial for achieving predictive maintenance, enhancing traffic efficiency, and supporting sustainable mobility strategies..This functionality is handy for urban planners and transport authorities trying to integrate their objectives with the United Nations (UN) Sustainable Development Goals (SDGs)—specifically SDG 9 (Industry, Innovation, and Infrastructure), SDG 11 (Sustainable Cities and Communities), and SDG 13 (Climate Action) (Wu et al. 2025).

A system based on information and communication technologies (ICTs) for the optimal management of infrastructure and transportation services aims to improve safety, reduce congestion, and minimize environmental impact (Chatti 2021). A cyber-physical system that mirrors the real-time status, working condition, or position of physical assets using data integration and simulation models. Transportation that meets society's mobility needs without compromising environmental, social, or economic sustainability (Ali et al., 2025d). Figure 4.2 visually represents how DTs function within smart transportation systems. The diagram illustrates the continuous feedback loop between the physical world (e.g., vehicles and infrastructure) and its digital counterpart through real-time data acquisition, simulation, and analytics. It shows key layers, including data sources, processing units, and decision-making modules. This conceptual flow enables predictive modeling, performance optimization, and adaptive control, supporting informed decision-making aligned with sustainability goals.

As urban populations continue to grow—expected to reach 68% of the global population by 2050, according to the United Nations—cities face mounting challenges in managing transportation systems that are efficient, inclusive, and environmentally sustainable (Adnan et al. 2024). Based on static

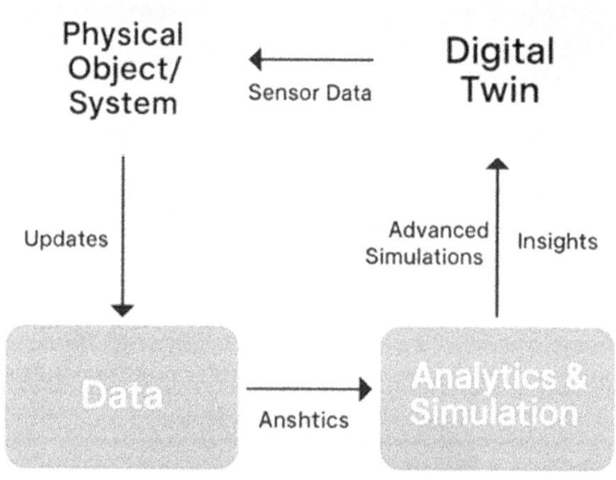

Figure 4.2 Conceptual diagram of DT technology.

models and historical information, the traditional transportation planning process no longer copes with the requirements of today's dynamic, complex urban environment. It has driven the evolution toward smart transportation systems that use real-time data, automation, IoT, and big data analytics to enhance mobility efficiency and quality of life in urban cities (Ali & Javaid 2023). DTs represent the transformative technology in this regard. DT is a digital emulation of a physical system or asset that is constantly updated using real-time data and simulations (Singh et al. 2021). In transportation, DTs are used for infrastructure (e.g., roads and bridges), vehicles, and complete mobility ecosystems to monitor performance, predict demand, and simulate system-wide reactions in real time (Ali & Javaid 2025).

In this research, we investigate the implementation of DT intelligence transport systems as a driver for sustainable development. By using the data-driven and real-time insights supported by DTs, cities can also plan adaptive and more sustainable transportation systems that increase access, decrease carbon footprints, diminish congestion, and contribute to the overall resilience of urban areas. The research also examines how DTs can help in decision-making processes, provide effective infrastructure management, and pave the way for transport systems' inclusion, safety, and sustainability. Through a comprehensive analysis of current applications, technological frameworks, and case studies, this research aims to establish a conceptual and practical foundation for the role of DTs in achieving sustainable transport objectives. Ultimately, the study seeks to contribute to the growing body of knowledge guiding the development of intelligent, responsive, and sustainable urban mobility systems in alignment with global sustainability agendas. The motivation for this study arises from the pressing global commitment to achieving the UN SDGs, especially SDG 9: Build resilient infrastructure, promote inclusive and sustainable industrialization, and foster innovation; SDG 11: Make cities and human settlements inclusive, safe, resilient, and sustainable; and SDG 13: Take urgent action to combat climate change and its impacts.

Although DT applications have been explored in manufacturing, aerospace, and healthcare industries, their application in transportation—specifically in sustainability and SDG alignment—remains underexplored. Existing studies focus on isolated use cases such as traffic monitoring or predictive maintenance. However, few have examined the integrated role of DTs in driving policy-level outcomes related to sustainable development. Moreover, the literature lacks a framework that links real-time data-driven transportation systems with strategic sustainability objectives. This chapter aims to bridge the gap between emerging DT technologies and the global sustainability agenda. The core contributions of this research include the following:

- A comprehensive review of DT applications in smart transportation systems across various global contexts.
- A proposed conceptual framework connecting DT-enabled transport solutions with relevant SDG targets and indicators.

- An exploration of real-world case studies that demonstrate the practical impact of DT integration in enhancing urban mobility and reducing carbon footprints.
- A set of strategic recommendations for city planners, policymakers, and transportation authorities to implement DT-based solutions aligned with sustainability goals.

The remainder of this chapter is organized as follows: Section 4.2 provides a detailed literature review covering smart transportation, DTs, and sustainable urban mobility. Section 4.3 introduces the proposed conceptual framework of the study. Section 4.4 presents the methodology. Section 4.5 discusses the analysis findings Finally, Section 4.6 concludes the discussion of the study and future research directions.

4.2 LITERATURE REVIEW

4.2.1 Evolution of smart transportation systems

Smart transportation systems have emerged as a central pillar of smart cities, leveraging ICT, IoT, and big data analytics to optimize mobility services. These systems aim to reduce congestion, improve safety, enhance public transit, and minimize environmental impacts (Elassy et al. 2024). According to Javaid et al. (2025), integrating intelligent transportation systems with real-time traffic data can significantly improve route optimization and travel efficiency. However, early smart transportation systems primarily focused on infrastructure-centric technologies—such as traffic light synchronization or GPS-based tracking—without system-level integration or predictive capabilities enabled by recent advances in cyber-physical systems (Zhu et al. 2018). Modern smart transportation moves beyond basic data collection to adaptive systems capable of simulation, feedback, and scenario planning, laying the foundation for DT applications. The focus on the evolution of smart transportation comes into play, given the increasing need for alignment between urban mobility solutions and the United Nations' SDGs. As cities grow and demand mobility increases, the need to address environmental degradation, decrease emissions, offer inclusive and resilient mobility systems, and do so urgently becomes paramount. Smart transportation systems using advanced technologies such as DTs can contribute to closing the sustainability gap by supporting more informed decisions, preventive maintenance, and system optimization in real time.

However, there is still a significant research gap about the full integration of the real-time DT models, which mimic, predict, and optimize the transportation ecosystems dynamically, despite the remarkable achievements that have been made in implementing intelligent transportation systems. Most of the literature covers detailed studies on specific technologies, such as IoT sensors or artificial intelligence in traffic management(AITM) rather than related to how a coherent DT framework can connect those various components so that they contribute to a synchronized ecosystem and intelligent mobility backdrop. Ali et al. (2024c) conducted cutting-edge research to illustrate the promise of DT-based transportation systems to predict the bottleneck, simulate the effects of interventions, and improve inter-modal integration. However, they also admit the need for more scalability and interoperability between city-level infrastructures. Bibri et al. (2024) studied AI-based edge computing embedded with DTs for real-time incident response in urban traffic systems. They also conducted a multi-city comparison and found that DT can further help achieve a 17% fuel consumption reduction in combination with eco-routing algorithms.

Werbińska-Wojciechowska et al. (2024) proposed predictive maintenance for public transit fleets via DT models, simulating the mechanical stress in diverse operational regimes to improve vehicle uptime and lifecycle performance. Zhu et al. (2018) investigated the integration of blockchain, used to secure the exchanged dataset in the DT-enabled transportation networks and to guarantee a transparent and verifiable data transaction, components required in an intersectoral trust and compliance perspective. They developed a hybrid DT model that integrated augmented reality (AR) interfaces with 3D-based traffic simulations, allowing planners and policymakers to visualize urban traffic interactively flows in real time. That adds to the ongoing discussion by combining recent trends in smart transportation research with an emphasis on DT technology. It also illustrates how integrating physical infrastructure with DTs can create a living urban mobility model, enabling ongoing learning, simulation-based policy testing, and predictive analytics. By surveying over 15 studies, this review forms the groundwork for considering technological trajectories and the potential to promote sustainable development with more innovative and holistic transportation systems.

4.2.2 DT technology: Concepts and capabilities

While initially developed for aerospace and industrial production, DT is now emerging into the urban mobility sector. Qian et al. (2022), a DT is characterized by a dynamic digital portrait of a physical entity or system that is to be constantly refreshed by real-time sensor data, advanced analytics, and simulation processes. At the heart of a DT is the potential to provide predictive (what will happen), prescriptive (what should be done), and descriptive (what is happening) insights. These capabilities fit DTs well, especially in complex, data-intensive environments like traffic systems. DTs provide hierarchical modeling of urban mobility networks in transportation that incorporates roads, vehicles, traffic signals, and pedestrian facilities. Through data ingestion from IoT sensors, GPSs, and smart cars, DTs are capable of (i) simulating traffic flows, (ii) forecasting congestion hotspots, and (iii) analyzing the impacts of different interventions (Aghaabbasi & Sabri, 2025). For example, the Virtual Singapore project has developed a full-fledged DT of the city-state, which is then employed for urban planning to model energy consumption, emergency egress pathways, and pedestrian-flow behavior. Also, in Helsinki, accessibility planning is improved through the 3D City Model Project, and support is given to carbon-neutral mobility transitions.

Fan et al. (2022) presented a hybrid DT model in a traffic real-time management system consisting of sensor networks, AI-based prediction, and edge computing to manage the dynamic traffic signal control action. Li et al. (2025) showed how Lisbon employed DTs in multimodal transportation optimization, achieving 12% savings in the average travel time and quantifiable emission reduction. These studies demonstrate how DTs may transcend static visualizations and become decision-support technologies actively influencing urban mobility. DTs have been successfully deployed for infrastructure monitoring and asset management. Armijo & Zamora-Sánchez (2024) demonstrated how in-traffic DTs (including stress, vibration, and load sensors) might monitor bridge structural health, leading toward predictive maintenance and cost savings. Chen et al. (2022) furthered this by utilizing machine learning in a DT setting to identify road surface deterioration and ensure timely resurfacing, enhancing road safety and minimizing resource wastage (Ali 2025).

Furthermore, despite these developments, the systematic integration of DT technology into the context of SDGs is still minimal in the literature and practice. Most applications are focused on technical optimization—vehicle flow and emissions—without explicitly considering broader sustainability considerations such as equity, resilience, or support of multiple modes. Armijo & Zamora-Sánchez's (2024) current cutting-edge mobility plans to steer society away from the social exclusion of low-income urban neighborhoods, while people-based planning tools like DT-based participatory planning tools are still recommended to simulate equitable mobility futures. Another difficulty is the absence of common paradigms to align DT case implementation with global sustainability metrics. Ali et al. (2024c) constructed a DT-based architecture to model the effects of policies on SDG-aligned targets such as zero carbon, public health, and access. However, they warned that inter-agency procedures and legal guidelines were necessary to extend such efforts. Tyagi et al. (2024) were able to make developing open data ecosystems and governance models key enablers of city-wide DT adoption with a balance in place between the advancement of technology and data privacy and citizen trust. DTs also provide an unparalleled opportunity for climate resilience and scenario planning.

Astarita et al. (2024) included climate models in an urban transportation twin to evaluate flood hazards and real-time evacuation procedures. They investigated the utility of artificial intelligence (AI)-driven DTs in predicting heat stress in public transportation systems and recommended service schedules and upgrading infrastructure. DT technology provides a breakthrough solution for controlling and maintaining the transportation system. It promisingly contributing to SDGs, as it can generate a real and dynamic urban mobility model. However, to fully realize this potential, four legacies must be addressed by future research around integration, inclusive design, and cross-sector collaboration frameworks that simultaneously make technology firm and fair.

4.2.3 Sustainable Development Goals and urban mobility

For anyone who does not know about it, the United Nations' 2030 Agenda for Sustainable Development offers a universal framework to guide us along a fairer, more inclusive, and environmentally healthier future. In this list of 17 goals, SDG 11 calls for inclusive, safe, resilient and sustainable cities, and human settlements, with a focus on safe and affordable access to transport systems (Yumnam et al. 2024). Critical, too, is SDG 13, which demands that we take action in handling climate change and its consequences, with transport still making up a large percentage of global greenhouse gas emissions. These twin motivations are interrelated concerning urban mobility, as new mobility options are crucial for tackling environmental and social equity imperatives.

For the last 20 years, literature has reported evidence that smart urban mobility systems can enable reaching the SDGs by ensuring better accessibility, limiting environmental footprints, and enhancing the quality of life. Ye et al. (2017) provided the theoretical ground for integrating transportation planning and sustainability by promoting transit-oriented development, non-motorized transportation, and demand management techniques. Alipour & Dia (2023) stress that integrated land use/transport planning is essential in meeting sustainability objectives by developing compact, walkable cities. Some researchers have investigated whether public transportation and SDG goals are related. Yumnam et al. (2024) analyzed policies in emerging countries. Their research showed that investment in bus rapid transit (BRT) and rail makes an important contribution to SDG 11 by improving equitable access to work and services. Moreover, they reinforce this argument by touting the positive environmental impact of mass transit and active mobility. They also demonstrated how collaborative mobility services can complement publicly owned infrastructure to decrease the system, enhance transport efficiency, and decrease vehicle kilometers traveled (VKTs), thus corresponding to SDG 11 and SDG 13.

However, these interventions continue to orient around what policies and technologies should be adopted rather than provide interactive tools that can be tracked, simulated, and made to respond to their impacts. This is where DT technology offers transformative potential. By enabling real-time modeling of urban mobility systems, DTs can assess the effectiveness of SDG-related initiatives under varying scenarios. Ali et al. (2025d) proposed a multi-domain DT architecture explicitly designed to align transportation strategies with SDG targets by simulating policy impacts across energy, emissions, equity, and efficiency. Berigüete et al. (2024) emphasized the role of digital governance and participatory platforms in supporting SDG-aligned urban DTs, highlighting the importance of data accessibility and citizen engagement. Beyond transportation planning, DTs are increasingly being explored for their capacity to support climate adaptation and disaster preparedness, both key components of SDG 13. Berigüete et al. (2024) demonstrated how transportation-focused DTs can model flood scenarios and extreme heat impacts to guide resilient infrastructure design and emergency service allocation. DTs hold significant potential to operationalize the SDGs within the transportation sector by bridging the gap between policy, planning, and real-time decision-making. As cities face mounting pressures from urbanization, climate change, and social inequality, the ability to simulate, test, and refine transportation interventions using DT platforms will become an essential asset for sustainable urban development (Ali et al. 2025b).

Despite the increasing adoption of smart transportation systems and the growing interest in DT technology, the existing literature reveals several critical gaps that must be addressed to advance sustainable urban mobility. One of the most significant gaps lies in the limited integration of DT frameworks with sustainability assessment models and SDG indicators. While numerous studies have explored the technical aspects of DTs in engineering and asset management, very few have connected these applications explicitly to environmental and social outcomes outlined in the SDGs. This disconnect hampers the ability of policymakers and city planners to quantify and track the broader impact of DT-enabled solutions on urban sustainability. Another gap pertains to the scarcity of real-world, system-level case studies demonstrating how DT technology can transform transportation decision-making and infrastructure planning. Most documented applications remain experimental or limited to specific components, such as vehicle monitoring or bridge maintenance. There is a clear need for comprehensive case studies that showcase end-to-end implementations of DTs in transportation systems, particularly in smart cities aiming to meet SDG targets such as reduced emissions, increased public transport use, and enhanced urban resilience.

Furthermore, little discussion is given to governance and policy structures that may facilitate widespread DT deployment, especially in low- and middle-income countries (LMICs). The deployment of DT systems typically demands substantial investment in digital infrastructure, a qualified labor force, and data governance systems, which are not yet sufficiently developed in developing countries. Unless these building blocks are laid, the upsides of DT technologies could be out of reach for the cities that most need them to meet the twin challenges of equitable and sustainable development. Third, interdisciplinary methods between transportation engineering, sustainability science, data science, and public policy are scarce. Dealing with complex urban challenges cannot rely on technological tools alone. It needs holistic models that include social, economic, and environmental factors. Future studies need to take a more comprehensive view that connects technical innovation with policy innovation and uses the full potential of DTs in fostering sustainable transportation systems. This chapter seeks to bridge these gaps by offering a coherent approach to linking DT technology with the strategic imperatives of the SDGs, backed by practical insights and takeaways from the field. To that end, doing so builds on the growing discussion of how new, advanced, digitally enabled technologies can construct smarter, greener, and more resilient urban mobility systems.

4.3 CONCEPTUAL FRAMEWORKS

4.3.1 Overview of DT functionality in transportation systems

The DT technique is a refined digital spinoff of physical systems that facilitates real-time emulation, reproduction, and control. A DT fuses information on infrastructure, vehicles, and users through a virtual dynamic digital transportation model. This "replica" of the world can update itself in real time with all the real data, and a complete understanding can be derived from how the actual system behaves The objective is to replicate the present condition and predict, simulate, and optimize operations for improved decision-making. The DT in mobility is based on three pillars: the physical part of the system (i.e., the road grid and public transportation), the virtual part (i.e., digital environment), and the data connectivity layer (sensors, IoT, and cloud). This triad ensures a smooth transition between the real and virtual worlds. Sophisticated modeling solutions, such as gazillion-model-based machine learning and AI-driven simulations, are integrated into the digital world to monitor system performance and user behaviors. Figure 4.3 shows how

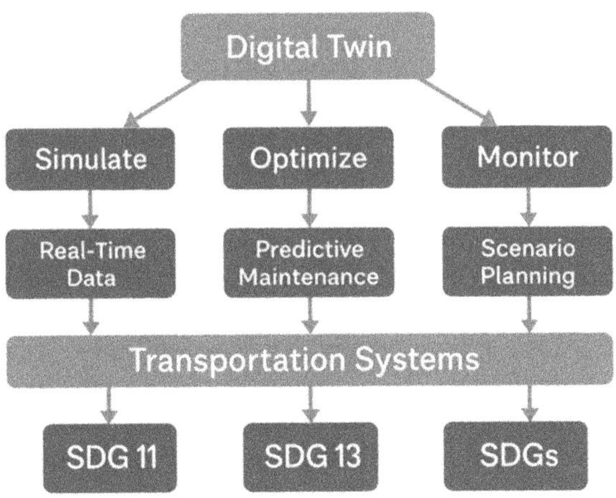

Figure 4.3 Conceptual framework linking DT capabilities to SDGs.

decision trees model, monitor, and optimize transportation systems. This figure depicts the primary operations—real-time data acquisition, predictive maintenance, and scenario planning—linked to the related SDGs such as SDG 11 (Sustainable Cities and Communities) and SDG 13 (Climate Action). This framework emphasizes the contribution of DTs to data-driven approaches for sustainable urban mobility.

DTs in transport have been engineered to work as an infinite loop. Live-field data is fed into the DT and processed for processing and analysis. The resulting insights are employed to suggest actions or to react in automation and fed back into the real-world system. This feedback mechanism guarantees that the conditions for traffic for each period will adjust to be consistent with optimal network performance measures. An interactive model of this sort moves beyond simple spatial query and data visualization; it is an inlet through which dynamic entities flow, allowing stakeholders to see what is happening in the world and make it happen. The DT is the control tower and the playground, where the policy impacts, design options, or infrastructure changes can be simulated before the physical implementation.

4.3.2 Simulation, optimization, and monitoring with DTs

4.3.2.1 *Simulation*

Simulation is an essential functionality of DTs, allowing planners to develop virtual environments as close to reality as possible and to use them to simulate transportation in the DT. From modeling traffic jams at rush hour to forecasting a new subway line's impact on city-wide commuting, DTs can perform many "what-if" analyses. These simulations consider meteorology, traffic flow, fuel usage, and user mode shifts. DTs also facilitate multiscale simulation ranging from individual vehicles to urban transit systems. Vehicle-to-infrastructure communication simulation can optimize traffic light timing, while multimodal transport simulation can decide bus-to-train connections or pedestrian walkability. The simulations are crucial for finding bottlenecks, estimating emissions, and planning for unexpected disruptions, such as natural disasters or protests. These simulations contribute to SDG 11 (Sustainable Cities and Communities) by providing evidence for addressing inclusive and resilient transportation system planning. They help achieve SDG 9 (Industry, Innovation, and Infrastructure) by maximizing the utilization and expansion of infrastructure. Urban development can be modeled to adhere to the long-range aims of sustainability without needing physical (trial-and-error) proofs. One also sees the forward simulation of scenarios (e.g., EV penetration, population growth, or travel demand) being of central importance for long-range planning. Planners can even evaluate the effects of proposed policy interventions or infrastructure investments long—even decades—before anything happens, allowing resources to be deployed more efficiently.

4.3.2.2 Optimization

Optimization in DT sequences consists of applying mathematical and computational methods to select the best operational policies under given constraints. This may mean reducing fuel consumption, increasing traffic flow, or decreasing passenger waiting time in transportation. These optimal plans are dynamic and updated based on real-time data and current situations. Real-time adaptation is one of the significant advantages of DT optimization. For example, when the DT receives information about a major accident from a sensor network in innovative city applications, it can instantly calculate an alternate traffic route and an optimized signal timing plan to reduce travel time. Logistics companies could employ DTs to optimize their delivery routes, lowering fuel consumption and the environmental impact.

These optimization features are consistent with SDG 13 (Climate Action) and SDG 12 (Responsible Consumption and Production). Optimized transport systems lead to less energy waste, reducing emissions and unnecessary moving of vehicles, which is suitable for the quality of the urban environment. Additionally, dynamic optimization makes it possible to immediately react to fleet management, which can be very important in emergencies (e.g., ambulances), public transport, and logistics. Further optimization is possible with the integration of AI and machine learning algorithms. These algorithms can be trained on historical data to forecast travel demand, schedule transit more efficiently, and minimize time when transit is unavailable. This results in operational efficiency, cost savings, and excellent environmental benefits—elements of a sustainable smart transportation plan.

4.3.2.3 Monitoring

Monitoring DT consists of continuously expressing the transport resources, the state of infrastructure, and the system's performance. It integrates with IoT, GPS, surveillance, and mobile applications. The information collected is then analyzed and presented on interactive dashboards, which will keep decision-makers up-to-date at all times. Real-time monitoring ensures real-time anomaly prevention (like car breakdowns, signal failures, bursts in congestion, etc.). During an emergency, early notification from the DT system can change vehicle rerouting and inform emergency services or notify commuters using digital signboards or mobile applications. This capability enhances the safety and reliability of urban mobility systems.

Environmental monitoring also contributes to SDG 3 (Good Health and Well-Being) concerning transportation systems' safe and accident-free functioning. It contributes to SDG 11 by increasing resistance to shocks, allowing for more predictive infrastructure maintenance, prolonging the lifespan, and cutting resources. Predictive maintenance, a key extension of real-time monitoring, uses the behavior of the data to predict when parts or systems will go down and to intervene before that happens. For instance, vibration sensors on a bridge or a metro line can warn the operators of structural fatigue, leading to some prevention measures and avoiding significant failures. At its core, real-time monitors reconfigure our transportation systems into responsive, adaptive systems of systems. Such ecosystems are more resilient and capable of responding effectively to the uncertainties of urban growth, climate change, and technological disruption, as well as delivering services sustainably and equitably.

4.3.3 Linking DT features to Sustainable Development Goals

DTs offer distinct features—real-time data collection, predictive analytics, simulation modeling, and system optimization—that align with the United Nations' SDGs. Each feature supports a specific aspect of sustainable development, from infrastructure resilience to environmental protection and public health. The linking of DT features to SDGs is represented in Table 4.1.

The conceptual framework outlined in this chapter demonstrates the transformative potential of DT technology in advancing innovative transportation systems that align with the SDGs. Through simulation, optimization, and continuous monitoring, DTs improve transportation networks' operational efficiency and resilience and serve broader sustainability objectives, such as reducing carbon emissions, enhancing safety, and promoting equitable mobility access. Integrating real-time data, predictive analytics, and scenario modeling enables stakeholders to make informed, future-proof decisions grounded in evidence. By mapping these technological capabilities directly to targeted SDGs—such as SDG 3, SDG 9, SDG 11, SDG 12, and SDG 13—this chapter highlights how digital innovation can be purposefully directed to solve pressing urban challenges. As cities grow and transportation demands increase, adopting DTs is a critical enabler of sustainable, innovative, and inclusive mobility ecosystems worldwide.

Table 4.1 Linking DT features to Sustainable Development Goals

DT feature	*Function in smart transport*	*Linked SDGs*
Real-Time Data Integration	Enables responsive traffic control, public transport scheduling	SDG 11, SDG 9
Predictive Maintenance	Anticipates infrastructure failures, extends asset lifespan	SDG 3, SDG 11
Scenario Planning and Simulation	Supports urban planning, emergency preparedness, and congestion modeling	SDG 11, SDG 13
Optimization Algorithms	Improves energy efficiency, reduces greenhouse gas emissions	SDG 12, SDG 13
User Behavior Analytics	Promotes sustainable transport choices like carpooling or cycling	SDG 3, SDG 11

4.4 RESEARCH METHODOLOGY

4.4.1 Research design

This study adopts a mixed-method approach, combining qualitative and quantitative methodologies to assess the role of DT technology in smart transportation systems and their contribution to achieving SDGs. The primary research design involves a case study analysis focusing on a smart city project implementing DT solutions for urban transportation management. This study incorporates simulation-based analysis, system modeling, and a meta-analytical review of existing research, offering a holistic understanding of the DT ecosystem within urban mobility frameworks. A methodology was selected due to its effectiveness in exploring real-world implementations in-depth, allowing for context-rich analysis of the challenges, innovations, and outcomes associated with DT integration. Alongside this, simulation modeling provides a replicable framework to test the performance and sustainability of DT-enhanced systems under various scenarios. Additionally, insights from stakeholder interviews—including urban planners, transportation engineers, IT specialists, and government officials—were gathered to assess practical challenges and institutional readiness.

4.4.2 Data collection

The data collection strategy consisted of three streams.

4.4.2.1 Primary qualitative data

In-depth interviews were conducted with 16 key stakeholders, including representatives from municipal transport authorities, DT software vendors, policymakers, and sustainability experts. Each interview lasted 30–45 minutes and followed a semi-structured format to balance consistency with response flexibility.

4.4.2.2 Quantitative system simulation

An innovative city project transport model was constructed using AnyLogic software to simulate key parameters—traffic flow, congestion rates, and energy consumption—before and after DT integration. Real-world data collected from IoT sensors and open-access city dashboards were input into the model.

4.4.2.3 Secondary data for meta-analysis

A meta-analysis was conducted on existing studies on DT applications in transportation systems. Using academic databases (Scopus, IEEE Xplore, ScienceDirect, and SpringerLink), 52 articles were initially identified based on keywords like "DT," "smart transportation," "urban mobility," and "sustainable development." After applying rigorous inclusion and exclusion criteria, 38 studies were retained for final analysis.

4.4.3 Inclusion and exclusion criteria

A rigorous set of inclusion and exclusion criteria was established to ensure the quality, relevance, and consistency of the studies selected for the meta-analysis. The inclusion criteria were defined to capture studies specifically addressing the intersection of DT technologies and sustainable smart transportation systems. Only peer-reviewed journal articles published between 2015 and 2024 were considered, ensuring recency and academic rigor. Furthermore, studies had to focus explicitly on applying DTs in transportation systems—such as public transit, traffic flow, or logistics networks—and include quantifiable outcomes relevant to sustainability, system efficiency, or urban mobility. Only English-language publications were included to maintain consistency in analysis and interpretation.

Unlike the latter, the exclusion criteria were defined to reject literature that did not meet the basic goals defined for the research. Only articles on product-level DTs (e.g., as employed in mechanical simulations unrelated to transport systems) were considered irrelevant. Studies that were not empirical or did not report measurable results (e.g., editorial, opinion, or pure-theory pieces) were also excluded. Articles were excluded if they were found to duplicate, have no full-text access, or were non-peer-reviewed, like conference posters, blogs, or white papers, to maintain the scientific integrity of the meta-analysis. A total of 52 studies were initially identified during a two-step screening with the title/abstract and then the complete text, from which 38 studies were finally included. This refined dataset served as the foundation for the meta-analytical component of the research, enabling a focused and evidence-driven exploration of how DT technologies contribute to sustainable urban transportation and support the achievement of the SDGs.

4.4.4 Data analysis and model estimation

The meta-analysis of the 38 studies was conducted using Hedges' g method because this method corrects for small sample bias and produces reliable effect size estimates from studies with different designs. This approach analyzed interventions on performance parameters such as fuel consumption, emission reduction, travel time, and user satisfaction. Results were also stratified by class of transportation system (car only, bus only, rail only, and mixed mode) and geographic region. Simulation results of the smart city case study were used to cross-verify the results from a meta-analysis. Variables such as average vehicle speed, public transit punctuality, and CO_2 emissions per trip were monitored across different policy scenarios (e.g., with vs. without predictive maintenance or real-time routing). These simulations were calibrated using baseline data from the city's transport control center and aligned with SDG indicators.

4.4.5 Ethical considerations

All stakeholder interviews were conducted with informed consent and complete anonymity. Institutional ethical approval was secured from the host university. Participants had the right to withdraw at any time, and all data was stored securely using encrypted digital platforms.

4.5 RESULTS AND DISCUSSION

The results of this study provide quantitative insights into the impact of DT technology on smart transportation systems with a specific focus on sustainability outcomes. The meta-analytical component, estimated using Hedges' method, evaluates fixed and random effects models. This dual modeling approach ensures that the findings account for variability across different studies while establishing a generalizable understanding of DT's role in sustainable transportation. Below, we discuss the interpretation of the omnibus and heterogeneity tests presented in Table 4.2.

The Omnibus Test of Model Coefficients reveals a statistically significant result ($Q = 42.786$, $df = 2$, $p < .001$), indicating that at least one of the predictors or effect sizes in the model contributes meaningfully to explaining the variability in outcomes across the included studies. This supports the inclusion of fixed effects, such as transportation type, region, or DT feature (e.g., predictive maintenance or real-time optimization), as significant contributors to the effectiveness of the intervention. The Test of Residual Heterogeneity ($Q = 96899.294$, $df = 38$, $p < .001$) indicates substantial residual variation not explained by the model's fixed effects alone. The high Q-statistic and significant p-value strongly support using a random effects model, as the estimated effects vary widely across studies. This level of heterogeneity is expected given the diverse contexts (geographical, technological, and operational) in which DTs have been applied. These findings suggest that while particular DT features consistently strongly impact sustainable transportation performance, the overall effect is also influenced by study-specific factors that should be accounted for in broader implementation strategies.

Table 4.2 Fixed and random effects

Test	*Q*	*df*	*p*
Omnibus Test of Model Coefficients	42.786	2	<.001
Test of Residual Heterogeneity	96899.294	38	<.001

Note: *p*-values are approximate.

Note: The model was estimated using the Hedges method.

Table 4.3 Coefficients

Parameters	*Estimate*	*Standard error*	*z*	*p*	*95% Confidence interval*	
					Lower	*Upper*
intercept	–234.269	78.483	–2.985	0.003	–388.093	–80.446
Standard deviation	5.029	0.964	5.216	<.001	3.139	6.919
Sample size	5.457	1.649	3.310	<.001	2.226	8.689

Note: Wald test.

Table 4.3 presents statistically significant coefficient estimates based on the Wald test, demonstrating that all included predictors—intercept, standard deviation, and sample size—meaningfully contribute to the model. The intercept value of –234.269 ($p = .003$) represents a baseline effect when other variables are held constant, though its practical interpretation is limited. The standard deviation (estimate = 5.029, $p < .001$) highlights considerable variability in DT effectiveness across studies, supporting the use of a random effects model. Additionally, the positive and significant sample size parameter (estimate = 5.457, $p < .001$) indicates that studies with larger sample sizes report greater DT impact, likely due to increased robustness and data quality in broader-scale implementations.

Table 4.4 presents essential model fit indices estimated using the Hedges estimation method, confirming the robustness of our analytical framework. The log-likelihood value of –193.441, while not directly interpretable alone, serves as a benchmark indicating the model's capacity to explain the observed data. The deviance of 359.627 further suggests a reasonable model-data alignment, underscoring the model's appropriateness. Fit indices such as the AIC (394.883), AICc (395.994), and BIC (401.737) reveal that our model maintains a sound balance between explanatory power and complexity. The closeness between AIC and AICc supports the model's suitability for smaller or moderately sized datasets, such as our sample of 38 studies. These metrics reinforce the validity of our findings and confirm that DT applications in smart transportation systems can be rigorously assessed using this modeling approach, supporting their role in advancing sustainable development objectives.

Table 4.4 Fit measures

Parameters	*HE*
Log-likelihood	–193.441
Deviance	359.627
AIC	394.883
BIC	401.737
Corrected Akaike Information Criterion (AICc)	395.994

Table 4.5 Residual heterogeneity estimates

		95% Confidence interval	
	Estimate	*Lower*	*Upper*
τ^2	788.610	526.145	1310.274
τ	28.082	22.938	36.198
I^2 (%)	99.999	99.998	99.999
H^2	88931.750	59333.905	147759.297

Table 4.5 provides a comprehensive assessment of residual heterogeneity in our meta-analysis evaluating the impact of DT technology in smart transportation systems on SDGs. These heterogeneity statistics offer crucial insight into the extent of variability in effect sizes that cannot be explained by the model's predictors alone. The τ^2 estimate of 788.610, with a 95% confidence interval ranging from 526.145 to 1310.274, quantifies the between-study variance in true effect sizes. This relatively high value indicates substantial differences across the studies, reflecting the diverse implementation contexts, transportation technologies, and DT applications considered in the analysis. The τ value, the square root of τ^2, is 28.082, with its confidence interval spanning from 22.938 to 36.198. Tau provides a more intuitive measure of heterogeneity on the same scale as the outcome and further underscores the variation between study-level effects. The I^2 value of 99.999% reveals that nearly all the observed variation in effect sizes is due to true heterogeneity rather than sampling error. Such an extremely high percentage strongly suggests that the impact of DT technology on transportation and SDGs is highly context-dependent, warranting detailed exploration of moderating factors. Lastly, the H^2 value of 88931.750 (confidence interval: 59333.905 to 147759.297) supports this conclusion, showing that the total variability is vastly greater than expected under a homogenous-effects model.

Table 4.6 presents the covariance estimates between the model parameters—intercept, standard deviation, and sample size. The diagonal elements show the variances of each parameter: the intercept has a variance of 6159.531, the standard deviation has a variance of 0.930, and the sample size has a variance of 2.719. Off-diagonal values represent the covariances between the parameters. The covariance between intercept and standard deviation is 3.448, suggesting a small positive relationship between these two parameters.

Table 4.6 Parameter covariances

Parameters	*Intercept*	*Standard deviation*	*Sample size*
intercept	6159.531	3.448	−128.869
Standard deviation	3.448	0.930	−0.186
Sample size	−128.869	−0.186	2.719

The covariance between intercept and sample size is –128.869, indicating a negative relationship between the intercept and sample size, while the covariance between standard deviation and sample size is –0.186, showing a very weak negative relationship. These covariances help understand the model parameters' interdependence and provide insights into how changes in one parameter might influence the others.

4.6 CONCLUSIONS AND FUTURE RESEARCH DIRECTIONS

This study investigates the disruptive effects of DT technology on the sustainability of transport systems. Simulation, optimization, and monitoring of transport networks are a few ways DTs can contribute to increased operational efficiency, lowered environmental impact, and enable more sustainable urban mobility. According to our results, adopting DT technologies into smart transport can have significant potential in predictive maintenance, real-time optimization, and scenario planning for achieving SDGs, especially sustainable cities, climate action, and efficient management of infrastructures. The meta-analysis determined that specific attributes of DTs, including real-time monitoring of data and predictive analytics, are their characteristics to enhance transportation system performance routinely. Nevertheless, the analysis also showed a high degree of residual heterogeneity, indicating that the effect of DTs varies widely based on local situation, technology infrastructure, and policy context. However, the strength of DT effects was determined by sample size, geographical location, and transportation modality, underscoring the need to employ place-specific interventions in applying technologies. DT technology has the potential to provide significant benefits. However, several challenges must be mitigated to ensure optimal effectiveness in promoting transport sustainability. One of the significant difficulties is the cost of this solution.

While the infrastructure (decisions based on fields) and out (sensors and communication protocols) are necessary to support and deploy DT systems, it can be costly for cities with small budgets. The technical difficulty and the requirement of special expertise to build and run DT systems can challenge the generalization. Another limitation is that the data source is arranged around DTs. The quality and consistency of the simulated results were significantly influenced by the accuracy and reliability of the data inputted into the models. Obtaining real-time, meso-level traffic data in a transportation network can be difficult and costly. Moreover, privacy and security concerns exist, especially regarding sensitive data on transportation habits, user actions, and the environment. People need to trust that the security protocols are in order to protect sensitive information.

Another challenge is the interoperability of DT systems with established transportation systems. Several cities still use legacy systems that may not be compatible with modern DT applications and may need to be upgraded or replaced. This may slow down the adoption chain and increase the cost of integration. There is also a need to work together across disparate stakeholders (government, private sector, and academia) to innovate and standardize frameworks to implement and scale technologies. In the future, DT technology will be used in sustainable transportation prospects. Future progress in AI, IoT, and machine learning will further improve DT systems' abilities to simulate reality, predict outcomes, and optimize transportation in real time. Additionally, when cities become more innovative and sustainable, combining DT with other smart city technologies, including energy management systems, waste management, and urban planning tools, will offer a more comprehensive and concerted solution for sustainability.

Furthermore, the future of DTs in transportation shows excellent potential for overcoming broader global issues, including climate issues and urban congestion. Using DTs for Sustainable Urban Mobility, cities are poised to design transportation systems that decrease carbon emissions and support public transportation and low-emission cars. With growing public sensitivity to environmental problems, there will be an increasingly higher demand for green transportation systems, increasing the popularity of DT technology. Despite the challenges, incorporating the DT into transportation can have a tangible sustainability benefit in urban mobility. By addressing the barriers to implementation, fostering cross-sector collaboration, and embracing future technological advancements, cities can pave the way for more sustainable, efficient, and resilient transportation systems.

REFERENCES

Adnan, N., M. F. Rashed, and W. Ali. 2024. "Embracing the Metaverse: Cultivating Sustainable Tourism Growth on a Global Scale." Current Issues in Tourism 28(20): 1–20.

Aghaabbasi, M., and S. Sabri. 2025. "Potentials of Digital Twin System for Analyzing Travel Behavior Decisions." Travel Behaviour and Society 38: 100902.

Ali, W. 2025. "Optimizing the E-Waste Management in India: A Sustainable Mathematical Modeling Approach to Circular Economy." Quality & Quantity:1–32.

Ali, W., and S. Javaid. 2023. "Enhancing Fermatean Fuzzy Transportation Problems: An Innovative Score Function-Based Optimization Strategy." International Research Journal of Economics and Management Studies (IRJEMS) 2 (4): 546–558.

Ali, W., and S. Javaid. 2025. "A Solution of Mathematical Multi-Objective Transportation Problems Using the Fermatean Fuzzy Programming Approach." International Journal of System Assurance Engineering and Management 16(2): 1–19.

Ali, W., S. Javaid, and U. Akbar. 2025d. "A Solution of Mathematical Multi-Objective Transportation Problems under Uncertain Environments." Palestine Journal of Mathematics,14 (1):987–1014

Ali, W., M. Khalid, N. A. Khan, and S. Javaid. 2025c. "Integrating Fermatean Fuzzy and Neutrosophic Goal Programming for Multi-Objective Healthcare Optimization under Uncertainty." Life Cycle Reliability and Safety Engineering: 1–21.

Ali, W., S. Sadia, F. Ahmad, S. Javaid, and N. Abidi. 2025a. "A Solution Programming Approaches on the Multi-Objective Capacitated Fractional Transportation Problem." International Journal of System Assurance Engineering and Management 16(2): 1–22.

Ali, W., S. Sadia, S. Javaid, N. Abidi, M. Nabeel, M. S. Javaid, and A. Nusrat. 2025b. "Solving the Multi-Objective Linear Plus Linear Fractional Programming Problem Using Different Mathematical Programming Approaches." Palestine Journal of Mathematics 14 (1): 1015–1040.

Alipour, D., and H. Dia. 2023. "A Systematic Review of the Role of Land Use, Transport, and Energy–Environment Integration in Shaping Sustainable Cities." Sustainability 15 (8): 6447.

Armijo, A., and D. Zamora-Sánchez. 2024. "Integration of Railway Bridge Structural Health Monitoring into the Internet of Things with a Digital Twin: A Case Study." Sensors 24 (7): 2115.

Astarita, V., G. Guido, S. S. Haghshenas, and S. S. Haghshenas. 2024. "Risk Reduction in Transportation Systems: The Role of Digital Twins According to a Bibliometric-Based Literature Review." Sustainability 16 (8): 3212.

Berigüete, F. E., J. S. Santos, and I. Rodriguez Cantalapiedra. 2024. "Digital Revolution: Emerging Technologies for Enhancing Citizen Engagement in Urban and Environmental Management." Land 13 (11): 1921.

Bibri, S. E., J. Huang, S. K. Jagatheesaperumal, and J. Krogstie. 2024. "The Synergistic Interplay of Artificial Intelligence and Digital Twin in Environmentally Planning Sustainable Smart Cities: A Comprehensive Systematic Review." Environmental Science and Ecotechnology:20 100433.

Chatti, W. 2021. "Moving Towards Environmental Sustainability: Information and Communication Technology (ICT), Freight Transport, and CO_2 Emissions." Heliyon 7 (10): e08123.

Chen, K., M. Eskandari Torbaghan, M. Chu, L. Zhang, and A. Garcia-Hernández. 2022. "Identifying the Most Suitable Machine Learning Approach for a Road Digital Twin." Proceedings of the Institution of Civil Engineers – Smart Infrastructure and Construction 174 (3): 88–101.

Duan, Z., L. Huang, Z. Zhu, S. Long, and Y. Liu. 2025. "Transformation and Inequity of Urban Green Space in Guangzhou: Drivers and Policy Implications under Rapid Urbanization." Sustainability 17 (5): 2217.

Elassy, M., M. Al-Hattab, M. Takruri, and S. Badawi. 2024. "Intelligent Transportation Systems for Sustainable Smart Cities." Transportation Engineering 16: 100252.

Fan, B., Z. Su, Y. Chen, Y. Wu, C. Xu, and T. Q. Quek. 2022. "Ubiquitous Control over Heterogeneous Vehicles: A Digital Twin Empowered Edge AI Approach." IEEE Wireless Communications 30 (1): 166–173.

Gilrein, E. J., T. M. Carvalhaes, S. A. Markolf, M. V. Chester, B. R. Allenby, and M. Garcia. 2021. "Concepts and Practices for Transforming Infrastructure from Rigid to Adaptable." Sustainable and Resilient Infrastructure 6 (3–4): 213–234.

Javaid, S., W. Ali, S. Sadia, and N. Abidi. 2025. "Optimizing the Electrical Vehicle Parking and Charging Assignments: A Balanced Approach Using Mathematical Modeling." Life Cycle Reliability and Safety Engineering 14(2): 1–16.

Javaid, M., A. Haleem, and R. Suman. 2023. "Digital Twin Applications Toward Industry 4.0: A Review." Cognitive Robotics 3: 71–92.

Jing Jin, Tong Liu, Yilei Wang, and Chunpeng Wang, "Isea: An Interesting Application of Chemistry Education Based on AR," in Proceedings of the 2019 International Conference on Virtual Reality and Visualization (ICVRV 2019) (2019), 283–285.

Li, W., B. Wang, R. Sun, L. Ai, and Z. Lin. 2025. "Energy-Efficient Multimodal Mobility Networks in Transportation Digital Twins: Strategies and Optimization." Energy 318: 134587.

Qian, C., X. Liu, C. Ripley, M. Qian, F. Liang, and W. Yu. 2022. "Digital Twin—Cyber Replica of Physical Things: Architecture, Applications and Future Research Directions." Future Internet 14 (2): 64.

Singh, M., E. Fuenmayor, E. P. Hinchy, Y. Qiao, N. Murray, and D. Devine. 2021. "Digital Twin: Origin to Future." Applied System Innovation 4 (2): 36.

Tyagi, A. K., S. Kumari, and T. Surve. 2024. "Integration of Digital Twin and Blockchain for Smart Cities." In Digital Twin and Blockchain for Smart Cities, 81–100.

Werbińska-Wojciechowska, S., R. Giel, and K. Winiarska. 2024. "Digital Twin Approach for Operation and Maintenance of Transportation System—Systematic Review." Sensors 24 (18): 6069.

Wu, D., A. Zheng, W. Yu, H. Cao, Q. Ling, J. Liu, and D. Zhou. 2025. "Digital Twin Technology in Transportation Infrastructure: A Comprehensive Survey of Current Applications, Challenges, and Future Directions." Applied Sciences 15 (4): 1911.

Ye, Y., C. Wang, Y. Zhang, K. Wu, Q. Wu, and Y. Su. 2017. "Low-Carbon Transportation Oriented Urban Spatial Structure: Theory, Model and Case Study." Sustainability 10 (1): 19.

Yumnam, G., Y. Gyanendra, and C. I. Singh. 2024. "A Systematic Bibliometric Review of the Global Research Dynamics of United Nations Sustainable Development Goals 2030." Sustainable Futures 7: 100192.

Zhu, L., F. R. Yu, Y. Wang, B. Ning, and T. Tang. 2018. "Big Data Analytics in Intelligent Transportation Systems: A Survey." IEEE Transactions on Intelligent Transportation Systems 20 (1): 383–398.

Chapter 5

Education 4.0

The integration of technology in learning and teaching

Namratha B.

5.1 INTRODUCTION TO EDUCATION 4.0

The advancement of human capital is attainable solely through scholarly pursuits, vocational training, and employment, which collectively lead to economic prosperity, personal satisfaction, and societal cohesion. The term 'learning' has emerged as the most prevalent concept in contemporary educational discourse. This is indicative of the meticulous monitoring of learning outcomes; students are now positioned as active learners, with educators assuming the role of facilitators. This shift has transformed adult education into a continuous learning journey, influenced by customized neoliberal policies. Learning is recognized as an individualized and complex progression, occurring often without conscious reflection on this nuanced and ineffable evolution. Research indicates that the learning processes during childhood and adolescence are minimally impacted by intentional and reflective pedagogical practices. Engagement with the world serves as the primary catalyst (De Souza and Debs 2024).

Economic and social imperatives are the principal factors that have facilitated the seamless transition witnessed during the digital revolution over the past five decades. Each revolutionary phase is characterized by advancements in energy and technology as shown in Figure 5.1. The initial wave of transformation was predicated on steam-driven industrialization. The subsequent wave was catalysed by the advent of industrial automation and electrical innovations. The integration of electronics and automated manufacturing delineates the third revolution. The contemporary fourth industrial revolution is characterized by the seamless interconnection between the physical and virtual realms. The term 'Industry 4.0' encapsulates the transformations that have taken place within the industrial sector since the mid-1990s, signifying the onset of the fourth industrial revolution (Hussin 2018).

A pedagogical figure is regarded as the central entity of knowledge from the initial stage of 1.0 to the advanced stage of 3.0 within the educational paradigm. The introduction of Smart Education 2.0 facilitated the utilization of portable calculators and correspondence-based remote learning

DOI: 10.1201/9781003586951-5

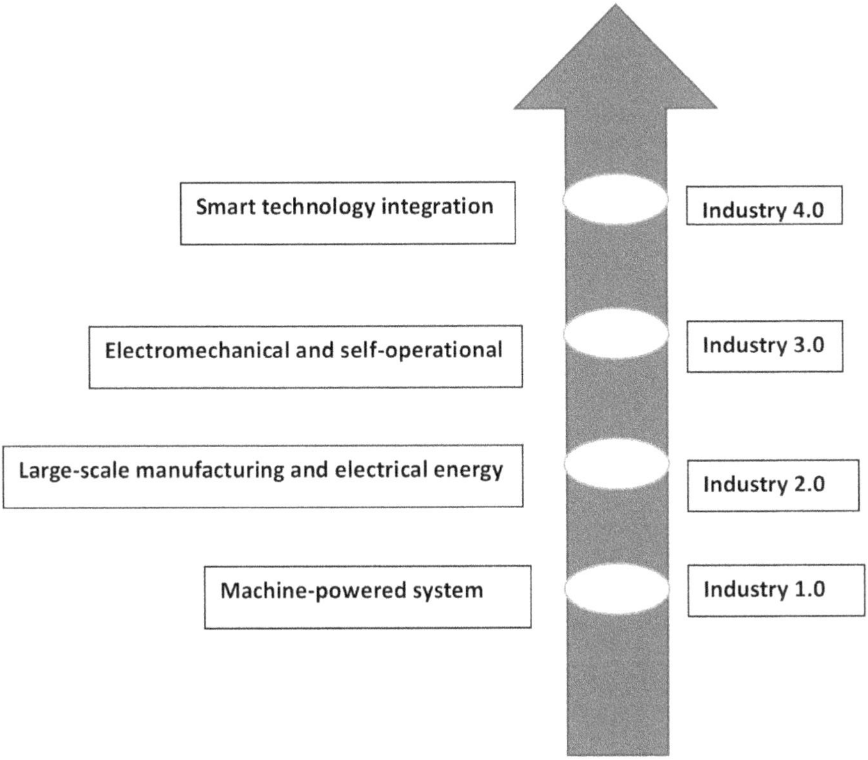

Figure 5.1 Major milestones of industrial revolution.

methodologies. The educational framework of Web 3.0 emphasized the integration of digital devices, online networks, digital pedagogy, and experiential learning methodologies. Subsequently, Education 4.0 emerged, which fundamentally relied on students for the learning process, with educational management systems playing crucial roles in the implementation of active methodologies for knowledge acquisition. Investigations into Education 4.0 underscore the significance of autonomous study, adaptive scheduling, practical skill development, and the incorporation of intelligent technologies within the educational process (Tikhonova and Raitskaya 2023).

The contemporary educational framework and modern competencies are strategically constructed through digital transformation. These elements serve as the connective tissue between digital technologies and both human and organizational dynamics. Innovative business models and strategic frameworks catalyse transformation within the organizational structure of work (Jugembayeva and Murzagaliyeva 2023).

Knowledge acquisition can occur in a myriad of contexts at any time, facilitated by self-directed e-learning tools in the first modality. The flipped classroom paradigm represents a compelling variant of interactive learning, whereby the theoretical components are engaged outside of designated classroom hours. The second modality of knowledge acquisition encompasses personalized learning tailored to individual students. Upon achieving proficiency in a specific subject area, students are presented with increasingly challenging tasks. This serves as a beneficial reinforcement, enhancing the overall learning experience and bolstering students' self-efficacy. In the third modality, learners are afforded the autonomy to select their preferred learning methodologies or tools, albeit the learning outcomes for a particular course are predetermined by academic institutions. In this scenario, instructors may implement hybrid educational models, inverted learning approaches, and technology-enhanced self-directed learning to facilitate the educational process (Gaines 2012; Moraes et al. 2022).

In the fourth modality, students are motivated to engage with subject matter through the completion of several short-term projects. This approach not only imparts knowledge but also cultivates essential skills. By engaging in such activities, students will develop competencies in deadline management, group dynamics, and systematic thinking, all of which are requisite for their future academic pursuits. In the fifth level, experiential learning opportunities, such as internships, mentorship initiatives, and collaborative projects, are advocated for knowledge acquisition. The progression of technology enables effective learning within specific domains, thereby creating additional opportunities for the acquisition of skills that necessitate human knowledge and interpersonal interaction (Pandey 2023).

In the sixth modality, students will confront data interpretation tasks, necessitating the application of their theoretical understanding to numerical data and the utilization of reasoning skills to draw logical inferences based on identified trends from provided datasets. The manual components of mathematical literacy are likely to become obsolete, as computational systems will undertake statistical analyses and forecast future trends. In the seventh modality, students will be evaluated using alternative assessment methods, rendering traditional assessment platforms potentially obsolete. Students' factual knowledge may be assessed throughout the educational process, while the practical application of that knowledge can be evaluated through their engagement in field-based projects (Dancsa et al. 2023).

According to the eighth modality, the perspectives of students will be duly acknowledged in the processes of curriculum design and revision. Their contributions assist curriculum developers in ensuring the curriculum remains relevant, current, and applicable. Ultimately, students will cultivate greater autonomy in their educational pursuits, thereby compelling educators to adopt a new role as facilitators who guide learners through their educational journeys. Education 4.0 represents a significant paradigm shift for educators. As they transition into the role of instructors, they are required to embrace multiple responsibilities during the transformation of the traditional teaching profession (De Souza and Debs 2024).

5.2 INTEGRATION OF ADVANCED TECHNOLOGIES IN EDUCATION

5.2.1 The role of digital platforms in pedagogy and evaluation

Learning technology enterprises are striving hard to find fresh approaches to provide education to folks who are devoid of proper academic infrastructure. Besides communication of information to one another, social web platforms play a vital role in job hunting and professional networking avenues. Standard lecture-based learning fails in rapid evaluation, extra involvement, and an instantaneous learning environment. On the contrary, e-learning resources fill the gaps created by such traditional pedagogical approaches (Haleem et al. 2022).

As wireless-enabled tech tools are gaining popularity day by day, learning centres can make use of this fact and put technology in the classroom. The utilization of projectors, computers, and other cutting-edge technical gear in the classroom may make studying fascinating and entertaining for students. Student learning can be more lively and involving when more and more technology is involved in classroom tasks such as team-based learning and seminar presentation. Participation can extend beyond verbal communication as well (Tian et al. 2024).

Digital platforms are good for the health of the environment by using less paper for information sheets and supplementary notes. E-learning also saves the time, helps in research, and cuts the expenses. This is a mode of sustainability in teaching and learning. In fact, as education is becoming globalized, integration of digital tools in teaching, learning, and evaluation has become mandatory (Nainggolan 2021).

During the time of COVID-19, pandemic developing countries struggled a lot and became reliant on digital technologies to save the education system of their country. Smart technologies helped in framing decision-making strategies, conceptual clarity, cognitive framework, and development of students. Moreover, there were bespoke learning plans designed to meet specific student goals (Ciarli et al. 2021).

As youngsters, nowadays, are so habituated for the use of electronic gadgets that learning in classrooms becomes livelier if technology is directly applied in learning. This can stimulate interest and foster active student participation without being distracted. Teachers should consider the widespread accessibility of mobile technology among students and make use of the same in study regimen (Alshammary and Alhalafawy 2023).

5.2.2 Learning instruments and digital tools

In the contemporary landscape characterized by rapid globalization and technological advancements, fundamental transformations in pedagogical processes and methodologies are imperative. A classroom enriched with digital rendering and elucidations will significantly conserve the lecturer's preparatory time and enhance the efficacy of subject matter dissemination. The widespread availability of augmented visualization prompts a resurgence of what several scholars deem a novel educational paradigm, referred to as the personal education framework (Dancsa et al. 2023).

Online resources within the educational setting can foster a more engaging and interactive atmosphere for learners. Digital media provides interactive audio-visual content that facilitates the dissemination of information to students through animation software and PowerPoint presentations in a dynamic manner. Digital technologies are instrumental in academic digital resources, particularly with the advent of e-books, which obviate the necessity of physically visiting libraries to procure texts (Bernacki, Greene, and Crompton 2019).

The emergence of innovative technologies such as smartphones, highly portable tablets, and notebooks necessitates that educators remain adept in integrating these new mediums into their instructional practices. Faculty members must possess comprehensive knowledge regarding the functionalities of contemporary digital mediums. Networked technologies enable learning to occur in diverse locations and at any time, transcending the confines of a traditional classroom setting limited to three hours per day. Computer-based tutorials provide visual learning advantages through animated videos, which are not typically afforded by classical education approaches. The capacity for visual comprehension among learners varies significantly, influenced by factors such as visual perception, socio-cultural background, heredity, visual skills, aesthetic sensibilities, and creativity. Digital devices fundamentally serve as adjuncts to the lecturer, designed to stimulate and enhance learners' engagement (García-Martínez et al. 2020).

Virtual laboratories present a viable resolution to the challenges posed by limited access to expensive and essential instruments required for technical practical studies. This approach enables laboratory equipment to be readily accessible online through an intuitive three-dimensional (3D) animated graphical interface, facilitating experiments analogous to those conducted in physical laboratories. By integrating virtual laboratories with a combination of web-based resources and diverse tools, including video clips, 3D animations, demonstrations, and simulators, faculty members can enhance interaction and conduct research effectively (Petko 2012).

E-learning encompasses all modalities of electronically facilitated education. It incorporates computer-based training (CBT), computer-based learning (CBL), Internet-based training (IBT), web-based training (WBT), online resource-based learning (ORBL), networked collaborative learning (NCL), and computer-supported collaborative learning (CSCL).

A Learning Management System/Learning Content Management System (LMS/LCMS) refers to software utilized for overseeing training programmes. It is also highly beneficial for disseminating course content via the Internet. The LCMS serves as software for the authoring and editing of educational materials. Macromedia Director is additionally advantageous in the authoring process (Sousa and Rocha 2018).

There exists a concept known as M-learning, which encompasses all forms of portable technologies, such as handheld computers, MP3 players, iPods, DVD players, talking books, personal digital assistants (PDAs), mobile phones, projector phones, handheld gaming consoles, and smartphones. Mobile learning involves the processes of authoring, content development, and delivery (Sousa and Rocha 2018).

A proficient educational instrument must prioritize a critical element known as learner motivation. Numerous studies exist within the domain of motivation and the learning process. The Web 2.0 technology directly addresses various dimensions of motivation, encompassing learner empowerment, agency, expression, novelty, collaboration, and interactivity. It adheres to the principles of motivation. Due to the emergence of social media platforms such as Flickr, Myspace, and YouTube, animation has increasingly become a significant factor within the educational milieu (Margaryan, Littlejohn, and Vojt 2010).

5.2.3 Flipped classroom: A paradigm shift in teaching

In the realm of higher education (HE), the instructional methodology referred to as the 'flipped classroom' (alternatively designated as the 'inverted classroom') has garnered significant scholarly attention. The underlying premise is that, instead of utilizing constrained classroom time for an educator to elucidate a concept (frequently via a lecture format), the educator may produce a video lecture, screencast, or vodcast that conveys the concept to students, thereby liberating precious class time for more interactive (and often collaborative) activities typically orchestrated by the educator. It is crucial to emphasize that this pedagogical strategy ought to extend beyond merely the 'take-home' video lecture (Strelan, Osborn, and Palmer 2020).

Research indicates that there are a few instructional methodologies that are less effective than traditional lecturing, and the continued prevalence of lectures as a foundational component of many curricula can primarily be attributed to two straightforward factors: tradition and financial considerations. Lectures are entrenched in our educational heritage and represent a relatively economical mode of content delivery (Chou, Chen, and Hung 2021).

One of the most significant challenges associated with conventional lecturing is the inherent misalignment of pacing. Students are more capable of comprehending the subject matter when their cognitive faculties are fully engaged. By delivering the content in the format of a video, podcast, or written text, students are afforded the opportunity to adhere to their individual learning rhythms, study at a pace that is conducive to their comprehension, and revisit the material as necessary (Joy et al. 2023).

The fundamental concept underlying the flipped classroom is a transformation in the pedagogical paradigm. In a conventional class—whether conducted in a physical or virtual environment—the instructor typically initiates the session by presenting the topic, usually through a lecture or by disseminating content via alternative means (such as guided reading, video presentation, or student experience sharing). Following this introduction, students may engage in practical application—applying the concepts imparted by the instructor to specific cases and problems, or by completing exercises or engaging in debates with their peers or the instructor. This experiential component of learning, and in many instances nearly all of it, transpires post-class, frequently in the form of assigned homework. Despite our accustomed familiarity with this instructional pattern and the purported rationale underpinning this approach, two principal issues arise (Campillo-Ferrer and Miralles-Martínez 2021).

Moreover, it is imperative that this approach incorporates both continuous and cumulative assessments, along with substantial live instructional sessions. Proponents of the reverse teaching model advocate for a plethora of advantages. Many of these advantages appear to be credible (e.g., increased time for more engaging instruction), particularly for educators operating within hybrid or blended educational environments that encompass a combination of on-site and remote teaching approaches; nonetheless, it is essential to acknowledge that the strategy is not without its constraints (Akçayır and Akçayır 2018).

There exist several disadvantages associated with this modality of learning. Firstly, the calibre of the video lecture may be significantly subpar; notwithstanding the possibility that an educator may excel in traditional classroom environments, he or she might not deliver a video that meets instructional and/or technical quality standards. Secondly, presuming that all students possess the capability to access the video lectures on their personal computing devices, the contexts within which they engage with the video may not be conducive to the effective assimilation of concepts. It can be contended that numerous distractions exist within physical classroom setup; however, the educator can at least gauge comprehension through various formative assessments. Thirdly, students may neglect to view or adequately comprehend the video, thereby rendering them unprepared or inadequately prepared for the more interactive activities that will transpire in an in-person format. Fourthly, students may require substantial scaffolding to facilitate their understanding of the material delivered in the video (Betihavas et al. 2015).

While proficient instructors are likely to incorporate effective scaffolding activities during the viewing process, such as 'stop, think, and answer' questions (and also permit rewinding if necessary), they may still fall short of offering sufficient scaffolding activities to accommodate diverse learner types. Fifthly, students are unable to pose inquiries to the instructor or their peers if they engage with the video in isolation. Consequently, critical just-in-time queries that would aid in their comprehension of the material cannot be addressed unless the instructor is accessible during the viewing, which poses a significant challenge (O'Flaherty and Phillips 2015).

Despite the myriad limitations inherent in the flipped classroom model and the absence of empirical evidence to validate its efficacy, subjective reports from numerous educators suggest that it can serve as an effective pedagogical approach across various educational tiers, contingent upon the characteristics of the learners, available resources, and temporal constraints. Furthermore, it appears to be particularly well suited for imparting procedural knowledge (Awidi and Paynter 2018).

Determining an appropriate instructional pace presents challenges across all educational settings; however, as the size of the cohort increases and the frequency of interactions between the educator and the students diminishes, the complexity of this endeavour escalates. Lectures remain, as previously noted, favoured by educational administrators in part due to their capacity to accommodate a large number of students simultaneously (Gillette et al. 2018).

While it is accurate that the implementation of the flipped classroom does not reduce the number of students within a cohort, it does proffer a potential solution for establishing an appropriate instructional tempo by permitting students to engage with the lesson content at their individual pace prior to the class, thereby reserving class time for interactions—both among students and between students and the educator—which is a feature that traditional lectures lack and even contemporary interactive lectures often fail to provide an adequate measure. It is important to acknowledge that finding the optimal teaching tempo may prove challenging even in smaller courses and group sessions, as variability in student pacing persists. Thus, contemplating the incorporation of flipped elements into even smaller courses holds merit and may indeed result in more significant enhancements in course efficacy (Gillette et al. 2018).

5.2.4 Gamification for enhanced student engagement

Learner involvement is recognized as a critical characteristic that significantly impacts scholarly achievement. The presence of varied learning preferences among students further influences their participation in activities orchestrated by educators. Maintaining students' interest and involvement presents a challenge that often places educators in a difficult position. This is due to the fact that knowledge immersion serves as a crucial determinant in their success and performance as assessed during either progress monitoring or cumulative assessment. Numerous studies have identified gamification as a strategy to enhance students' accomplishment (Reed et al. 2016).

Gamification refers to the application of game design elements within non-game contexts to engage participants and foster desired behaviours. Advances in technology facilitate the incorporation of game elements into pedagogical models, thereby expanding the methodologies available to educators for constructing lesson plans. However, not all educators possess the creativity requisite for integrating gamification into their instructional designs. Consequently, online platforms such as Kahoot!, Quizizz, Socrative, and Quizalize offer commendable alternatives for educators, enabling the diversification of lesson plans and activities that can effectively engage and motivate students, thereby enhancing their participation during classroom instruction (Balalle 2024).

The assurance of student satisfaction and favourable outcomes through 'academic engagement' has emerged as a paramount political and economic concern within the contemporary landscape of HE. The implementation of gamification as an instructional strategy to augment learner motivation and engagement is still an evolving concept. Research in this domain has predominantly concentrated on elucidating the perceived correlation between heightened motivation, time and effort invested 'on task', and academic achievement. The concept of gamification has existed for a considerable duration, with educators employing related practices for some time. Although the term has gained popularity relatively recently, the core principle associated with gamification offers a valuable framework to enhance learning design (Smiderle et al. 2020).

Gamification is distinct from game-based learning and serious games, as it incorporates certain elements of games without transforming the educational experience into a full-fledged game. The term 'gamification' became prevalent in mainstream discourse around 2010, and various definitions of game mechanics in education are currently in circulation. Research regarding the effects of gamification and gaming on learning remains constrained, primarily focusing on motivation, which represents merely one aspect of student engagement (Welbers et al. 2019).

Numerous theoretical frameworks exist regarding gamification, with Landers' theory of gamified learning being particularly noteworthy as it represents the most robust conceptualization currently available. This theory integrates research pertaining to serious games and gamification through the commonality of game attributes and posits a psychological framework for gamified learning, albeit primarily emphasizing behaviour modification. However, this theory exhibits limitations, as it neglects to consider the social, cognitive, and affective dimensions of student engagement that are pertinent to HE, as well as the broader implications of engagement, such as student satisfaction or overall wellbeing (Da Rocha Seixas, Gomes, and De Melo Filho 2015).

Nonetheless, evidence suggests that games and playful learning can positively affect other dimensions of the student experience, such as levels of interest, intellectual engagement, and intrinsic motivation by providing opportunities for autonomy, relatedness, and competence. It is feasible to select game attributes that are suitable for facilitating the attainment of specific learning objectives categorized within the three domains of learning: cognitive, affective, and psychomotor. Moreover, it is possible to identify a game attribute for inclusion in an interactive learning design by recognizing the psychological domain that aligns the learning outcome or educational objective with the desired modifications to the student experience of engagement (Rajcsanyi-Molnar, Andras, and Czifra 2025).

Finally, it is imperative to articulate that the game-inspired learning is constrained by ethical boundaries and poses the potential danger of engendering scenarios wherein the contextual essence of an experience is transformed to such an extent that it obfuscates the authentic nature of a task, thereby potentially nullifying the informed consent of participants. When attributes characteristic of games are incorporated with the intention to mislead, coerce, or placate participants regarding the essence of the task, playful learning transcends ethical boundaries. Consequently, endeavours to gamify a task must be accompanied by ethical safeguards that address these concerns, ensuring the transparency and accessibility of the methodologies and objectives underlying the gamification of a task (Saxena and Mishra 2021).

5.3 LEARNING MANAGEMENT SYSTEMS AND MOOCS

LMS plays a critical role in enabling students to access educational resources through course outlines, submission of assignments, retrieval of grades, fostering interactive exchanges between students and instructors, promoting peer interactions, facilitating engagement with educational tools, knowledge sharing, and participation in digital competency tests (Bradley 2020).

Similarly, the Massive Open Online Course (MOOC) has profoundly impacted adult learning paradigms. Open-source courses are predicated on the principle of collaborative authorship, where knowledge is generated and disseminated by a vast array of learners (Castaño-Muñoz and Rodrigues 2021).

5.3.1 Impact of LMS in online learning

The rapid evolution of informatics has prompted a transformation within the educational sector, especially concerning the sphere of learning. Once confined to in-person interactions, education is now predominantly delivered through online modalities utilizing an array of devices and platforms. A notable platform that has gained significant traction in online education across schools and universities is the LMS (Hernández-García et al. 2024). The LMS operates as an all-encompassing platform that enhances educational interaction between instructors and students within the framework of remote instruction as shown in Figure 5.2.

Virtual learning environment empowers educators to facilitate and exemplify discussions, orchestrate online activities, establish learning objectives, present learners with choices, and aid in problem-solving through systematic decision-making processes. The presence of an instructor within an LMS fosters an interactive learning atmosphere. The utilization of LMS allows students to maintain their independence, enthusiasm, and impetus. Educational stakeholders should seek empirical research to substantiate their contributions to LMS platforms for enhancing students' learning in mathematics and other academic disciplines (Hu and Xiao 2025).

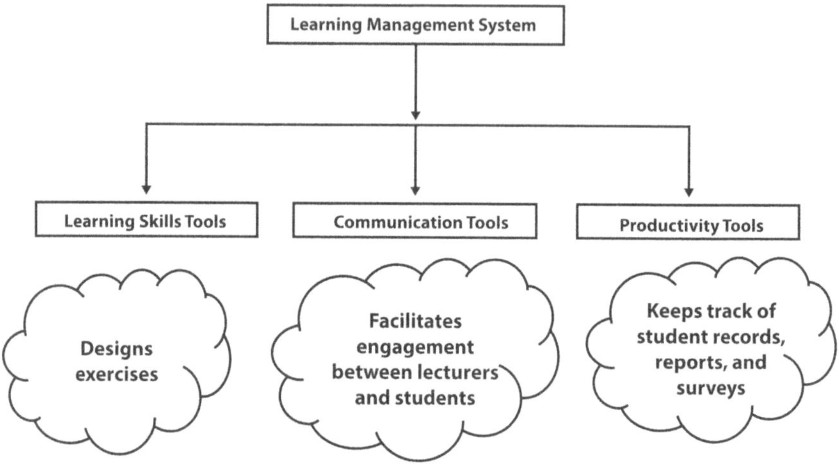

Figure 5.2 Tools in the learning management system.

Computer designers from the 1950s posited the application of an LMS for educational objectives as both feasible and essential. Various methodologies were identified for employing an LMS as an educational tool, encompassing a diverse lexicon pertinent to computer utilization. Over the years, the technology and tools that underpin online learning frameworks have evolved significantly since the inception of online education in the mid to late 1990s (Bradley 2021).

An LMS establishes a framework for asynchronous delivery methods that encompass emails, discussion forums, audio presentations, and academic bulletin to foster constructive interactions. Asynchronous delivery mechanisms enable learners to engage with one another without the impediments imposed by distance and temporal constraints. Such environments permit learners to harmonize their educational pursuits with other commitments and responsibilities. Furthermore, within asynchronous frameworks, learners are required to navigate the LMS to access course materials, partake in meaningful communication, and manage the technological components of the course (Huang 2022).

An LMS also delineates a structure for synchronous delivery methods through the implementation of video and online conferencing. Learners can interact with their instructors via video, online discussions, and live chat, in addition to presentations and document files. Digital conferencing within synchronous frameworks enhances interaction between learners and their instructors (Turnbull, Chugh, and Luck 2020).

The virtual learning environment nurtures a setting conducive to engagement and learner success, enabling students to enrol in courses, monitor their academic performance, and stay informed on updates and announcements related to their courses. Users of the LMS gain access to resources and information disseminated by instructors in both synchronous and asynchronous contexts (Agrawal et al. 2021).

LCMS is a modern version of virtual learning environment. An LCMS provides instructional designers who are equipped with methodologies that facilitate the systematic creation of e-learning content. This instructional content management enables the reuse of content through the capabilities to create, store, assemble, and deliver customized e-learning materials in the format of learning objects (LOs) (Al-Dhief et al. 2024). In a parallel manner, LOs constitute pivotal elements within an LCMS or an LMS, offering viable alternatives for users across diverse educational contexts. The integration of LOs enhances contemporary educational settings by providing tools that can be tailored to address the particular requirements of learners, as well as operational strategies that accommodate both large and small audience sizes, which may facilitate changes in resource allocation (Bradley 2020).

Educational institutions are confronted with the critical choice of adopting either a licensed or free LMS. This decision is primarily influenced by the available resources and the expertise level of users within the educational institutions. Private learning systems utilize exclusive code, necessitating that educational organizations acquire a subscription to access the functionalities of the LMS. Notable examples of custom LMS solutions include Blackboard, PowerSchool, SchoolWires, Edline, eSchoolView, and SchoolPointe. Conversely, open-source systems operate under a free license, allowing users unrestricted access and utilization of the system. Illustrative examples of open-source systems encompass Moodle, WordPress, and Drupal (Furqon et al. 2023).

A significant challenge facing instructors utilizing virtual learning environment is the necessity to implement both managerial and pedagogical modifications, transitioning from a uniform state of inadequacy to a framework of advancement that supports all learners. The academic progression of students utilizing an LMS exhibits variability, wherein low-achieving students may encounter difficulties in adhering to assignment deadlines, while high-achieving students may find limitations in their ability to surpass mere deadline compliance (Pacheco et al. 2025).

5.3.2 The growth and influence of MOOCs

The implementation of online educational innovations has increasingly proliferated in recent years, with the expectation that such advances will augment human capital and mitigate disparities. Among these educational innovations, MOOCs have attained significant prominence within the educational sphere. These courses are characterized by their virtually limitless enrolment capacity and the absence of stringent enrolment prerequisites, aside from access to the Internet. Typically, they are provided at no cost to participants; however, certain ancillary services, such as the issuance of certificates, may incur fees (Berde et al. 2024).

Numerous institutions of HE are currently implementing free online learning programmes as a means to augment their conventional mission and provide succinct and adaptable lifelong learning opportunities through non-degree programmes that are frequently associated with professional advancement. Furthermore, empirical evidence suggests that MOOCs do not supplant traditional HE; rather, they serve as a strategic marketing instrument aimed at enhancing enrolment figures in conventional educational settings. Additionally, a recent investigation indicates that the introduction of a more affordable online master's programme in computer science does not diminish the demand for the equivalent programme in a traditional format, but rather broadens the overall demand, with MOOCs capable of fulfilling a comparable function (Czerniewicz et al. 2016).

Integrating virtual open learning platforms into the framework of professional development constitutes the most auspicious application of these educational modalities in the foreseeable future. In contrast to traditional learners, adult and employed individuals possess distinct requirements and anticipations, valuing the flexible and self-directed educational opportunities that MOOCs are adept at providing. The ingenuity inherent in open access online courses represents a digital publication for educational materials covering high-quality courses offered at no cost (Czerniewicz et al. 2016).

MOOCs are characterized by unique attributes that differentiate them from conventional online courses. The subsequent discussion underscores the characteristics that distinguish the learning experience within MOOCs. It is evident that the attributes of MOOC innovations have profoundly transformed the educational learning process through: firstly, multifaceted platforms, enabling courses to engage a substantial number of students and participants. Secondly, these courses are accessible to any interested individual, provided free of charge, without obligation or prior conditions, at any time and from any location. Thirdly, participants originate from varied cultures, backgrounds, and possess a multitude of motivations (Kwiek 2001).

It is widely acknowledged that the primary role of higher educational institutions is to equip individuals with essential skills, knowledge, and a commitment to lifelong learning. To fulfil this mandate, HE has been significantly shaped by the rapid advancements in information and communication technologies. This evolution has facilitated the development of novel technological resources, such as MOOCs, aimed at imparting the skills and methodologies evident in contemporary scientific inquiry, thereby enabling learners to establish a foothold in a competitive landscape (Kwiek 2001).

Digital open learning courses furnish a platform for complimentary online courses accessible to extensive audiences globally, with content curated and pre-recorded by educational authorities. These experts provide a range of educational materials, including text, audio, and video resources. They also engage with social networking platforms to exchange experiences and foster learning from peers. This collaborative educational approach distinctly delineates MOOCs from other learning platforms, offering a more participatory and interactive experience than traditional methods (Liyanagunawardena, Adams, and Williams 2013).

The design of MOOCs must transcend the mere provision of standard online content. Whether delivered in a virtual or mixed mode, MOOCs should incorporate both passive and declarative elements to enhance students' problem-solving capabilities. Video lectures and readings markedly influence participants' studies regarding the application and direction of their MOOC learning within their professional practices. While MOOCs themselves may not be inherently innovative, they do present opportunities for novel ideation and operational methodologies. Furthermore, by offering courses in unconventional formats, MOOCs remain accessible to all individuals, irrespective of location, with potential connections to traditional degree-granting programmes of institutions (Ma and Lee 2018).

The significance of MOOCs lies in their provision of open-ended online course experiences, which have enabled participants from diverse nations to engage with international universities and benefit from their distinct programmes and offerings. MOOCs have alleviated the apprehension that accompanied the initial emergence of online courses, which appeared to pose a threat to the sustainability of academic institutions. This apprehension stemmed from the belief that if students could secure a degree or enrol in courses offered by a prestigious online university, they would opt out of incurring the substantial financial burdens associated with traditional on-campus education. Certain scholars contend that online education represents the most viable solution for economically disadvantaged students who would otherwise lack access to HE (Ma and Lee 2018).

The physical campus will persist in attracting students who seek interaction with their peers and professors while acquiring a degree from esteemed universities. Conversely, should MOOCs be utilized to establish a framework that acknowledges demonstrable proficiency, they may further diminish the significance of campus-based networking. When leveraged to directly connect talent with prospective employers, large-scale online education has the potential to bypass one of the few remaining justifications for pursuing a conventional college experience (Maphosa and Maphosa 2023).

The adaptable nature of MOOCs has garnered significant appreciation from students. Learners possess a preference for exploring new subjects or augmenting contemporary knowledge, and they have expressed curiosity regarding MOOCs, motivated by personal challenges and the aspiration to amass numerous completion certificates (Maphosa and Maphosa 2023).

The administration of the university can facilitate faculty incentives for those interested in developing MOOCs to explore and assess various learning modalities. Consequently, this support can help mitigate the challenges that hinder faculty members from engaging in teaching through MOOCs. By streamlining the administrative processes instituted within universities, faculty members will be empowered to embrace creativity and innovation in order to fulfil their pedagogical objectives (Liyanagunawardena, Adams, and Williams 2013).

5.4 PERSONALIZED LEARNING THROUGH AI AND ML

In the framework of the Fourth Industrial Revolution, the customization of educational experiences has attained paramount importance, as it epitomizes the escalating synthesis of human competencies with technological innovations. This phenomenon is propelled by transformative technologies such as artificial intelligence (AI), the Internet of Things, and robotics. AI, characterized as the capacity of machines to analyse data through algorithms, discern patterns, and employ that knowledge in decision-making akin to human cognition, has witnessed substantial progress. From the initial intelligent tutoring systems to individualized learning environments, AI has enabled the delivery of resources and assessments specifically tailored to each learner, adapting to their unique profiles, preferred learning modalities, and cognitive capacities. Furthermore, biometric and contextual monitoring technologies have surfaced, facilitating the recognition of emotions and preferences, thereby allowing systems to be fine-tuned with greater precision to address students' requirements (Castro et al. 2025).

Recent advancements in technology have significantly revolutionized the educational landscape. The convergence of AI and machine learning (ML) technologies represents one of the most promising and impactful advancements within this domain. These formidable instruments have transformed numerous sectors, and their influence on education has been profoundly substantial. Educators can discern the strengths, weaknesses, and learning trajectories of individual students through the application of data analytics and ML algorithms. Armed with this insight, they can formulate tailored learning pathways, establish targeted intervention objectives, and furnish timely feedback. Ultimately, this approach fosters a more effective learning environment (Essa, Celik, and Human-Hendricks 2023).

Generative AI, exhibiting human-like characteristics and social behavioural capabilities, is beginning to augment the roles of educators, particularly in the facilitation of practical tasks and the enhancement of prior knowledge mastery. In this context, AI emerges as a transformative agent that provides unparalleled capabilities for student profiling, adaptive content delivery, and real-time feedback (Wang et al. 2024).

One of the principal motivations behind the escalating demand for AI in various domains, including education, is its capacity to process and analyse data with greater speed and efficiency than human beings. Nevertheless, it is imperative to acknowledge the existing challenges and barriers associated with the incorporation of AI into the curriculum. Concerns regarding data security and student privacy represent a significant issue. Additionally, there are technological constraints, as well as ethical and moral dilemmas that warrant careful consideration (Okagbue et al. 2023).

The notion of personalized learning is not a novel concept within the educational domain; however, the emergence of AI and big data analytics is facilitating the advent of innovative methodologies for its implementation.

Policymakers are compelled to prioritize the allocation of funding and the development of infrastructure to guarantee that all educational institutions possess access to the technological resources requisite for the efficacious integration of AI. Furthermore, the formulation of transparent regulatory frameworks that address data privacy, ethical considerations in AI, and the dynamics of teacher-student interaction within AI-mediated learning environments is imperative for the protection of the interests of both educators and learners (Wang et al. 2024).

There exist platforms that amalgamate specialized knowledge with cutting-edge ML and AI technologies to assist educators in evaluating student advancement. These platforms facilitate the conservation of time that is typically allocated to academic evaluations and assessments.

At a global scale, the amalgamation of AI with other digital innovations has commenced but is still in the early stages. For example, in Germany, AI is combined with augmented reality through the utilization of smart glasses, enabling students to receive instructions instantly before their visual field. Similarly, in Canada, investigations are underway regarding the potential application of simulations and virtual reality in conjunction with AI to enhance the educational experience. Within these simulated environments, students receive practical guidance from an AI, after which they can promptly engage with a specific problem (Munir, Vogel, and Jacobsson 2022).

In China, AI is employed not solely for educational purposes and assessments but also for monitoring students through AI systems that utilize cameras to observe attentiveness levels. Additionally, AI-powered robotics is currently being designed to require students to impart new information or skills, thereby facilitating a deeper understanding of the lesson material as students elucidate it. This phenomenon aligns with the protégé effect, which posits that learners acquire knowledge more swiftly when they have the opportunity to explain their understanding to peers. This approach operates under the premise that a teachable robot can receive 'infinite' explanations, whereas a human might be inclined to feign comprehension to preserve their reputation (Munir, Vogel, and Jacobsson 2022).

Despite the potential of AI-driven technologies to furnish robust instruments for adaptive instruction, educators encounter substantial pedagogical and psychological impediments. The incorporation of AI within educational contexts necessitates that educators modify their pedagogical methodologies and assimilate emerging technologies into their instructional practices. Nevertheless, resistance to these novel technologies frequently emerges due to apprehensions regarding workload, motivational factors, and the perceived intricacy of AI-augmented systems. Moreover, a deficiency of professional development initiatives specifically designed to address the exigencies of educators exacerbates the challenges related to implementation (Munir, Vogel, and Jacobsson 2022).

A subset of pedagogical responsibilities may be subject to automation through AI, certain existing duties are likely to gain heightened significance, and novel responsibilities will emerge as a consequence of AI integration within educational frameworks. Among the responsibilities susceptible to automation, the curation of instructional materials and the processes of assessment and grading prominently figure. The provision of evaluative feedback to learners may similarly be undertaken by AI systems. Given that certain responsibilities are amenable to automation via AI, this facilitates opportunities for educators to concentrate on tasks that are challenging or unfeasible to automate using AI, such as social responsibilities including mentorship and coaching. Furthermore, the integration of AI will engender the emergence of new responsibilities. For educators to implement AI in a responsible manner, they must possess the capacity to interpret AI outputs and translate these into actionable measures (Castro et al. 2025).

Facilitators retain the autonomy to determine the organization of their lessons. Consequently, the design of AI must be approached through a design thinking framework. In the context of design thinking, the needs of the user are paramount in the development of technology. For the effective implementation of AI in educational settings, the AI must be adaptable in its application and align with the various learning modalities exhibited by educators (Essa, Celik, and Human-Hendricks 2023).

5.5 CHALLENGES AND CONSIDERATIONS IN IMPLEMENTING EDUCATION 4.0

Although personalized learning has been the focus of considerable scholarly discourse, the existing literature delineates numerous impediments that remain to be surmounted. These obstacles can be categorized into five distinct dimensions as follows (Mian et al. 2020).

Perceptual Barriers: These pertain to the heterogeneity in the comprehension and execution of concepts related to customized learning.

Organizational Barriers: These encompass resistance to transformation, the absence of technological infrastructure, inadequate temporal resources, insufficient institutional backing, and challenges related to funding.

Emotional Obstacles: Psychological barriers are associated with the perception of increased workload, scepticism regarding technological competencies, and a deficit of motivation towards self-paced learning.

Technology Constraints: These consist of restricted access to technological resources, inadequacies in digital competencies, insufficient technical assistance, difficulties in integration, and ethical considerations.

Curriculum Limitations: These relate to the frameworks employed for personalized instruction, the deficiency in pedagogical acumen necessary for the adaptation of content and learning activities, inadequate pedagogical support for educators, and the intricacy of assessing and monitoring students' individual progress.

Technology-enhanced learning is individualized and empowers students to cultivate greater autonomy, thereby enriching their educational experiences. With the implementation of AI, the generation of instructional materials and media is rendered more accessible, alleviating the necessity for educators to possess an in-depth understanding of technology. Educators merely need to select from the plethora of platforms and applications available. The educational paradigm must also evolve to maintain relevance and equip forthcoming generations to confront technological advancements and increasingly intricate global challenges. Through this technological integration, the educational curriculum can be transformed into a dynamic framework that is attuned to individual needs and emphasizes the development of skills pertinent to contemporary society (Costan et al. 2021).

The conditions for the successful implementation of AI applications extend beyond mere technical considerations. Educators, school administrators, and leaders must also recognize the intrinsic value of AI applications and deploy them judiciously. To achieve this, it is imperative that personnel within the educational domain possess adequate digital competencies. Subsequently, subject-specific AI applications can be integrated into the instructional process. Ultimately, once users gain sufficient experience with the responsible application of AI, educational institutions can progress towards the integration of diverse AI systems (Costan et al. 2021).

Irrespective of the technological feasibility or legal permissibility of specific AI applications, it is imperative to ascertain whether particular applications are deemed desirable or acceptable. Within the educational milieu, there exists a prevailing belief that the responsibility for this ethical assessment ultimately resides with educators and educational institutions. An AI application, notwithstanding its efficacy, must invariably adhere to ethical norms. Even in instances where pupils exhibit significantly accelerated learning rates within an adaptive learning environment, the ethical justification for its application remains tenuous due to the lack of transparency regarding the personalization of learning materials (Luthra and Mangla 2018).

From a utilitarian perspective, the benefits and drawbacks must be meticulously evaluated against one another. In the context of the adaptive learning system, it becomes essential to ascertain whether the enhancements in learning outcomes surpass the deficiencies in transparency. When a consensus is reached among all relevant stakeholders that the utilization of AI yields greater advantages compared to a scenario devoid of AI, the ethical justification for the application of AI can be established. Objectives are pivotal within the framework of virtue ethics. Concerning the opaque adaptive learning system, one may question whether it is employed for a virtuous purpose. Does an educational institution utilize the system to enhance student performance, or is it primarily aimed at reducing the teaching workforce? In the former instance, one could argue for the validity of the objective, whereas in the latter case, one might query whether the deployment of systems with ambiguous operational mechanics for the purpose of cost reduction constitutes a justifiable aim (Luthra and Mangla 2018).

In a scenario where learning is entirely automated, resulting in students navigating the educational process devoid of human intervention, one may ponder the potential inhumanity of such an educational model. This risk is exacerbated when simulations are employed excessively within pedagogical contexts. There exists a possibility that knowledge and skills acquired through simulations may not be transferable to real-world applications (Moktadir et al. 2018).

There exists a potential concern that facilitators may require the acquisition of novel competencies to utilize AI in a responsible manner. At present, data-centric pedagogical approaches are not incorporated into teacher education programmes; whereas the utilization of AI necessitates, those users possess at least fundamental digital proficiency. Should educators fail to adapt to this evolving landscape, there is a possibility that future educators will emerge as data scientists lacking a robust pedagogical foundation. Nevertheless, the likelihood that the role of the facilitator will be supplanted by automation is exceedingly minimal. For an AI system to assume all responsibilities typically held by an educator, a variant of artificial general intelligence would be essential, which current assessments deem implausible in the foreseeable future (Moktadir et al. 2018).

Both educators and AI must undergo transformation to mitigate this potential risk. Both the elucidation of ML algorithms and the competency of educators (pertaining to AI) must be augmented to successfully address this risk. Greater emphasis should be allocated to digital competencies within teacher training programmes. The incorporation of digital skills should be mandated, akin to didactic and pedagogical competencies. Encourage educators possessing expertise in data science or AI. This initiative has the potential to catalyse a ripple effect within an educational institution that can facilitate the integration of AI (Costan et al. 2021).

5.6 FUTURE PROSPECTS OF EDUCATION 4.0

The incorporation of emerging Information and Communication Technology (ICT) content, subjects, and disciplines into educational curricula generates novel curricular requirements.

Decision-makers within HE are confronted with the imperative to harmonize the promotion of student learning and the facilitation of faculty research with the exigencies of keeping pace with advancements in technology. Educational technology (EdTech) can serve these objectives by enhancing access to content, fostering collaborative opportunities, augmenting interactivity in pedagogical practices, enabling the customization of instructional methods, and yielding vast quantities of data for analytical scrutiny. Concurrently, this raises significant concerns regarding data security and privacy. Numerous HE administrators are endeavouring to regulate unrestricted acquisitions across campuses, which result in the proliferation of EdTech resources (Rodriguez-Segura 2021).

The prospective effectiveness of online instruction can be delineated in terms of competencies, including the knowledge, skills, and motivational factors that collectively underpin successful performance, or alternatively, in terms of competence, which refers to the behaviours indicative of one's ability to perform within an online educational environment. Additionally, the success of future faculty can be conceptualized comprehensively in terms of competence, reflecting the behaviours associated with effective performance, such as the integration of EdTech within an online instructional framework (Bonfield et al. 2020).

Within the Smart Classroom, facilitators are afforded the opportunity to employ diverse natural modalities while engaging with remote learners, thereby replicating the effects of traditional instruction in a physical classroom setting with local students. In the Smart Classroom, educators can navigate freely, utilizing conventional pedagogical strategies to instruct remote students. Given their presence in a tangible classroom environment, they can simultaneously accommodate local learners. The concurrent instruction of both local and remote students necessitates a reduced workforce compared to distinct on-campus and tele-education operations. Furthermore, the lecture, recorded as hypermedia coursework, remains accessible for playback following the class session. Such a smart classroom paradigm can be actualized via mobile learning methodologies aimed at modernizing and globalizing educational framework as innovative pedagogical strategies (Min-Allah and Alrashed 2020).

Smart campuses, characterized as digitally transformed university environments, are frequently regarded as scaled-down versions of smart cities and function as living laboratories for the research, development, and implementation of smart technologies, alongside their conventional roles in knowledge dissemination, skill development, and scientific inquiry. Furthermore, a community centred on human values and user-driven value creation is established to enhance the overall university experience, facilitated by a mobile application based on Wi-Fi hotspot connectivity to foster social networking. A knowledge economy is cultivated through the expanding global market for smart education, which generates income. Presently, there exists a limited comprehension of the conceptualization and practical application of the smart campus (Polin et al. 2023).

5.7 CONCLUSION

In contemporary discourse, the plethora of technological advancements available within the educational realm has increasingly relegated traditional pedagogical approaches (notably the chalk-and-talk method) to a marginalized status. This conventional method of instruction is characterized as being predominantly teacher centred and is progressively waning in prevalence among contemporary educators.

In recent years, the notion of gamification within educational settings has emerged as a burgeoning field of inquiry among researchers. Scholars characterize playful learning as the integration of game elements into non-gaming contexts, which can serve as a mechanism to enhance student engagement within the classroom. Students exhibit a pronounced preference for game-based learning platforms, which stimulate a competitive ethos among peers and ultimately culminate in improved scholastic achievement.

The flipped classroom paradigm enables educators to allocate increased attention to its more challenging components. The instructional material is delivered through asynchronous multimedia and reading materials that learners can revisit at their convenience and employ during their examination preparations; comprehension is subsequently assessed through self-administered weekly quizzes, while students are afforded the opportunity to articulate their inquiries during synchronous sessions, thereby facilitating a more manageable engagement with the foundational elements of a subject.

LMS signifies a progression from the mechanisms and systems previously established by specific institutions for the registration of students in designated courses and the maintenance of records pertaining to students' activities. A variety of learning options have been developed to facilitate students' participation in online courses, which may either be integrated into the formal curriculum or arise from the necessity for institutional certification. Learners engaging with LMS functionalities are likely to initiate self-regulatory practices concerning their academic progress.

Conventional educational frameworks frequently struggle to accommodate the diverse needs and learning preferences of students, culminating in suboptimal outcomes. Nevertheless, with the emergence of AI and ML, novel opportunities have arisen to customize and personalize the learning experiences of each student. Despite the progressive developments in contemporary AI, it remains in the nascent phase of its integration with educational methodologies and pedagogical practices, and has yet to reach a stage of maturation. Owing to the finite number of educators and the associated financial implications, personalized one-on-one instruction is not generally feasible from a societal perspective.

The rapid technological transformation is reshaping comprehension and tutoring paradigms, progressing from fundamental e-learning modalities to the emergence of smart campuses. This evolution permits the delivery of educational experiences at any location and at any time through distance learning, thereby expanding and creating innovative programmes tailored for implementation. Additionally, educational opportunities have become accessible to all societal segments, as geographical constraints are no longer an impediment. Moreover, the implementation of Smart Classrooms through mobile learning is advancing the modernization and internationalization of educational networks, as new learning methodologies are developed through the integration of information technology.

REFERENCES

Agrawal, Shruti, Nidhi Sharma, and Sumedha Bhatnagar. 2021. "Education 4.0 to Industry 4.0 Vision: Current Trends and Overview." In *Lecture Notes in Mechanical Engineering*, 475–85. doi:10.1007/978-981-16-3033-0_45.

Akçayır, Gökçe, and Murat Akçayır. 2018. "The Flipped Classroom: A Review of Its Advantages and Challenges." *Computers & Education* 126: 334–45. doi:10.1016/j.compedu.2018.07.021.

Al-Dhief, Fahad Taha, Ali Al Nasser, Shafazawana Mohamed Tharikh, Hassan Al Nasser, Ali Abdul Ghaffar Al-Mosleh, Musatafa Abbas Abbood Albadr, and Majid Razaq Mohamed Alsemawi. 2024. "Review of Learning Management Systems: History, Types, Advantages, and Challenges." *Indonesian Journal of Electrical Engineering and Computer Science* 33 (1): 350–60. doi:10.11591/ijeecs.v33.i1.

Alshammary, Farhan Mohammed, and Waleed Salim Alhalafawy. 2023. "Digital Platforms and the Improvement of Learning Outcomes: Evidence Extracted from Meta-Analysis." *Sustainability* 15 (2): 1305. doi:10.3390/su15021305.

Awidi, Isaiah T., and Mark Paynter. 2018. "The Impact of a Flipped Classroom Approach on Student Learning Experience." *Computers & Education* 128: 269–83. doi:10.1016/j.compedu.2018.09.013.

Balalle, Himendra. 2024. "Exploring Student Engagement in Technology-Based Education in Relation to Gamification, Online/Distance Learning, and Other Factors: A Systematic Literature Review." *Social Sciences & Humanities Open* 9: 100870. doi:10.1016/j.ssaho.2024.100870.

Berde, Éva, Seyoum Teffera Mengesha, Belaynew Asrie Mola, and Sándor Remsei. 2024. "Massive Open Online Courses (MOOCs) for Ethiopian Educators: Evaluating MOOCs as a Global Development Tool." *Research in Globalization* 9: 100263. doi:10.1016/j.resglo.2024.100263.

Bernacki, Matthew L., Jeffrey A. Greene, and Helen Crompton. 2019. "Mobile Technology, Learning, and Achievement: Advances in Understanding and Measuring the Role of Mobile Technology in Education." *Contemporary Educational Psychology* 60: 101827. doi:10.1016/j.cedpsych.2019.101827.

Betihavas, Vasiliki, Heather Bridgman, Rachel Kornhaber, and Merylin Cross. 2015. "The Evidence for 'Flipping out': A Systematic Review of the Flipped Classroom in Nursing Education." *Nurse Education Today* 38: 15–21. doi:10.1016/j.nedt.2015.12.010.

Bonfield, Christopher Alan, Marie Salter, Alan Longmuir, Matthew Benson, and Chie Adachi. 2020. "Transformation or Evolution?: Education 4.0, Teaching and Learning in the Digital Age." *Higher Education Pedagogies* 5 (1): 223–46. doi:10.1080/23752696.2020.1816847.

Bradley, Vaughn Malcolm. 2021. "Learning Management System (LMS) Use with Online Instruction." *International Journal of Technology in Education* 4 (1): 68. doi:10.46328/ijte.36.

Campillo-Ferrer, José María, and Pedro Miralles-Martínez. 2021. "Effectiveness of the Flipped Classroom Model on Students' Self-Reported Motivation and Learning during the COVID-19 Pandemic." *Humanities and Social Sciences Communications* 8 (1): 176. doi:10.1057/s41599-021-00860-4.

Castaño-Muñoz, Jonatan, and Margarida Rodrigues. 2021. "Open to MOOCs? Evidence of Their Impact on Labour Market Outcomes." *Computers & Education* 173: 104289. doi:10.1016/j.compedu.2021.104289.

Castro, Gina Paola Barrera, Andrés Chiappe, María Soledad Ramírez-Montoya, and Carolina Alcántar Nieblas. 2025. "Key Barriers to Personalized Learning in Times of Artificial Intelligence: A Literature Review." *Applied Sciences* 15 (6): 3103. doi:10.3390/app15063103.

Chou, Chi-Pu, Kuo-Wei Chen, and Chia-Jen Hung. 2021. "A Study on Flipped Learning Concerning Learning Motivation and Learning Attitude in Language Learning." *Frontiers in Psychology* 12: 753463. doi:10.3389/fpsyg.2021.753463.
Ciarli, Tommaso, Martin Kenney, Silvia Massini, and Lucia Piscitello. 2021. "Digital Technologies, Innovation, and Skills: Emerging Trajectories and Challenges." *Research Policy* 50 (7): 104289. doi:10.1016/j.respol.2021.104289.
Costan, Emily, Gamaliel Gonzales, Roselyn Gonzales, Lynne Enriquez, Felix Costan, Decem Suladay, Nadine May Atibing, et al. 2021. "Education 4.0 in Developing Economies: A Systematic Literature Review of Implementation Barriers and Future Research Agenda." *Sustainability* 13 (22): 12763. doi:10.3390/su132212763.
Czerniewicz, Laura, Andrew Deacon, Michael Glover, and Sukaina Walji. 2016. "MOOC—Making and Open Educational Practices." *Journal of Computing in Higher Education* 29 (1): 81–97. doi:10.1007/s12528-016-9128-7.
Dancsa, Daniel, Iveta Štempeľová, Ondrej Takáč, and Norbert Annuš. 2023. "Digital Tools in Education." *International Journal of Advanced Natural Sciences and Engineering Researches* 7 (4): 289–94. doi:10.59287/ijanser.717.
Da Rocha Seixas, Luma, Alex Sandro Gomes, and Ivanildo José De Melo Filho. 2015. "Effectiveness of Gamification in the Engagement of Students." *Computers in Human Behavior* 58: 48–63. doi:10.1016/j.chb.2015.11.021.
De Souza, Alex Sander Clemente, and Luciana Debs. 2024. "Concepts, Innovative Technologies, Learning Approaches and Trend Topics in Education 4.0: A Scoping Literature Review." *Social Sciences & Humanities Open* 9: 100902. doi:10.1016/j.ssaho.2024.100902.
Essa, Saadia Gutta, Turgay Celik, and Nadia Emelia Human-Hendricks. 2023. "Personalized Adaptive Learning Technologies Based on Machine Learning Techniques to Identify Learning Styles: A Systematic Literature Review." *IEEE Access* 11: 48392–409. doi:10.1109/access.2023.3276439.
Furqon, Muhammad, Parlindungan Sinaga, Liliasari Liliasari, and Lala Septem Riza. 2023. "The Impact of Learning Management System (LMS) Usage on Students." *TEM Journal* 12(2):1082–89. doi:10.18421/tem122-54.
Gaines, Brian R. 2012. "Knowledge Acquisition: Past, Present and Future." *International Journal of Human-Computer Studies* 71 (2): 135–56. doi:10.1016/j.ijhcs.2012.10.010.
García-Martínez, José-Antonio, Francisco-César Rosa-Napal, Isabel Romero-Tabeayo, Sara López-Calvo, and Eduardo-José Fuentes-Abeledo. 2020. "Digital Tools and Personal Learning Environments: An Analysis in Higher Education." *Sustainability* 12 (19): 8180. doi:10.3390/su12198180.
Gillette, Chris, Michael Rudolph, Craig Kimble, Nicole Rockich-Winston, Lisa Smith, and Kimberly Broedel-Zaugg. 2018. "A Meta-Analysis of Outcomes Comparing Flipped Classroom and Lecture." *American Journal of Pharmaceutical Education* 82 (5): 6898. doi:10.5688/ajpe6898.
Haleem, Abid, Mohd Javaid, Mohd Asim Qadri, and Rajiv Suman. 2022. "Understanding the Role of Digital Technologies in Education: A Review." *Sustainable Operations and Computers* 3: 275–85. doi:10.1016/j.susoc.2022.05.004.
Hernández-García, Ángel, Carlos Cuenca-Enrique, Laura Del-Río-Carazo, and Santiago Iglesias-Pradas. 2024. "Exploring the Relationship between LMS Interactions and Academic Performance: A Learning Cycle Approach." *Computers in Human Behavior* 155: 108183. doi:10.1016/j.chb.2024.108183.
Hu, Juan, and Wen Xiao. 2025. "What Are the Influencing Factors of Online Learning Engagement? A Systematic Literature Review." *Frontiers in Psychology* 16: 1542652. doi:10.3389/fpsyg.2025.1542652.
Huang, Qiaoyi. 2022. "Teachers' Intention to Use an Electronic Learning Management System in the Long Term." *Interactive Learning Environments* 31 (10): 7182–95. doi:10.1080/10494820.2022.2062607.

Hussin, Anealka Aziz. 2018. "Education 4.0 Made Simple: Ideas for Teaching." *International Journal of Education and Literacy Studies* 6 (3): 92. doi:10.7575/aiac.ijels.v.6n.3p.92.

Joy, Praisy, Rajeev Panwar, R. Azhagiri, Asha Krishnamurthy, and Mallikarjun Adibatti. 2023. "Flipped Classroom – A Student Perspective of an Innovative Teaching Method during the Times of Pandemic." *Educación Médica* 24 (2): 100790. doi:10.1016/j.edumed.2022.100790.

Jugembayeva, Bakytgul, and Aliya Murzagaliyeva. 2023. "Innovation Readiness for Digital Learning within the University 4.0 Model." *Asia Pacific Education Review* 25 (5): 1363–77. doi:10.1007/s12564-023-09909-2.

Kwiek, Marek. 2001. "Globalization and Higher Education." *Higher Education in Europe* 26 (1): 27–38. doi:10.1080/03797720120054157.

Liyanagunawardena, Tharindu Rekha, Andrew Alexandar Adams, and Shirley Ann Williams. 2013. "MOOCs: A Systematic Study of the Published Literature 2008–2012." *The International Review of Research in Open and Distributed Learning* 14 (3): 202. doi:10.19173/irrodl.v14i3.1455.

Luthra, Sunil, and Sachin Kumar Mangla. 2018. "Evaluating Challenges to Industry 4.0 Initiatives for Supply Chain Sustainability in Emerging Economies." *Process Safety and Environmental Protection* 117: 168–79. doi:10.1016/j.psep.2018.04.018.

Ma, Long, and Chei Sian Lee. 2018. "Investigating the Adoption of MOOCs: A Technology–User–Environment Perspective." *Journal of Computer Assisted Learning* 35 (1): 89–98. doi:10.1111/jcal.12314.

Maphosa, Vusumuzi, and Mfowabo Maphosa. 2023. "Opportunities and Challenges of Adopting MOOCs in Africa: A Systematic Literature Review." In IntechOpen eBooks. doi:10.5772/intechopen.1001298.

Margaryan, Anoush, Allison Littlejohn, and Gabrielle Vojt. 2010. "Are Digital Natives a Myth or Reality? University Students' Use of Digital Technologies." *Computers & Education* 56 (2): 429–40. doi:10.1016/j.compedu.2010.09.004.

Mian, Syed Hammad, Bashir Salah, Wadea Ameen, Khaja Moiduddin, and Hisham Alkhalefah. 2020. "Adapting Universities for Sustainability Education in Industry 4.0: Channel of Challenges and Opportunities." *Sustainability* 12 (15): 6100. doi:10.3390/su12156100.

Min-Allah, Nasro, and Saleh Alrashed. 2020. "Smart Campus—A Sketch." *Sustainable Cities and Society* 59: 102231. doi:10.1016/j.scs.2020.102231.

Moktadir, Md. Abdul, Syed Mithun Ali, Simonov Kusi-Sarpong, and Md. Aftab Ali Shaikh. 2018. "Assessing Challenges for Implementing Industry 4.0: Implications for Process Safety and Environmental Protection." *Process Safety and Environmental Protection* 117: 730–41. doi:10.1016/j.psep.2018.04.020.

Moraes, Eduardo Baldo, Liane Mahlmann Kipper, Ana Clara Hackenhaar Kellermann, Leonardo Austria, Pedro Leivas, Jorge André Ribas Moraes, and Marcus Witczak. 2022. "Integration of Industry 4.0 Technologies with Education 4.0: Advantages for Improvements in Learning." *Interactive Technology and Smart Education* 20 (2): 271–87. doi:10.1108/itse-11-2021-0201.

Munir, Hussan, Bahtijar Vogel, and Andreas Jacobsson. 2022. "Artificial Intelligence and Machine Learning Approaches in Digital Education: A Systematic Revision." *Information* 13 (4): 203. doi:10.3390/info13040203.

Nainggolan, Sudung. 2021. "Evaluating of Digital Platforms Related Online Learning during COVID-19 Pandemic: Students' Satisfaction View." *AL-ISHLAH Jurnal Pendidikan* 13 (2): 1358–65. doi:10.35445/alishlah.v13i2.912.

O'Flaherty, Jacqueline, and Craig Phillips. 2015. "The Use of Flipped Classrooms in Higher Education: A Scoping Review." *The Internet and Higher Education* 25: 85–95. doi:10.1016/j.iheduc.2015.02.002.

Okagbue, Ekene Francis, Ujunwa Perpetua Ezeachikulo, Tosin Yinka Akintunde, Mustapha Bala Tsakuwa, Samuel Nchekwubemchukwu Ilokanulo, Kosiso Modest Obiasoanya, Chidiebere Emeka Ilodibe, and Cheick Amadou Tidiane Ouattara. 2023. "A Comprehensive Overview of Artificial Intelligence and Machine Learning in Education Pedagogy: 21 Years (2000–2021) of Research Indexed in the Scopus Database." *Social Sciences & Humanities Open* 8 (1): 100655. doi:10.1016/j.ssaho.2023.100655.

Pacheco, Alex, Renato Yupanqui, Diana Mogrovejo, Joel Garay, and Yrene Uribe-Hernández. 2025. "Impact of Digitization on Educational Management: Results of the Introduction of a Learning Management System in a Traditional School Context." *Computers in Human Behavior Reports* 17: 100592. doi:10.1016/j.chbr.2025.100592.

Pandey, Anushree. 2023. "E-Learning and Education 4.0: Revolution in Education of 21st Century." In *Lecture Notes in Networks and Systems*, 431–38. doi:10.1007/978-3-031-29860-8_44.

Petko, Dominik. 2012. "Teachers' Pedagogical Beliefs and Their Use of Digital Media in Classrooms: Sharpening the Focus of the 'Will, Skill, Tool' Model and Integrating Teachers' Constructivist Orientations." *Computers & Education* 58 (4): 1351–59. doi:10.1016/j.compedu.2011.12.013.

Polin, Ken, Tan Yigitcanlar, Mark Limb, and Tracy Washington. 2023. "The Making of Smart Campus: A Review and Conceptual Framework." *Buildings* 13 (4): 891. doi:10.3390/buildings13040891.

Rajcsanyi-Molnar, Monika, Istvan Andras, and Sandor Czifra. 2025. "Integrating Serious Games and Gamification for Diverse Learner Groups: Lessons from the 'GeoGecko' Project." *Education Sciences* 15 (4): 440. doi:10.3390/educsci15040440.

Reed, Lauren A., Richard M. Tolman, L. Monique Ward, and Paige Safyer. 2016. "Keeping Tabs: Attachment Anxiety and Electronic Intrusion in High School Dating Relationships." *Computers in Human Behavior* 58: 259–68. doi:10.1016/j.chb.2015.12.019.

Rodriguez-Segura, Daniel. 2021. "EdTech in Developing Countries: A Review of the Evidence." *The World Bank Research Observer* 37 (2): 171–203. doi:10.1093/wbro/lkab011.

Saxena, Manisha, and Dharmesh K. Mishra. 2021. "Gamification and Gen Z in Higher Education." *International Journal of Information and Communication Technology Education* 17 (4): 1–22. doi:10.4018/ijicte.20211001.oa10.

Smiderle, Rodrigo, Sandro José Rigo, Leonardo B. Marques, Jorge Arthur Peçanha De Miranda Coelho, and Patricia A. Jaques. 2020. "The Impact of Gamification on Students' Learning, Engagement and Behavior Based on Their Personality Traits." *Smart Learning Environments* 7 (1): 3. doi:10.1186/s40561-019-0098-x.

Sousa, Maria José, and Álvaro Rocha. 2018. "Digital Learning: Developing Skills for Digital Transformation of Organizations." *Future Generation Computer Systems* 91: 327–34. doi:10.1016/j.future.2018.08.048.

Strelan, Peter, Amanda Osborn, and Edward Palmer. 2020. "The Flipped Classroom: A Meta-Analysis of Effects on Student Performance across Disciplines and Education Levels." *Educational Research Review* 30: 100314. doi:10.1016/j.edurev.2020.100314.

Tian, Shuang, Ashutosh Sharma, Lin Wu, and Kulwant S. Pawar. 2024. "A Systematic Literature Review on the Digital Platform and Its Role in the Circular Economy: State-of-the-Art and Future Research Directions." *Journal of Digital Economy* 3: 100122. doi:10.1016/j.jdec.2024.12.001.

Tikhonova, Elena, and Lilia Raitskaya. 2023. "Education 4.0: The Concept, Skills, and Research." *Journal of Language and Education* 9 (1): 5–11. doi:10.17323/jle.2023.17001.

Turnbull, Darren, Ritesh Chugh, and Jo Luck. 2020. "Learning Management Systems: A Review of the Research Methodology Literature in Australia and China." *International Journal of Research & Method in Education* 44 (2): 164–78. doi:10.1080/1743727x.2020.1737002.

Wang, Shan, Fang Wang, Zhen Zhu, Jingxuan Wang, Tam Tran, and Zhao Du. 2024. "Artificial Intelligence in Education: A Systematic Literature Review." *Expert Systems with Applications* 252: 124167. doi:10.1016/j.eswa.2024.124167.

Welbers, Kasper, Elly A. Konijn, Christian Burgers, Anna Bij De Vaate, Allison Eden, and Britta C. Brugman. 2019. "Gamification as a Tool for Engaging Student Learning: A Field Experiment with a Gamified App." *E-Learning and Digital Media* 16 (2): 92–109. doi:10.1177/2042753018818342.

Chapter 6

Leveraging technology for smart climate surveillance

An educational approach to environmental sustainability

A. Leela Glory and M. Balasubramaniam

6.1 INTRODUCTION

As quoted by Robert Swan – The thought that someone else will save our planet is the greatest threat to it. We all must do our part to sustain planet Earth. The technique of persisting without exhausting natural resources, keeping in mind the requirement for future generations, is termed sustainability. By doing so, we protect and conserve our natural resources like our surroundings, raw natural minerals, and forms of energy, which is termed Environmental Sustainability. There are many types of sustainability, like resource sustainability, economic sustainability, social sustainability, and environmental sustainability, but the most important of these is environmental sustainability because if nature is not cared for, then the very existence of mankind will be put in danger. It is important to develop sustainable practices and train the younger generation to do so too, as it is the need of the hour, considering all the climatic changes that are happening around us.

In an era of today which is undergoing a technology advancement at a rapid phase, it is essential to integrate technology into these sustainable practices. A global collaboration to sustain our environment is a prerequisite to ensure a healthy atmosphere. In today's technological era, the quality of the air, water, and soil can be constantly monitored to ensure a lack of pollution, and this is carried out by using smart technologies like remote sensing via satellites, which give us valuable data that, in turn, aids in managing the ecosystem. Artificial intelligence (AI) programs can also be used to detect pollution, air quality check and monitor any type of alteration in the ecosystem. The use of AI in weather surveillance can forecast disasters, giving us enough time to react to the upcoming situation, like evacuating people from the area and taking precautionary measures.

To use AI and smart technologies to carry out these complex monitoring processes, it is the paramount duty of the education field to educate and train individuals who can handle these procedures in their respective fields. It is also important to bring about environmental awareness among people related to climate change and educate them on how they can do their part to save nature.

 DOI: 10.1201/9781003586951-6

6.2 WHAT IS SMART SYSTEM TECHNOLOGY?

Sensing, analyzing, monitoring, and detecting are functions performed by smart technology using a computer system. Technological methods are used to acquire data based on the function being carried out (Admin 2020), and decisions are made based on this data to bring about smart changes that will assist in controlling the adverse effects of environmental changes due to human exploitation.

6.3 WHAT ARE ENVIRONMENTAL CHANGES?

Environmental or climate changes are the effects that arise in our environment due to anthropogenic activities that create an imbalance in nature. The excessive misuse of resources is making them deplete at a rapid rate. These effects must be constantly monitored to detect shifts in temperature levels, the decline of biodiversity, changes in weather patterns, and a sea level rise due to global warming. The data collected and the results received after analysis will aid in creating a sense of responsibility among individuals to make conscious decisions to safeguard our environment from these effects. Conserving resources is a major need of the hour today, keeping in mind the environmental changes that are occurring.

6.4 EFFECTS OF ENVIRONMENTAL CHANGES

As a result of anthropogenic activities, the environment is facing many adverse effects, like:

6.4.1 Global warming

In recent times, the Earth's overall temperature has increased drastically, referring to Global warming. A surge in human activities has disrupted the natural processes occurring in nature ("Global Warming," n.d.). The greenhouse effect, which warms up planet Earth's atmosphere, is amplified due to global warming, creating massive climate change.

6.4.2 Increase in the greenhouse effect

The heat from the sun is trapped in the Earth's atmosphere by a natural process called the greenhouse effect, which maintains a temperature that is suitable to sustain life on Earth. The five major greenhouse gases are water vapor, nitrous oxide, methane, carbon dioxide, and chlorofluorocarbons (US EPA 2024). The increase in these greenhouse gases due to the excessive burning of fossil fuels, due to human activity, has caused global warming, which creates an adverse effect on the atmospheric temperature (Cermak 2024).

6.4.3 Rise in sea level

Global warming creates a rise in the sea level, which is a repercussion of the burning of fossil fuels, which releases a large amount of greenhouse gases, increasing the heat level in the atmosphere. The ocean absorbs this excessive heat, and when water gets heated up, it expands, creating a rise in sea level. Global warming also causes the glaciers and ice caps to melt, raising ocean water levels. This rise in the sea level worldwide poses a threat to coastal areas due to storms and floods. If not monitored, this would lead to dwellers of coastal areas being relocated. Consequences include increased intensity of storm surges ("Sea Level Rise" n.d.).

6.4.4 Ozone layer depletion

The layer above the troposphere of the Earth's atmosphere is the stratosphere, which extends for approximately 10–50 kilometers above the surface of the Earth. The major role of the stratosphere, which has a high concentration of ozone (O_3), is to absorb and filter the harmful UV rays of the sun to protect existence on Earth. Global warming has caused a large depletion in the ozone layer, which is now causing the harmful UV rays of the sun to reach the Earth's atmosphere, causing serious health conditions like skin cancer to occur among livestock on Earth. These harmful gases that are emitted into the stratosphere cause the breakdown of ozone (O_3) in the layer, leading to ozone layer depletion (Contributors 2025).

The above conditions must be dealt with immediately as they are causing drastic effects to the environment, like climate change, floods, droughts, earthquakes, and the melting of ice caps.

6.5 SMART CLIMATE SURVEILLANCE – PROMOTING TECHNOLOGICAL INNOVATION

In this fast-growing technological era, using innovative technological methods to carry out smart climate surveillance has become mandatory, leading to environmental monitoring. Developing countries' climate alteration responses are supported by inculcating innovative technologies by the Climate Technology Centre and Network (CTCN) (Lee and Mwebaza 2021). The use of modern AI algorithms, drones, remote sensing, AI and machine learning, Internet of Things (IoT), and data analytics for monitoring ecological changes aids in genuine data collection, which is then analyzed to check shifts in climate conditions. In real-world situations, students must be encouraged to develop smart, innovative technological tools that will contribute to data collection and surveillance.

6.6 HARNESSING SMART TECHNOLOGY FOR ENVIRONMENTAL MONITORING

In a scenario today, where we are facing a constant and rapid change in the climate conditions due to anthropogenic activities, monitoring these changes becomes pivotal to control the effects caused and avoid damage to the environment. In a technological era, smart technological digital tools can be used to observe the Earth and give data-driven insights, monitor these climatic crises, and find appropriate solutions. These smart digital tools provide us with more accurate data, which assists authorities in taking advanced corrective measures to mitigate disasters caused by climate change. To possess a sustainable planet, it is important to foster digital transformation. Using data-driven technological tools for data analysis assists in making the right decisions, informed choices, enhances efficiency and accuracy, and displays results based on real-time perceptions (Khanh and Khanh 2024). Data-driven tools aid in improving resilience regarding change in climate conditions by giving better insights about the environment. Some smart technologies that can be harnessed to aid in climate change monitoring are:

6.6.1 Climate change adaptation using AI

AI is used to develop the latest sophisticated climate change monitoring tools that can give professionals prior knowledge about climate imbalance. Weather monitoring done using these AI tools provides us with precise and perfect data. One example of a smart AI technological tool in which AI algorithms are included is the Sea Surface Temperature Ocean models, which can speed up the analysis of huge amounts of data, improving weather forecasts to get a deeper understanding of climate change and ocean current patterns. In the field of agriculture, AI is also used to analyze the genetic data of plants to come up with species that are resistant to drought and can survive with the least amount of water available. AI works smartly when integrated into sewer systems, which, in times of heavy rainfall, prevents flooding (Jain et al. 2023).

6.6.2 Climate adaptation using AI drones

Drones are also called Unmanned Aerial Vehicles, which are equipped with high-end cameras and can hover over vast distances and gather required data. These are controlled using remote controls and can be equipped with sensors that can scan areas and gather precise data. These drones can also possess GPS to be more exact while tracking a location. Alterations in climate can be monitored using drones that can collect visual data that forecasts prominent risks due to climate change. Drones can also be used to reach remote affected areas that are hard to get to for search and rescue operations in case of any natural disaster (World Economic Forum 2024).

6.6.3 Climate adaptation using earth observation

The changes in the environment can be stored and recorded by satellites, which transmit this data to ground antennas; thereafter, the data reaches the weather monitoring stations for weather forecasting ("Imagine the

Universe!" n.d.). Remote sensing is used to monitor and gather data about natural resources, oceanography, disaster control, and agricultural practices from a distance without being physically present by using satellites or sensors. This data is used to check on the changing trends in the environment. For example, information about Earth's climate change, water resources, and melting ice caps is gathered by the European Space Agency (Smith 2025). Earth observation also acts as an early warning signal regarding natural disasters like tornadoes, floods, hurricanes, etc. Satellites are capable of advanced multispectral imaging and thermal imaging that can detect changes in temperature, ecosystem monitoring, and track the lifestyle of wildlife, so we can be aware if they are in any trouble due to climatic conditions.

6.6.4 Climate adaptation using advanced computing

Computers equipped with the capacity for high accuracy and speed are called supercomputers. Quantum computers use photons, which are subatomic particles of light that can carry out many calculations at a single time and carry quantum information known as qubits ("What Is Quantum Computing?" n.d.). They can improve climate surveillance and use techniques like fluid dynamics to enhance weather forecasting. These processes cannot be carried out by traditional equipment. Supercomputing ability aids in accurate weather forecasting and climate modeling.

6.6.5 Climate adaptation using the Internet of Things (IoT)

Handheld devices or sensors are sophisticated technological tools with excellent network connectivity, aiding them in collecting data from other devices using the internet. IoT can act as a monitoring system and help in collecting and sharing data. IoT uses sensors to gather environmental data and pass it on to devices to be processed and analyzed, based on which the environmental changes can be studied. IoT can function without human-to-human or human-to-computer interactions by using networks to transfer data ("Internet of Things (IoT)" n.d.). The sensors in the IoT can sense and gather data related to air temperature fluctuations and air quality. Natural disasters like wildfires can be detected using the sensors and information sent as mobile phone warnings and alerts to people in the vicinity of the disaster. For example, California, which is known for its wildfires, uses the PanoAI, designed by a California-based company with the help of precise cameras and complex data feeds, to detect wildfires by scanning more than five million acres of land area and provide information to the emergency services and the Fire Department professionals. PanoAI is empowered by rapid data analysis, which is supported by AI (YourStory 2023).

6.6.6 Climate adaptation using augmented reality and virtual reality

Users can get an experience of superimposed digital features to get a real-time feeling of being in a particular environment, using headsets to give a fully involved and immersed escapade. Using AR and virtual reality (VR) can give a simulated 3D model exposure of the Earth's surface to individuals and how the world can be impacted due to today's drastic climate change, transforming weather patterns, and biodiversity effects. These experiences that are created can bring about awareness among the public, making them more responsible about the nature around them, safeguarding it to reduce their carbon footprint and protect the environment for future generations (AleksandraSzlacheta 2023). AR can provide experts with real-time data on carbon emissions, air quality, and weather conditions, assisting scientists and environmentalists to take steps to curtail the adverse effects. AR and VR can also help farmers with tools and knowledge to make informed choices to escalate their crop yield by reducing the harm done to the environment due to farming (Letrange 2023). For example, the Climate Tipping Points Hub is a real-time experience designed by a Company and its Co, which utilizes the VR simulators' 3D technology to provide people with an immersive experience of climate fluctuations and their impacts, like drought, Arctic Sea ice that is melting and disappearing, etc. This gives us an overview and visualized form of the impacts in the future due to climate transformation on the global ecology and polar zones (World Economic Forum 2023).

6.6.7 Environmental data integrity using Blockchain

Data once entered cannot be altered or deleted is called an immutable ledger, which is integrated in Blockchain technology. These stores' environmental data regarding carbon emissions, weather changes, etc., and blockchains can check whether organizations are abiding by the rules of sustainability, thus meeting their goals. Sustainable practices can be tracked using these blockchains to verify whether resources are being used responsibly, reducing environmental harm and damage due to misuse.

6.7 BENEFITS OF SMART TECHNOLOGY IN CLIMATE CHANGE MONITORING

Utilizing smart technologies in climate change surveillance has many benefits, as follows:

6.7.1 Data-driven smart decisions

Strategies can be decided based on the timely and accurate data provided by smart surveillance technologies. Policymakers, scientists, and businesses can make quick decisions as smart technology gives data at a rapid speed.

Smart tools also aid in emission tracking, monitoring conservation efforts, and checking environmental programs' effectiveness (Emergent Africa 2024). These smart devices are also used to predict climate change much in advance by displaying data that can be analyzed by professionals. By studying the data, stakeholders can also decide on which place can be used to install a particular resource like solar panels or wind turbines, to use a natural resource to its optimum use, and to improve climatic conditions.

6.7.2 Availability of real-time data

Environmental data that is continuously monitored and collected by smart technology is immediately available, precise, and accurate in real time. This access to data at high speed enables authorities and professionals to take quick action in case of an upcoming environmental disaster like a flood, wildfire, tsunami, a rise in climate temperature, etc. Some technological tools that gather and analyze data in real time are IoT, SOS (the Sensor Observation Service), GIS (Geographic Information Systems), Telemetry Systems, and Machine Learning techniques (Rohan et al. 2025).

6.7.3 Cost-Effectiveness of smart technology

Traditional data tracking and collection methods use costly equipment, huge infrastructure investment, and large-scale manpower. On the other hand, smart technologies used like remote sensing, use of IoT, etc. to perform the same data collection function are more accurate, efficient, and cost-effective. Data can be collected over a large area at tremendous speeds, thus speeding up the data analyzing process and giving ample time to react in case of an environmental threat situation ("How Internet of Things (IoT) Technology Helps Address Climate Change and Environmental Challenges" n.d.).

6.7.4 Awareness and engagement among the public

Data from the environment that is gathered by these smart technologies, once analyzed, can be accessed on phones. People can track and get weather forecasts on their mobile phones using smart technological platforms. Environmental changes like rainfall, air humidity levels, water pollution levels, etc., can be accessed by individuals on their mobile phones. Warnings can also be forwarded to people on their phones by professionals and authorities to alert them about urgent, immediate evacuation plans due to a disaster threat (Chakravarty 2017).

6.7.5 Future climate scenarios predicted

The use of modern and advanced technological tools in climate surveillance can help us to predict future climate scenarios. Data collected from weather ground stations, satellites, sensors, drones, etc., aids in predicting the impact of climate change, like global warming effects on the ecosystem and human habitats. Getting future warnings about natural disasters like floods, droughts, tornadoes, wildfires, etc., gives us enough reaction time to adapt to changes and mitigate the effects. Sustainable methods like proper planning, awareness, resource management, and change in agricultural practices can bring about a better, healthier environment (Emergent Africa 2024).

6.8 IDENTIFYING COMMUNITY-LEVEL CLIMATE CHALLENGES – TARGETING SOLUTIONS

Every locality or area is vulnerable to different events like heat waves, floods, droughts, wildfires, etc., which are identified at the community level. Risks that communities face depend on their geographic location, infrastructure, socio-economic conditions, resources available, etc., which must be considered when data is collected using AI drones, sensors, and surveys of patterns of the surrounding area for climate surveillance. The residents can best give proper insights about the locality they live in. Thus, collecting insights from the residents and analyzing the environmental data collected will lay the groundwork for climate resilience planning and avoiding disasters in the future by coming up with quick and smart solutions.

6.9 ROLE OF EDUCATION INSTITUTIONS – SHAPING A SUSTAINABLE FUTURE

With a vision for a sustainable future, the role of educational institutions is very important in educating students and the public about climate change and how to integrate the latest technologies in climate surveillance, as education can act as a stimulant for resilience toward climate change. Training the students about the latest technological methods and upskilling them to be professionals of tomorrow is the role of educational institutions.

6.9.1 EdTech in climate and environmental awareness

EdTech means integrating the latest and modern technology in the field of education into the classroom environment, by utilizing engaging methods like VR, gamified methods in teaching, robots, the IoT, etc. to educate the students about climate change and its effects, climate surveillance, and environmental changes ("Education Technology: What Is Edtech? A Guide" n.d.). It also brings about awareness among students about their surroundings and environment, and how they can do their part to reduce these effects. EdTech in education also prepares students to be tech-savvy to meet the growing needs in their professional field in the future.

6.9.2 Climate education integrated into the curriculum

Future generations can be made aware of environmental sustainability, weather fluctuations, climate change, and the science behind them, by integrating environment and related topics into the school and university curriculum. Topics related to smart technology like AI, satellites, drones, and sensors should be included in the syllabus so that students gain knowledge about their functions and how these tools aid environmentalists, professionals, and scientists to carry out surveillance, helping them to address the various climate change issues. When students grasp these concepts, they are equipped with skills so that in the future they can be a part of this field of environmental technology ("Climate Education and Leadership".). Training individuals about climate surveillance at the education and learning stage makes leaders of tomorrow, who will help to mitigate environmental issues and come up with smart solutions.

6.9.3 Technology integration in education

The leverage of technology for climate education gives students an idea of the latest trends in the field of technology related to climate surveillance. Integrating complex technological concepts like data analysis, sensor technology, satellite technology, quantum computing, remote sensing, and IoT in the field of education will enhance the learning and understanding of students about how these modern complex technologies are used for climate change surveillance. Courses can be organized by educational institutions related to advanced technological tools that educate students on how to analyze and interpret data related to the climate, thus enabling them to make the right choices and decisions to reduce the impacts of climate change. Universities and schools can integrate topics like climate change surveillance into the STEM curriculum. An amalgamation of hands-on practical exposure and theoretical content knowledge can enable students to get educated about how to innovatively develop and implement these smart technologies.

6.9.4 Collaborations with tech start-ups

Educational institutions must collaborate with tech start-ups that are developing innovative smart solutions for climate conditions. These partnerships will provide the students with ideas and insights about advanced research, giving them access to modern technology like machine learning, satellites, IoT, etc. Students can learn cutting-edge technological methods from these start-ups, which will aid them in becoming experts in their field when they become professionals and will motivate them to be alert about environmental alterations.

6.9.5 Promoting citizen science: Training future leaders

Students are also motivated to participate in citizen science programs that collect data on weather changing patterns, air quality, water pollution, contamination, etc. They can utilize smart technological tools like drones, sensors, and IoT to collect and analyze data, thus doing their part to monitor climate change and seek smart solutions for these changes. Communities and non-professionals can be encouraged to participate in collecting data by offering citizen science projects, which enable individuals to contribute directly to monitoring climate change (Citizen Science 2024). People can be trained on data collection methods like checking air quality, pollution levels in water, local temperature measurement, and humidity levels. The public must be educated on how to contribute to climate monitoring by using smart technologies like mobile apps, community sensor networks, and other advanced tools. They can then report any climatic changes in the local environment to the environmental authorities for quick mitigation of effects. This creates a wide awareness among the public, and they can also become a part of monitoring the environment of the local area where they live.

6.9.6 Hands-on practical learning for students

Educational institutions encourage students to participate in educational, practical hands-on projects and fieldwork (Khanna 2024). Universities can provide advanced programs on the science of climate, sustainability, environmental policies, pollution control rules, etc. The focus of these programs is to create a practical learning session so that students can learn to analyze and monitor climate-related data in a real-time scenario, and come up with smart ideas to mitigate the impact climate change brings about in the environment. Students can be encouraged to take part in research related to the topic of climate, thus fostering problem-solving and innovative ideas for sustaining the environment.

6.9.7 Entertaining cross-disciplinary collaboration

Climate change monitoring is a complex process that involves professionals from various fields like engineering, policymaking, environmental science, and data science, which creates an interdisciplinary collaboration (NIFU 2023). This collaboration in education helps students understand how these multiple fields merge to promote environmental sustainability. For example, students in the engineering field can design a low-cost drone or sensor that can collect data related to climate monitoring, and students in the data science field can attempt to analyze the data collected to recognize changes in patterns in the environment.

6.9.8 Promoting innovation in technology

New and modern methods of environmental monitoring can be discovered when educational institutions collaborate with high-tech companies, and give opportunities to students to invent AI algorithms that can carry out advanced data analysis. Students can create modern drones or sensors that are capable of remote sensing to monitor the ecosystem (World Economic Forum 2024). Such collaborations with real-time applications aid in enhanced learning among students, thus creating an advanced climate monitoring system.

6.9.9 Education on sustainable living

Educating the public and students about how to lead a sustainable lifestyle (earning2gether 2025), like opting for energy resources that can be renewed, decreasing energy consumption, and utilizing products that can be recycled or eco-friendly, will reduce ecological imbalance, which is created by communities harming the environment due to misusing nature's resources. Educational programs and workshops can be organized to educate society about how disturbances caused by climate conditions could be a repercussion of their actions and can lead to people becoming aware and dealing with the issue at its core.

6.9.10 Interactive learning platforms

Students can be taught about climate change issues, like rising sea levels and rise in atmospheric temperature, and their effects by using climate models shown online, virtual simulators, and environmental video games that are interactive to give a real-time experience to the student ("Discover Thousands of Collaborative Articles on 2500+ Skills," n.d.). By exposing them to this smart technology, students will learn to monitor the environment and make decisions to mitigate the impacts of these climatic changes. Students get an experience even before they experience it in real life, and will be ready to face the problem when it comes. These interactive platforms also help to create in them a mindset of how to avoid this situation from happening in the environment.

6.9.11 Educational field trips and outdoor educational activities

Real-time learning can be experienced by the students when educational institutions organize field trips and visits to the local monitoring stations that monitor the climate, botanical gardens, energy sites that are based on renewable resources, and smart farms. This practical exposure will enhance and deepen their knowledge and understanding of environmental issues, smart technologies, and environmental sustainability. After the field trip visit, the institutions can create clubs wherein the students can discuss and present a report on what they have learned during the visit (Sph-Super-Admin 2025).

6.9.12 Training programs and workshops organized

Professional training programs and workshops should be organized by educational institutions for scientists, students, educators, policymakers, and the public to keep them up-to-date with the latest and modern technologies. Advanced workshop on AI prediction of climate, tools for climate surveillance, technologies for remote sensing, etc., should be conducted by educational institutions.

6.9.13 Creating sustainable campuses through education

Environmental sustainability is one of the most important and revered topics today, owing to the climatic changes that are happening around us, due to human exploitation of resources. The education field must include topics related to climate monitoring and sustainable methods to overcome these effects. The public should be educated on how to recycle waste, consider carpooling or public transport, switch over to electric vehicles (EVs), conserve natural resources like water, and replace plastic with paper or cloth bags (Pugh 2025). By switching over to sustainable methods, we can make sure that we will reduce the carbon footprint and thus reduce the harm that is done to our Earth and its environment.

6.9.14 Environmental sustainability and community engagement

Community engagement as a whole and educating them in sustainable and nature-friendly methods is very important to reduce the impact of anthropogenic exploitation of natural resources. The society must be taught about how to monitor their environment using a smart app that can be installed on their mobile, to be aware of climatic changes and become a part of rectifying the issue.

6.9.15 Teachers' role: Champions to initiate a change

While education is vital in shaping individuals, it cannot succeed without teachers, who are essential to the education system. Educators' primary focus should be to train students with skills and knowledge about current environmental issues and ways to address and mitigate them. Additionally, preparing these individuals to become the professionals of tomorrow is a key responsibility of teachers (Kaur 2013). Students must be taught and exposed to the latest smart technologies during their learning and training period to implement them when they embark on their professional journeys. Implementing AI as part of the curriculum will also aid educators in training students to meet the demanding needs of tomorrow.

6.9.16 Advocacy and policy education

In the context of environmental change, educational institutions should educate students about advocacy and policies that should be followed during climate data monitoring. Students should be made aware of policy decisions that are governed by climate surveillance, data monitoring, environmental safety rules, equity, use of resources, and climate change. They should also be upgraded with the existing sustainable rules and practices to bring about a positive alteration at the local, national, and global levels.

6.10 TECHNIQUES OF PROMOTING SUSTAINABLE PRACTICES

Some of the techniques used to promote sustainable practices are as follows:

6.10.1 Agriculture aided with climate smart technologies

Agricultural practices used should be such that they reduce climate change and the impact of environmental issues. Some methods that use modern technologies to do so are using smart crop watering methods, smart agricultural practices, and climate change resilient crops to enhance crop productivity, thus bringing down the environmental change impact.

6.10.2 Shifting to renewable energy

To reduce the carbon footprint, it is mandatory to shift over to renewable resources like solar energy with the use of solar panels, grid-smart technology to generate power, and other alternative smart energy methods that reduce climate change effects.

6.10.3 Sustainable smart urban planning

An urban sustainable environment should be created by utilizing smart technological means like smart waste management, shifting to alternative transport systems that reduce air pollution, reducing consumption of energy, recycling, and reusing. Modern cities should integrate high-end technologies like AI, the IoT, and advanced data analysis methods to increase sustainability and monitor the climate (Bhoda 2024). These latest technologies can also monitor waste generated and management, water utility and consumption, air pollution and quality, and energy levels utilized, thus keeping track of the environment and carbon footprint. They also help society to adapt and find ways to mitigate changes in weather and rising levels of the temperature of the atmosphere.

6.10.4 Creating green buildings

The use of building materials that are eco-friendly, home smart systems that utilize less power, integration of energy that is renewable energy, and smart home devices reduces the carbon footprint. By integrating such methods, the emissions and impact on the environment are drastically reduced. The CO_2 emissions should be reduced by using technologies that mitigate their effects, hence protecting the environment.

6.10.5 Daily sustainable practices

Individuals should be encouraged and educated on how to carry out these sustainable practices daily to reduce climate change, and not use them once in a while.

6.10.6 Raising awareness about climate change

Awareness education is a very important and crucial part where people should be made aware and educated on climate issues and their effects. Checks should be made to ensure that policymakers are meeting their goals on emission tracking, projects on conservation, forest condition monitoring, like deforestation checks, etc. Data collected should be analyzed properly, and people should be kept aware of the results obtained and methods used to eradicate these effects. The effectiveness of initiatives taken to meet sustainability should be constantly monitored for awareness among everyone.

6.11 CASE STUDIES: SUSTAINABILITY AND DIGITAL TECHNOLOGY MERGED SUCCESS STORIES

6.11.1 Canon Global: Sustainability at Canon

The headquarters of Canon are in Shimomaruko, Tokyo, Japan, which carries out many global initiatives. Since 1988, this company, Canon, has strived to live up to its motto of "Kyosei," which means – for a common good cause, living and working together to promote worldwide happiness and prosperity. In April 2024, Canon started its sustainable committee with a CEO and a group of the board of directors who would discuss sustainability-related issues and upcoming activities planned. A society must live in harmony in mutual coexistence with the nature around them to conserve natural resources for future generations to come, irrespective of their caste, creed, religion, language, region, etc. To bring about the existence of such a type of society, Canon has come up with new innovative ideas, merging modern present-day technology and creating products that will bring a solution to the problems the world is facing today due to environmental changes. It promotes a change in lifestyle by using resources wisely to protect our environment and coming up with smart choices. Canon works toward maintaining a sustainable society through its corporate and collaborative activities. Some of the global activities that Canon carries out are creating sustainable policies, organize educational workshops and programs for the community at large, initiating awards for sustainable activities carried out by people, environmental education programs, initiatives to target science innovation for smart climate surveillance, collaborative global activities initiatives, start-up of the sustainability channel by Canon entertaining sustainable practices among people, food drives encouraged, creating a harmony between nature and humans through nature saving initiatives, etc. ("Approach to Sustainability | Canon Global," n.d.)

6.11.2 Offshore wind power and digital technology: A greener world

To create a world that runs completely on Green Energy is our vision, says Michael Biermann, CIO of Orsted Company. He added that they have transformed from a coal and oil-based company completely into a wind energy company. With the environmental change we are experiencing today, it is a priority to shift and embrace smart technological techniques that can create a greener and less polluting atmosphere to live in. Orsted has reduced its consumption of coal by 82% and hopes to reduce its consumption further in the future. To accomplish its mission, Orsted has shifted to data analytics with the use of advanced AI, in collaboration with Microsoft's latest technology, which together work for a greener tomorrow. This initiative by the company can smartly transform the data from 1300 offshore wind turbines into useful green energy that will conserve natural resources and save precious time, too. Every single turbine is installed with sensitive sensors that can provide a large amount of data that can be analyzed using smart technological tools. Integration of present cloud-based modern tools has aided Orsted's company in collaboratively working with 5900 employees effectively daily. Cloud technology has also created smart climate surveillance methods and has reduced the computation time to set up a new windmill from weeks to four to six hours ("Ørsted Builds a Greener World With Offshore Wind Power and Digital Technology - Source" 2023).

6.12 CONCLUSION

Thus, it is paramount to carry out smart climate surveillance today to track the climatic changes that have been brought about mainly due to anthropogenic exploitation. Smart digital tools must be utilized to determine the impact of climate change in real time so professionals can figure out methods to mitigate these effects, creating a bridge between data collection, analysis, and action carried out. It is pivotal to be aware of environmental changes at the Global level and not just at the local level, as climate change has become a change that is affecting our environment all around the globe. It has become of utmost importance for educational institutions to educate students with skills and knowledge about environmental changes and climate change so that they can come up with innovative ideas and solutions to aid in climate surveillance. They must educate them on the advanced technology of today so that they can have this knowledge at their fingertips and be able to use it when they become professionals of tomorrow, to not just be aware but put into action the knowledge they have learned. Integrating smart technology today strengthens the resilience needed to face today's issues of climate change. The inclusion of smart tools as mobile apps helps to make climate monitoring user-friendly and accessible to all so that climate monitoring can happen at a community level, where locals can monitor the climate in their locality and inform professionals about any upcoming climatic disturbances in advance, thus boosting citizen science activities. Collaboration among various sectors and an interdisciplinary approach are very important to come up with smart methods to tackle environmental issues and mitigate their effects at the earliest. Hence, leveraging the smart technology of this modern era for climate surveillance has made it easy to collect, analyze, and come up with solutions. Educational institutions must thus upskill students and the entire society toward sustainable practices so that we can create a safer and cleaner climate for the upcoming generations, and rectify our mistakes for a better, greener planet Earth.

REFERENCES

Admin. "What Are Smart Systems?" *SSIs*. 2020. https://ssi-master.eu/what-are-smart-systems/

AleksandraSzlacheta. 2023. "How Can We Mitigate Climate Change With VR and AR?" GL. April 7, 2023. https://giantlazer.com/how-can-we-reduce-climate-change-with-vr-and-ar/

"Approach to Sustainability | Canon Global." n.d. Canon Global. https://global.canon/en/sustainability/policy/

Basics of Climate Change | US EPA." 2025. US EPA. August 12, 2025. https://www.epa.gov/climatechange-science/basics-climate-change

"Better Understanding of Interdisciplinary Collaboration in Climate Change Research - BICC (12820330)." 2023. NIFU. March 24, 2023. https://www.nifu.no/prosjekter/better-understanding-of-interdisciplinary-collaboration-in-climate-change-research-bicc-12820330/

Bhoda, Santosh Kumar. 2024. "Smart Cities' Role in Combating Climate Change." January 20, 2024. https://www.linkedin.com/pulse/smart-cities-role-combating-climate-change-santosh-kumar-bhoda-a9wic/

Built In. "Education Technology: What Is Edtech? A Guide." *Built In*. Accessed July 4, 2025. https://builtin.com/edtech

Caltech Science Exchange. "What Is Quantum Computing?" *Caltech Science Exchange*. Accessed October 15, 2025. https://scienceexchange.caltech.edu/topics/quantum-science-explained/quantum-computing-computers

Cermak, Alicia. 2024. "Causes - NASA Science." NASA Science. October 23, 2024. https://science.nasa.gov/climate-change/causes/

Chakravarty, Debalina. "Awareness About Climate Change Adaptation Through Mobile Applications." *ResearchGate*. 2017. https://www.researchgate.net/publication/322982185_Awareness_about_Climate_Change_Adaptation_through_Mobile_Applications?utm_source=chatgpt.com

"Citizen Science." *A STAP Background Note*. 2024. https://stapgef.org/sites/default/files/2024-01/Citizen%20Science_0.pdf.

"Climate Education and Leadership." *Teach For All*. Accessed October 15, 2025. https://teachforall.org/climate-education-and-leadership

Contributors, Next Ias. 2025. "Ozone Layer Depletion: Causes, Effects &Amp; Control Measures." NEXT IAS Blog. July 10, 2025. https://www.nextias.com/blog/ozone-layer-depletion/

Discover Thousands of Collaborative Articles on 2500+ Skills." n.d. https://www.linkedin.com/advice/3/how-can-you-use-vr-increase-environmental-awareness-qsqhc

earning2gether. "Education for Sustainable Living." *Learning Together*. 2025. https://www.learning-together.eu/courses-training/education-for-sustainable-living/.

Emergent Africa. "Data-Driven Climate Change Mitigation." *LinkedIn*. 2024. https://www.linkedin.com/pulse/data-driven-climate-change-mitigation-emergent-africa-fqxcf/.

"Global Warming." n.d. https://education.nationalgeographic.org/resource/global-warming/

"How Internet of Things (IoT) Technology Helps Address Climate Change and Environmental Challenges." n.d. Viasat.Com. https://www.viasat.com/perspectives/enterprise/2024/how-internet-of-things-iot-technology-helps-address-climate-change-and-environmental-challenges/

"Imagine the Universe!" n.d. https://imagine.gsfc.nasa.gov/observatories/data/data_ground.html. Accessed Oct 15, 2025.

Jain, Harshita, Renu Dhupper, Anamika Shrivastava, Deepak Kumar, and Maya Kumari. 2023. "AI-enabled Strategies for Climate Change Adaptation: Protecting Communities, Infrastructure, and Businesses From the Impacts of Climate Change." *Computational Urban Science* 3 (1). https://doi.org/10.1007/s43762-023-00100-2

Kaur, Sukhwinder. 2013. "Role of Teachers in Imparting Environmental Education for Sustainable Development." *International Educational E-Journal* 2(2, April–June): 10–15. https://www.oiirj.org/ejournal/apr-may-june2013/02.pdf. Accessed October 15, 2025.

Khanh, Linh, and Linh Khanh. 2024. "What Is Data-Driven Analysis? Methods and Examples." *Thoughts about Product Adoption, User Onboarding and Good UX | Userpilot Blog* (blog). September 24, 2024. https://userpilot.com/blog/what-is-data-driven-analysis/

Khanna, Rishita. "Practical Skills Are Crucial to Prepare Kids for Climate Change." *The New Indian Express*. 2024. https://www.newindianexpress.com/cities/bengaluru/2024/Dec/01/practical-skills-crucial-to-prepare-kids-for-climate-change

Lee, Woo-Jin, and Rose Mwebaza. 2021. "Digitalization to Achieve Technology Innovation in Climate Technology Transfer." *Sustainability* 14 (1): 63. https://doi.org/10.3390/su14010063

Letrange, Caroline. 2023. "Sustainability in Focus: The Power of AR & VR in Climate Data Visualisation." March 3, 2023. https://www.linkedin.com/pulse/sustainability-focus-power-ar-vr-climate-data-caroline-letrange/

Ørsted Builds a Greener World With Offshore Wind Power and Digital Technology – Source." 2023. Source. May 12, 2023. https://news.microsoft.com/source/features/digital-transformation/videos/orsted-greener-world-offshore-wind-digital-technology/

Pugh, John. "Top 55 Sustainability Ideas for Businesses & Green Campuses in 2025." *Sunbolt.* 2025. https://gosunbolt.com/green-campus-sustainability-ideas/

Rohan, Md., Mst. Khatun, Md. Rahman, and Abhijit Pathak. 2025. "Enhancing Flood Disaster Response Through Real-Time Monitoring and IoT: The Case of SentryLeaf." *Internet of Things and Cloud Computing* 13 (1): 1–14. https://doi.org/10.11648/j.iotcc.20251301.11

Sea Level Rise." n.d. https://education.nationalgeographic.org/resource/sea-level-rise/?utm_source=chatgpt.com

Smith, Angela. 2025. "Earth Observation at the Frontline of Climate Action." ITU. March 5, 2025. https://www.itu.int/hub/2024/03/earth-observation-at-the-frontline-of-climate-action/

Sph-Super-Admin. 2025. "9 Tips to Cultivate Sustainability and Environmental Awareness for Students." Sekolah Pelita Harapan | International School, Jakarta, Tangerang, Sentul, Cikarang. May 6, 2025. https://sph.edu/blogs/environmental-awareness-for-students/

World Economic Forum. "How Metaverse Technology Is Catalysing Action on Polar Ice Tipping Points." *World Economic Forum.* 2023. https://www.weforum.org/stories/2023/09/metaverse-technology-catalysing-action-polar-ice-tipping-points/

World Economic Forum. "6 Technologies to Help the World Adapt to Climate Change." *World Economic Forum.* 2024. https://www.weforum.org/stories/2024/02/ai-climate-adaptation-technologies/

World Economic Forum. "World Economic Forum Annual Meeting 2024." *World Economic Forum.* Accessed October 15, 2025. https://www.weforum.org/meetings/world-economic-forum-annual-meeting-2024/

Yasar, Kazim, and Adam S. Gillis. Internet of Things (IoT)." *IoT Agenda, TechTarget.* Accessed October 15, 2025. https://www.techtarget.com/iotagenda/definition/Internet-of-Things-IoT

YourStory. 2023. "Pano AI Wildfire Detection AI Technology." *YourStory.* Accessed October 15, 2025. https://yourstory.com/2023/07/pano-ai-wildfire-detection-ai-technology

Chapter 7

Smart hospitals and healthcare facilities impact of artificial intelligence in healthcare management

Divyedeep Bishnoi, Divanshu Bansal, Dimple Dhiman, Bhawna, Ujjwal Kaushik, and Neha Kukreti

7.1 INTRODUCTION

The uses of artificial intelligence (AI) technology in healthcare are emerging at a rapid pace, and there is several AI solutions aimed to solving the most pressing problems that healthcare organizations currently face. Healthcare operations and management are being revolutionized by AI through improved decision-making, streamlined processes, and enhanced patient care; it becomes more precise, non-disruptive to the person, and streamlined by the employment of natural language processing (NLP), predictive analytics, and machine learning (ML; Samajdar et al., 2025). Industry leaders need to appreciate the current developments of AI technologies to ensure value-based care and how they could improve the efficiency, safety, and availability of healthcare service delivery (Chen & Decary, 2020). Growing evidence supports the use of big data management and handling in the healthcare sector to enhance therapeutic services. In an effort to advance precision medicine, numerous corporate and public organizations have created, saved, and examined various forms of big healthcare data, including clinical, electronic, personal, and sensory data as well as omics data (Gupta & Kumar, 2023). It is clear that AI and related technologies are developing quickly, enabling healthcare providers to increase patient value and operational effectiveness (Mudgal et al., 2022). Few of the impacts of AI on healthcare management have been depicted in Figure 7.1.

7.2 AI-POWERED PATIENT MONITORING

The integration of AI technology with distant patient monitoring represents a novel turning point in the evolution of healthcare. AI is producing a paradigm shift in the healthcare system that might significantly change many features of medical research, patient care, and healthcare management. Remote patient monitoring is one of the few prestigious disciplines (Tsvetanov, 2024). Through the use of AI, this new method tracks vital signs continuously beyond the confines of a hospital (Topol, 2019). These innovations

DOI: 10.1201/9781003586951-7

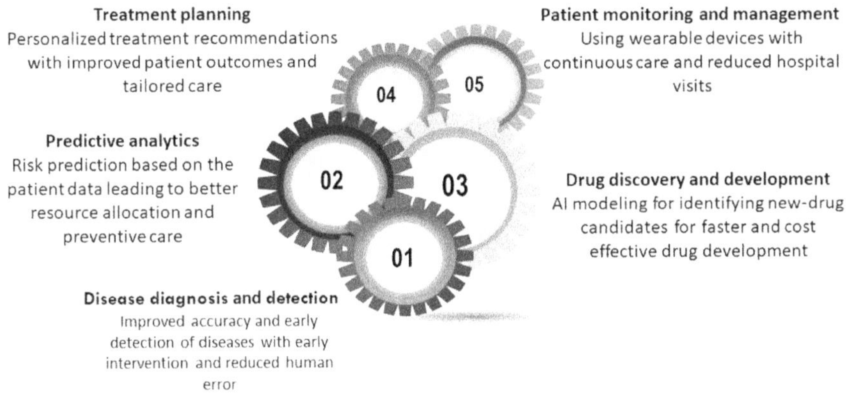

Figure 7.1 Role if AI in healthcare system.

serve to enhance the unparalleled potential breakthroughs in care – innovations that augment healthcare outcomes while easing the system burden, encourage new patient autonomy movements, and enabling sophisticated multi-dimensional interventions. AI-powered RPM systems make it possible to collect, process, and communicate information efficiently. Such systems are complex ecosystems comprising wearable technologies and sensors along with data acquisition and transmission systems, AI algorithms and analytics, as well as user interfaces (Shajari et al., 2023). The use of advanced analytical techniques and data-driven insights has led to the development of predictive models with intelligent alerts which serve as more efficient approaches for offering preventative care. Predictive models enable healthcare practitioners to pay special attention to preventative medicine for those individuals who statistically are more prone to develop particular diseases in the future (Dixon et al., 2024). Further, smart warnings by predictive models inform healthcare practitioners about individuals exhibiting initial warning signals or risk indicators of particular diseases. This supports early and effective treatment, as well as ensures quick preventative intervention. These methods, which utilize the approaches to compare simulated and real outcomes, offer helpful information for model improvement and treatment adjustment, hence promoting accuracy in forecasts and preventive treatments implemented (Dogheim & Hussain, 2023). For a more holistic evaluation of a patient's status, AI-driven RMP systems combine data from wearables, smart home systems, and mobile health apps. RPM systems also offer extensive patient monitoring through monitoring of vital parameters, physical activity levels, medication compliance, and mental health information (Yang et al., 2019). AI can assist RPM systems in reducing hospital and emergency department readmissions through early diagnosis and treatment of patients, thus presenting a gigantic cost savings to healthcare systems (De Guzman et al., 2022) shown in Table 7.1.

Table 7.1 Real-time applications of AI-based monitoring systems

AI-based patient monitoring system	*Area of application*	*Details*	*Reference*
Video-based monitoring – Telehealth monitoring using sensors, RGB camera	Able to track vital indicators in patients, including heart rate, blood pressure, oxygen saturation (SpO_2), cough analysis, and respiratory rate.	To estimate the vital indicators, e.g., breathing rate and heart rate, image and video processing algorithms helped determine a region of interest on the patient, e.g., facial benchmarks. Depending on remote photoplethysmography (rPPG) found in a video frame captured by an ordinary smartphone camera, heart rate, blood volume pulse, and SpO_2 were calculated.	Bousefsaf et al. (2019), Khalid et al. (2022), Rivella et al. (2023)
IoT-enabled wireless networks and Bluetooth for patient monitoring	Wearable vital signs sensing needs can take many different forms, ranging from physiological measurements like blood glucose, blood oxygen saturation, electrocardiograms, capnography, and skin sweat to cardiac implantable and mobility evaluation devices.	A subdomain of the Internet of Things called wireless body area networks (WBANs) links wireless sensors on a patient's body to the network. RPM sensors limit the daily activities of people with chronic illnesses by requiring skin contact in order to get vital indicators.	El-Rashidy (2021)
Continuous glucose monitoring (CGM) using CGM-derived metrics, such as the glucose management indicator (GMI)	—	Allows continuous real-time glucose monitoring, which provides crucial information that has an immediate effect on the management of hyperglycemia and the molecular processes that support it, such as insulin resistance and beta-cell dysfunction.	Klupa et al. (2023), Uhl et al. (2024), Jospe et al. (2024)
Advancing pain management with AI-enhanced RPM such as the NXTSTIM EcoAI™ and neuromodulation therapies	—	Real-time data analysis by these technologies can identify physiological and biochemical alterations that could point to changes in pain pathways or the effectiveness of treatment. In order to restore therapeutic advantages, AI can suggest changes to neuromodulation parameters, such as altering pulse width, amplitude or occurrence.	Tran and Crawford (2020), Mekhail et al. (2024) Brandao et al. (2024)
AI-powered teleassessment technologies such as TeleFE and a remote smart home healthcare support system (ShHeS)		Because it allows for remote patient monitoring, patients can receive medical advice without putting a strain on hospitals during a pandemic. Along with support for health information on a smartphone that sends notifications, it also has smart home features to manage appliances at home.	Taiwo and Ezugwu (2020)

(Continued)

Non-contact screening system	For COVID-19 patients, this system employed logistic regression and XGBoost methods. Ultra-wideband radio-impulse radar was used to track physiological markers like respiration, heart rate, body movements, and sleep quality.	It contrasted 144 detector nursing data from healthy patients with 140 sensor nursing data from 23 COVID-19 patients. XGBoost + LR outperformed other machine learning algorithms in terms of discriminating.	Huang et al. (2020)
Cloud-based remote patient monitoring system (CloudDTH)	Using this approach, wearable technology was utilized to track, diagnose, and forecast people's health in an effort to improve it, especially as they age. The issue of accurate crisis warning and real-time supervision for senior citizens in healthcare services was addressed by the proposal of digital twin healthcare (DTH).	By identifying and warning employees of potentially concerning condition shifts in a tenth of a second, this system, which combines the real and virtual worlds, can reduce the end-to-end latency. It would also be beneficial to test and validate the system in a real-world setting.	Rahim et al. (2024)
RFID (radio frequency identification) is a non-invasive digital technology used for remote patient monitoring	In a mental health hospital, it makes contactless monitoring of critically ill patients possible. Additionally, it opens a predictive environment where patients' vital signs can be predicted for the future.	Using its built-in NCS mechanism, an AI-enabled RPM system framework with a non-invasive digital technology RFID retrieves patients' vital signs and physical movements. Determine the patients' vital signs for the next three hours based on the time series data that was recovered, and categorize their physical activities into ten designated physical activities. This paradigm aids in preventing unanticipated clinical catastrophes and implementing preventative measures with timely medical intervention.	Tao et al. (2021)
NodeMCU microcontroller system	Using sensors attached to the patient's body, this gathers the patient's temperature, heart rate, SpO_2, ECG, and other vital signs. It then sends the patient's doctor detailed medical information and, if the parameter threshold is reached, the current emergency status.	To measure data in real time, all of the sensors are attached to the patient's body. The microcontroller then receives the measured results. The ESP32 internet module is connected to the NodeMCU microcontroller. All of the data from this controller is gathered by an IoT platform (Blynk) and shown to the physician. An emergency email would be sent to the doctor if any of the parameter values deviated from the threshold. A physician can use live calling to speak with a patient.	Nooruddin et al. (2020)

7.3 AI-ENABLED SMART DEVICES IN PATIENT HEALTHCARE

The traditional healthcare system has been replaced by smart healthcare due to recent developments in the Internet of Things (IoT), AI, and cloud computing. Important technologies like IoT and AI can be used to enhance medical services (Mansour et al., 2021). By assisting physicians in gathering objective data on patients' chores and progress, AI systems have the possible to progress healthcare services (Zafar et al., 2024). Continuous monitoring of medication adherence, lifestyle factors, and vital signs is made possible by wearable technology, remote monitoring, and AI algorithms. Healthcare professionals may quickly adjust treatment regimens and give patients immediate assistance thanks to these real-time insights (Manickam et al., 2022). Numerous sensors collect patient data, which is then transmitted to a fog, edge, or cloud server for processing over a LoRa or Wi-Fi gateway. To monitor their patients' health, doctors, and their staff use a range of user interface technologies, including web or mobile applications (Shiwlani et al., 2025).

7.3.1 Wrist mount devices

Watches and bands are examples of wrist-mounted gadgets. Wearable technology comes in a variety of forms, including smart watches, smart bands, smart glasses, and bright clothing.

7.3.2 Smart watches

Smartwatches (SWs) are among the most substantial wearable technology in the modern era. In addition to serving as a fitness following device that helps handlers to keep track of their day-to-day activities, such as heart rate, workout times, calories burned, and step counts. Smart watches also monitor convinced human biological indications and biomechanics. They gather data and store it on a cloud server for later use (Mohankumar & Jayaramaiah, 2021). For some physicians, the volume of information that patients offer them via wearable health monitoring devices is becoming too much to handle. However, the wearables industry's explosive growth presents a challenge to the medical sector. Not all information gathered by wearable technology is useful to doctors, and not all wearable technology is made equal. Doctors cannot be sure that the data they get is accurate because the FDA does not currently regulate the majority of wearable technology (Sharma et al., 2021). SWs were initially designed to track messages and alarms without requiring a phone to be removed. When SWs are worn on the wrist, it facilitates health and wellbeing monitoring. For example, SWs can use their sensing abilities to determine whether a doctor's appointment is necessary. Additionally, SWs help identify those who have sleep apnea. People in this condition start breathing again while they're asleep. If treatment is not received, it may lead to elevated blood pressure, a persistent problem with concentration, and a higher risk of stroke (Bilge et al., 2024).

7.3.3 Smart glasses

In addition to serving as a computer, smart glasses are a kind of wearable technology that may be worn on the face and achieve the initial goal of improving vision. Numerous studies on smart glasses have been carried out in a variety of sectors, and a number of review papers have assessed the devices' usefulness for study (Gao et al., 2024). Dickey et al. created a Google Glass application to train urology residents using augmented reality during surgery, specifically for inflatable penile prosthesis placement. Google Glass successfully demonstrated the potential for future use in the training process and fostering interaction between trainees and specialists (Dickey et al., 2016).

Dr. Homero Rivas from Stanford University implemented augmented reality in surgical labs to teach anatomy in 3D using HoloLens by Microsoft. The device enables students to see and interact with anatomical structures, providing better understanding than traditional two-dimensional graphics in books. Pilot studies were conducted at Case Western Reserve University and Cleveland Clinic (Wrzesińska, 2015). Researchers also evaluated the advantages of smart glasses for Parkinson's disease including motion tracking, voice and visual cues, and object recognition, with patients in the Netherlands enthusiastic about using this technology (Zhao et al., 2015). Smart glasses can be also used for inspection, therapist observations, coaching, and patient care, particularly for visually impaired, dementia, autism, and obese individuals, providing quick access to electronic health records (EHRs; Zuidhof et al., 2024). Smart glasses are gaining interest in healthcare for remote instruction, procedure documentation, patient empowerment, signal data reading, instructional films, wound care, education, and monitoring and simulation (Klein et al., 2015; Wüller et al., 2019).

7.3.4 Wrist patches

For the purpose of analyzing sweat samples in real time, a flexible patch system based on microfluidics was created. This sensor is built using an exclusive spiral-channel microfluidic entrenched with ion-selective devices on a stretchy plastic substratum. The sensor may be able to track sweat rate and ion concentrations (H^+, Na^+, K^+, and Cl^-), which would make it easier to track physiological and pathophysiological conditions in humans using sweat parameters (Mohankumar & Jayaramaiah, 2021). This study insights a pioneering wearable wristband with microneedles (MNs) for portable health intensive care, addressing trials in incessant, precise finding of small analyte capacities and concentrations. The MN patch is united with a multilayer device using agarose, $Mn_3[Fe(CN)_6]2$, and cotton fiber, which reacts with TMB (3,3′,5,5′-tetramethylbenzidine) to show the oxalate levels over color changes from green to light or dark blue, depending on concentration (Amourizi et al., 2025).

7.3.4.1 Chest-mounted devices add case studies

The key concerns of caretakers or health professionals are the monitoring of individuals' postural disabilities and falls. There are two commercially available alert systems for safety monitoring: the Alter One medical alert system and the Life Alert Classic from Life Alert Emergency Response Inc (Junaid et al., 2022). These gadgets have a push button built into a necklace, and pressing it sends a wireless message to a distant place. In parallel, the Well core system uses accelerometer units and sophisticated microprocessors to track postural movement. This technology can further communicate with a remote center and distinguish between normal and fall-related bodily movements. Another chest-worn gadget for tracking temperature, sleep patterns, and heart rate is MyHaloTM by Halo MonitroingTM. In conclusion, a mobile device with an integrated system that has a balance sensor that initiates automated emergency contact dialing in the event of a fall will be beneficial in a significant way (Mohankumar & Jayaramaiah, 2021). A low-power wearable device patch is anticipated for IoT-connected remote long-term healthcare applications, allowing wearable health monitoring of body temperature, ECG, heart rate, and PPG (photoplethysmography) (Wu et al., 2020). Participants used VitalPatch and gold standard devices to perform seven consecutive movements: tapping, rest, drinking, sit-to-stand, turning pages, using a tablet, and rubbing. Participants underwent hypoxia, reducing oxygen saturation to 80%. Accuracy was measured using VitalPatch estimates compared to gold standards, with clinically suitable rates within 3 respirations per minute for RR and 5 beats per minute for HR. The study demonstrated that VitalPatch, despite some limitations, is highly accurate for early detection of cardiorespiratory worsening (Morgado Areia et al., 2021).

7.3.5 Smart lenses

Tears are complex fluid consists of more than 98% water, along with various proteins, electrolytes, and metabolites. Some components in tears might help diagnose human metabolites. To gather tears without harming the eye, contact lenses offer a good option as they touch the basal tears (Park et al., 2024). Smart contact lenses (SCLs) differ from typical contact lenses because of their sophisticated functionalities that do more than correct vision issues. Such adaptable devices are capable of monitoring health since they contain small electronic components embedded within them. Such components enable SCLs to monitor alterations in tear fluid composition examining parameters such as glucose lactic acid, and other indicators to monitor various eye diseases (Kazanskiy et al., 2023). SCLs contain different parts like power source, sensors, biomaterial, and circuits making them capable of measuring and communicating with users or physicians. Humans use SCLs to provide drugs, test IOP, and measure glucose and lactic acid concentrations (Wu et al., 2024). The devices include minute technologies that can capture physical features of eyes and biomarkers to offer relevant information for the observation, detection, and therapy of diseases (Seo et al., 2023). Jang et al. introduced an SCL in March 2021 to diagnose and manage chronic ocular surface inflammation (OSI) diseases. OSI diseases such as dry eye syndrome and meibomian gland dysfunction are progressively widespread, but physicians still struggle to diagnose them because of atypical clinical signs and indications (Jang et al., 2021). Pan et al. developed a graphene field-effect transistor biosensor integrated into a contact lens based on the specific antigen-antibody recognition mechanism. The biosensor, which has a metalloproteinase-9 (MMP-9)-specific antibody attached to it, can detect OSI concentration in tears by changing its electrical signal (Pan et al., 2025). Smart contact lenses are easy to use and work well with eye disorders. These lenses help more of the medicine stay in the eye and less of it wash away. This makes it easier for people with long-term eye issues to stick to their treatment. New electronics and tiny tech solve many of the problems these lenses used to have. Now companies can make and sell them, which means they can help even more people (Jang et al., 2021). Scientists have come up with a soft clever contact lens that can measure how much cortisol is in tears without wires. This lens has a special sensor, a device for near-field communication, and an antenna. It is battery-free and wire-free. Experiments on humans and animals indicate that the lens is safe to use and could assist doctors in detecting diseases through computer programs that learn patterns (Ku et al., 2020).

7.4 AI IN IMAGING AND DIAGNOSTICS

AI, the technology forerunner in data processing, has revolutionized medical imaging through its myriads of applications enhancing patient care, diagnostic accuracy, and efficiency (Kufel et al., 2023). Diagnostic imaging in healthcare is profoundly being changed by AI. The interpretation and use of medical images, including MRIs, CT scans, and X-rays have become much better with this instrument, which incorporates complex ML and algorithms (Najjar, 2023). Extraction and quantification of numerical features of medical images are the aim of "radiomics" research area. Intensity, shape, texture, etc., properties integrated within medical information analyze to furnish features (Mayerhoefer et al., 2020). Texture analysis is a valuable method that employs images of several modes to provide ML models with parameters that are meaningful for diagnosis. For example, in contrast to deep learning methods, texture analysis of individual vertebrae in CT scans provided a higher accurate prediction of the age of the patient (Obuchowicz et al., 2021).

A few of the most significant benefits of AI in medical imaging include the ability of AI to accelerate the analysis of medical images. The traditional image interpretation procedures are timewasters and prone to human mistakes.

The time it would take to treat a patient can be shortened considerably, however, with the use of AI, which can see and study images much quicker (Tang, 2019). With the seconds ticking in a life-or-death scenario, this level of speed is extremely crucial. Further, AI increases diagnostic accuracy. ML algorithms in AI can detect patterns and deviations that the human eye might miss by learning from enormous databases of medical images. This improved precision is very important in cutting the rates of misdiagnosis and ensuring that patients are given proper treatment as quickly as possible (Srivastav et al., 2023).

Acquisition, feature extraction, preprocessing, classification, registration, segmentation, visualization, and object localization are the eight broad categories into which AI-based imaging techniques can be divided. Comprehending the complicated processes of the human brain and diagnosis of neurological diseases rely considerably on the procedures adopted in brain imaging and diagnosis of brain tumors, an area of great interest for AI in medical imaging and diagnosis (Pinto-Coelho, 2023). Techniques such as positron emission tomography (PET), electroencephalography signals (EEG), magnetic resonance imaging (MRI), and functional MRI (fMRI) yield significant information about the brain's structure, connectivity, and function. With increased understanding of the complex nature of the human brain, the integration of AI with brain imaging has the capability to transform neuroscience and patient treatment. Brain tumors are detected and classified by a potent combination of deep learning methods and the sine–cosine fitness grey wolf optimization algorithm (Yen et al., 2023). It treats the necessity of precise detection and classification of tumors and the challenges that come with it. Convolutional neural networks (CNNs) tackle complexity and ambiguity by automatically learning and obtaining relevant characteristics for tumor research. Due to its importance in the diagnosis and monitoring of COVID-19 disease, lung imaging has also attracted a lot of interest from researchers. Identification and assessment of lung pathology rely significantly on accurate lung area segmentation in medical images (Thanoon et al., 2023). It focused on using encoder–decoder networks for segmentation with super-pixel scaling to segment lung regions in chest X-ray images. Metrics of evaluation such as the Jaccard index, Dice similarity coefficient, specificity, accuracy, and sensitivity illustrate that the proposed approach, being a combination of super-pixel resizing and an encoder–decoder network for segmentation, is very efficient compared to other methods (Lee et al., 2022).

7.5 AI-POWERED DRUG DELIVERY AND CLINICAL SUPPORT

7.5.1 Precision medicine and clinical decision support

Precision medicine, also known as personalized medicine or customized therapy, is a way of delivering medical treatment that is tailored to every patient according to their specific characteristics, such as environment, genetics, biomarkers, and lifestyle (Quazi et al., 2022). One of the emerging fields that demonstrate the potential of precision medication on a large scale is personalized treatment (Johnson et al., 2021). Prediction of the response to chemotherapy was well established in research by Huang et al., employing patient gene-expression information to train a provision ML model (Huang et al., 2018).

The authors used the gene-expression profiles of 175 cancer patients to forecast the manner in which the patients would respond to certain standard-of-care chemotherapy drugs. Importantly, the study was positive in results, with predictive accuracy above 80% for a number of drugs. The results show how AI can be used to forecast response to treatment. In another study, Sheu et al. utilized AI and 17,556 patients' EHRs to forecast the response of individual patients to various antidepressant classes. The study showed that real-world EHR data with AI modeling can be used to predict antidepressant response well, opening up the possibility of creating clinical decision support systems for better medication selection (Sheu et al., 2023).

7.5.2 Dose optimization and therapeutic drug monitoring

With the help of AI algorithms, doctors can optimize dosages of medicine for specific patients and forecast future adverse drug events. Creating an AI prediction model for prothrombin time international normalized ratio (PT/INR) and a decision support system for warfarin maintenance dose optimization was the aim of the project (Gan et al., 2025). The algorithm performed better than experienced clinicians with substantial differences in the prediction of future PT/INRs, and the derived individualized warfarin dose was accurate, according to authors' analysis of data from 19,719 inpatients in three centers (Lee et al., 2021).

7.5.3 AI assistance in population health management

To enhance patient outcomes, decrease expenses, and facilitate predictive analytics and risk assessment, ML algorithms and other technology are utilized to examine data and create prognostic models. Finding individuals at risk of developing long-lasting diseases like endocrine or heart disorders is one area where predictive analytics can be of immense help. Predictive models can identify who is likely to develop the disorders and treat or prevent them by gathering information such as demographics, lifestyle, and medical history (Nelson et al., 2020). AI is able to improve the quality of care and lower costs while enhancing population health. It can identify patients eligible for pre-emptive health management or at high risk, recognizing patterns in free text, imaging, and health system interactions, augmenting patient identification. AI-powered technologies are able to improve population health services by managing nonclinical tasks such as appointment scheduling and enabling the human workforce to work on patient-facing tasks (Maleki Varnosfaderani & Forouzanfar, 2024). They are also able to anticipate clinical deterioration risk, extract patient data, and detect risk factors. Health systems can also benefit from AI's ability to inform payers of risk factors through overcoming the drawbacks of manual chart review (Mutharasan & Walradt, 2024). AI use in disease forecasting greatly enhances public health through anticipating infectious illness propagation and guiding-related health measures accordingly. AI also has the vital role to play in public health by being able to aid fast outbreak control and response by applying huge analysis of data, spotting trends, and projecting forward outcomes (Noorbakhsh-Sabet et al., 2019). AI can promote risk prediction for public health through big data analyses, including huge electronic health data, and viewing complex data like genomics and images. The combination of AI with other technologies, like wearable devices and genomics, can make more precise predictions. Spatial modeling, where interventions are localized to disease-burdened areas, can also be aided by the efficacy and accuracy of AI (Olawade et al., 2023). AI algorithms analyze EHR data, yielding information for disease pattern detection, personalized medicine, and early outbreak identification. They enhance diagnostic procedures, improve individual and community health, and reduce lab costs, though they have drawbacks (Fitzpatrick et al., 2020; Gunasekeran et al., 2021).

7.5.4 AI in drug information and consultation

ML and AI in the pharmaceutical industry have the potential to add almost $100 billion a year to the US healthcare system, according to the McKinsey Global Institute. According to researchers, these technologies improve decision-making, foster innovation, boost the efficacy of clinical trials and research, and create new tools that help patients, insurers, regulators, and healthcare professionals (Alowais et al., 2023).

AI can improve healthcare efficiency, patient flow, and caregiver experience by reducing inefficiency and enhancing caregiver safety. It can be applied to remote patient monitoring, identifying and providing timely care, and integrating AI chatbots with wearable devices for improved behavior, sleep, and wellness. AI has shown to outperform human experts in image-based analyses in various medical specialties, including dermatology, radiology, cardiology, and pathology (Dailah et al., 2024). Researchers are discovering the cellular and molecular basis of disease, assembling multimodal datasets for biological and digital biomarkers for severity, progression, and diagnosis. Forthcoming AI applications include immunomics/synthetic biology and drug discovery (Bajwa et al., 2021).

7.6 AI IN DRUG RESEARCH

The current state of drug-target interaction prediction research can be broadly categorized into three groups: chemical genome-based methods, molecular docking-based methods, and ligand-similarity-based methods. By evaluating the chemical structural similarities between medicines and targets, ligand-similarity-based approaches generate predictions (Zhang et al., 2023). The molecule-docking method simulates the binding process between the medication and the target in order to assess the possibility of binding. Chemogenome-based methods predict a drug's target by examining how it affects the genome (Huang et al., 2024).

Pun et al. used deep learning techniques to predict drug pharmacological effects in amyotrophic lateral sclerosis (ALS) by identifying 18 druggable genes based on disease-specific data (Pun et al., 2023). Deep learning was used to identify genes linked to age-related diseases and adult cells to uncover a novel target (COX7A1) in embryonic-fetal transition, providing insights into normal development and cancer. AI can transform biomedical research through insights and solutions, especially target identification for drug discovery. BIOVIA Generative Therapeutics Design application utilizes 3D structural models to precisely represent. AI technology can make clinical trial design better, improve success rates, and facilitate hypothesis generation, drug discovery, cohort assembly, monitoring, compliance, and endpoint selection, all needing collaboration to collect and organize data (Davkhar et al., 2024). AI is important in drug discovery as it detects drug-drug interactions, which can result in modified effects or drug adverse reactions. This is done by digging through large sets of data and identifying patterns. New AI algorithms precisely predict new drug pairs, showing its potential in drug discovery (Gupta et al., 2021). The ML algorithm effectively identified new inhibitors for proteins such as beta-secretase (BACE1) and for Alzheimer's disease, showing its potential in drug discovery (Dhamodharan & Mohan, 2022).

Supervised and unsupervised learning are AI methods applied in drug discovery. Supervised learning using input-labeled data to train models for classification or prediction while unsupervised learning methods employ unlabeled data to discover patterns and cluster input without prior knowledge. These methods involve classification and regression algorithms, and open-source libraries such as Scikit-learn, PyTorch, and Keras are employed to practice these algorithms (Chen et al., 2023).

7.7 CHALLENGES WITH ARTIFICIAL INTELLIGENCE IN HEALTHCARE

Although there are numerous benefits to using AI-based patient monitoring, there are also strong disadvantages, including the necessity to safeguard sensitive healthcare information from cyber-attacks and unauthorized access (Khalid et al., 2023). In addition, these systems are potentially resource-consuming and complex, requiring large-scale technological investments along with training of healthcare professionals. Adoption is financially hindered by the immense cost of incorporating and sustaining AI technologies within RPM systems, particularly for disadvantaged populations and smaller health centers (McGlynn & Asch, 2020). However, challenges are usually managed with due consideration by practitioners, and whenever collaboration is realized, scaling can always be managed.

7.8 DISCUSSION

A paradigm shift in the provision, planning, and optimization of medical services has been brought about by the use of AI into healthcare management. AI is transforming healthcare systems with the goals of increasing productivity, cutting costs, and improving patient outcomes through anything from administrative automation to predictive analytics. AI algorithms can help clinicians make rapid and correct judgments by evaluating large datasets from EHRs, diagnostic images, and patient histories. This improves patient safety and care quality by increasing diagnostic accuracy and lowering the possibility of medical errors. AI-powered solutions assist hospital managers with resource management by predicting patient admissions, optimizing staff schedules, and controlling bed occupancy rates. These predictive models help hospitals make better use of their infrastructure and guarantee greater readiness for spikes in demand, like those that occur during pandemics. AI is essential to personalized treatment as well. Healthcare professionals can use ML models to customize treatment regimens based on a patient's genetic composition, lifestyle, and current medical data. Therapeutic results and patient satisfaction are enhanced by this change from one-size-fits-all to patient-centric therapy. Administratively, AI tools like robotic process automation (RPA) are making regular operations like arranging appointments, processing claims, and billing more efficient. In addition to lowering operating expenses, this frees up healthcare professionals to concentrate more on patient care and engagement. Additionally, NLP-powered chatbots and virtual health aides are revolutionizing patient relationships. By offering round-the-clock assistance, prescription reminders, symptom monitoring, and mental health interventions, these tools improve the responsiveness and accessibility of healthcare. The use of AI in healthcare administration presents ethical, legal, and regulatory obstacles despite its revolutionary promise. Strong governance frameworks and interdisciplinary cooperation are required due to worries about algorithmic bias, data privacy, and the requirement for transparent AI decision-making.

7.9 FUTURE SCOPE

AI in healthcare management has a bright future ahead of it, with the potential to completely transform the way medical services are provided, optimized, and customized. It is anticipated that as AI technologies develop, their incorporation into healthcare systems will become more impactful, intelligent, and seamless. Predictive analytics will advance in the upcoming years to accurately predict disease outbreaks, readmissions, and the need for emergency treatment. AI will provide evidence-based insights during diagnosis, surgery, and treatment planning, assisting in real-time clinical decision-making more and more. AI-powered chatbots and virtual health aides are probably going to grow more perceptive and sympathetic, allowing for ongoing, individualized care outside of hospital settings. AI in administrative activities will improve operational efficiency and lessen systemic pressures, from personnel management to supply chain optimization. Healthcare management systems will be further strengthened by the integration of AI with other cutting-edge technologies such as blockchain, augmented reality, and the Internet of Medical Things (IoMT), which will guarantee data security, interoperability, and improved training through simulations. However, the future will require strong ethical frameworks, ongoing AI algorithm validation, and proactive regulation to address algorithmic fairness, accountability, and data privacy in order to realize this potential. In summary, AI will not only assist but also serve as a fundamental component of intelligent, effective, and just healthcare management systems across the globe.

7.10 CONCLUSION

The conclusions show that most hospitals usage of AI-based instrument to improve skills and knowledge of their healthcare specialists for patient diagnosis and treatment. AI systems have also been shown to expand the efficacy and organization of hospitals' nursing and management functions. Healthcare professionals are definitely accepting AI in several arenas. Though, its applications offer both the utopian (new prospects) as well as the dystopian (challenges). Dissimilar cynics, AI should not be seen a probable source of progression.

REFERENCES

Alowais, S. A., S. S. Alghamdi, N. Alsuhebany, T. Alqahtani, A. I. Alshaya, S. N. Almohareb, A. Aldairem, M. Alrashed, K. Bin Saleh, H. A. Badreldin, M. S. Al Yami, S. Al Harbi, and A. M. Albekairy. "Revolutionizing Healthcare: The Role of Artificial Intelligence in Clinical Practice." BMC Medical Education 23, no. 1 (2023): 689.

Amourizi, F., K. Dashtian, R. Zare-Dorabei, and S. Hajati. "Transformative Personalized Oxalate Biomarker Analysis through a Wrist-Worn Microneedle Device Integrated with Duplex Nanozyme Toolbox." Sensors and Actuators B: Chemical 423 (2025): 136731.

Bajwa, J., U. Munir, A. Nori, and B. Williams. "Artificial Intelligence in Healthcare: Transforming the Practice of Medicine." Future Healthcare Journal 8, no. 2 (2021): e188–e194.

Bilge, P. C., E. K. Fatma, S. Cansu, S. Haydar, K. Deniz, K. Alisher, and O. Ibrahim. "Scanning of Obstructive Sleep Apnea Syndrome Using Smartwatch: A Comparison of Smartwatch and Polysomnography." Journal of Clinical Neuroscience 119 (2024): 212–219.

Bousefsaf, F., A. Pruski, and C. Maaoui. "3D Convolutional Neural Networks for Remote Pulse Rate Measurement and Mapping from Facial Video." Applied Sciences 9, no. 20 (2019): 4364.

Brandao, M., M. Talih, S. Holden, F. Fernandes, T. Graven-Nielsen, and R. Lucas. "Pain History and Experimental Pressure Pain Responses in Adolescents: Results from a Population-Based Birth Cohort." European Journal of Pain 28, no. 1 (2024): 70–82.

Chen, M., and M. Decary. "Artificial Intelligence in Healthcare: An Essential Guide for Health Leaders." Healthcare Management Forum 33, no. 1 (2020): 10–18.

Chen, W., X. Liu, S. Zhang, and S. Chen. "Artificial Intelligence for Drug Discovery: Resources, Methods, and Applications." Molecular Therapy – Nucleic Acids 31 (2023): 691–702.

Dailah, H. G., M. Koriri, A. Sabei, T. Kriry, and M. Zakri. "Artificial Intelligence in Nursing: Technological Benefits to Nurse's Mental Health and Patient Care Quality." Healthcare (Basel, Switzerland) 12, no. 24 (2024): 2555. https://doi.org/10.3390/healthcare12242555

Davkhar, S. S., S. D. Sanika, and N. B. Khedkar. "A Review of Artificial Intelligence in Drug Discovery and Development." International Journal of Pharmaceutical Sciences 2, no. 3 (2024): 7–22.

De Guzman, K. R., C. L. Snoswell, M. L. Taylor, L. C. Gray, and L. J. Caffery. "Economic Evaluations of Remote Patient Monitoring for Chronic Disease: A Systematic Review." Value in Health 25, no. 6 (2022): 897.

Dhamodharan, G., and C. G. Mohan. "Machine Learning Models for Predicting the Activity of AChE and BACE1 Dual Inhibitors for the Treatment of Alzheimer's Disease." Molecular Diversity (2022): 1–17.

Dickey, R. M., N. Srikishen, L. I. Lipshultz, P. E. Spiess, R. E. Carrion, and T. S. Hakky. "Augmented Reality Assisted Surgery: A Urologic Training Tool." Asian Journal of Andrology 18, no. 5 (2016): 732–734.

Dixon, D., H. Sattar, N. Moros, S. R. Kesireddy, H. Ahsan, M. Lakkimsetti, M. Fatima, D. Doshi, K. Sadhu, and M. Junaid Hassan.(2024) "Unveiling the Influence of AI Predictive Analytics on Patient Outcomes: A Comprehensive Narrative Review." Cureus 16, no. 5 (2024): e59954. https://doi.org/10.7759/cureus.59954

Dogheim, G. M., and A. Hussain. "Patient Care through AI-driven Remote Monitoring: Analyzing the Role of Predictive Models and Intelligent Alerts in Preventive Medicine." Journal of Contemporary Healthcare Analytics 7, no. 1 (2023): 94–110.

El-Rashidy, N., S. El-Sappagh, S. M. R. Islam, H. M. El-Bakry, and S. Abdelrazek. "Mobile Health in Remote Patient Monitoring for Chronic Diseases: Principles, Trends, and Challenges." Diagnostics 11, no. 4 (2021): 607.

Fitzpatrick, F., A. Doherty, and G. Lacey. "Using Artificial Intelligence in Infection Prevention." Current Treatment Options in Infectious Diseases 12, no. 2 (2020): 135–144.

Gan, T. R. X., L. W. J. Tan, M. Egermark, A. Truong, T. L. Kumar, K. Tan, S. B. Tang, S. Blasiak, A. Goh, B. C. Ngiam, and K. Y. Ho, D. "AI-assisted Warfarin Dose Optimisation with CURATE.AI for Clinical Impact: Retrospective Data Analysis." Bioengineering & Translational Medicine 10, no. 3 (2025): e10757. https://doi.org/10.1002/btm2.10757

Gao, L., C. Wang, and G. Wu. "Wearable Biosensor Smart Glasses Based on Augmented Reality and Eye Tracking." Sensors 24 (2024): 6740.

Gunasekeran, D. V., R. M. W. W. Tseng, Y. C. Tham, and T. Y. Wong. "Applications of Digital Health for Public Health Responses to COVID-19: A Systematic Scoping Review of Artificial Intelligence, Telehealth and Related Technologies." NPJ Digital Medicine 4, no. 1 (2021): 40.

Gupta, N. S., and P. Kumar. "Perspective of Artificial Intelligence in Healthcare Data Management: A Journey towards Precision Medicine." Computers in Biology and Medicine 162 (2023): 107051.

Gupta, R., D. Srivastava, M. Sahu, S. Tiwari, R. K. Ambasta, and P. Kumar. "Artificial Intelligence to Deep Learning: Machine Intelligence Approach for Drug Discovery." Molecular Diversity 25 (2021): 1315–1360.

Huang, C., Y. Wang, X. Li, L. Ren, J. Zhao, Y. Hu, L. Zhang, G. Fan, J. Xu, X. Gu, Z. Cheng, T. Yu, J. Xia, Y. Wei, W. Wu, X. Xie, W. Yin, H. Li, M. Liu, Y. Xiao, H. Gao, L. Guo, J. Xie, G. Wang, R. Jiang, Z. Gao, Q. Jin, J. Wang, and B. Cao. "Clinical Features of Patients Infected With 2019 Novel Coronavirus in Wuhan, China." The Lancet 395, no. 10223 (2020): 497–506.

Huang, D., M. Yang, X. Wen, S. Xia, and B. Yuan. "AI-Driven Drug Discovery: Accelerating the Development of Novel Therapeutics in Biopharmaceuticals." Journal of Knowledge Learning and Science Technology 3, no. 3 (2024): 206–222.

Huang, X., H. Liu, X. Li, L. Guan, J. Li, A. M. Tellier, and J. Zhang. "Revealing Alzheimer's Disease Genes Spectrum in the Whole-Genome by Machine Learning." BMC Neurology 18, no. 1 (2018): 8.

Jang, J., J. Kim, H. Shin, Y. G. Park, B. J. Joo, H. Seo, and J. U. Park. "Smart Contact Lens and Transparent Heat Patch for Remote Monitoring and Therapy of Chronic Ocular Surface Inflammation Using Mobiles." Science Advances 7, no. 14 (2021): eabf7194.

Johnson, K. B., W. Q. Wei, D. Weeraratne, M. E. Frisse, K. Misulis, K. Rhee, and J. L. Snowdon. "Precision Medicine, AI, and the Future of Personalized Health Care." Clinical and Translational Science 14, no. 1 (2021): 86–93.

Jospe, M. R., K. M. Richardson, A. A. Saleh, L. C. Bohlen, J. Crawshaw, Y. Liao, K. Konnyu, and S. M. Schembre. "Leveraging Continuous Glucose Monitoring as a Catalyst for Behaviour Change: A Scoping Review." International Journal of Behavioral Nutrition and Physical Activity 21, no. 1 (2024): 74.

Junaid, S. B., A. A. Imam, A. O. Balogun, L. C. De Silva, Y. A. Surakat, G. Kumar, M. Abdulkarim, A. N. Shuaibu, A. Garba, Y. Sahalu, A. Mohammed, T. Y. Mohammed, B. A. Abdulkadir, A. A. Abba, N. A. I. Kakumi, and S. Mahamad. "Recent Advancements in Emerging Technologies for Healthcare Management Systems: A Survey." Healthcare (Basel, Switzerland) 10, no 10 (2022): 1940. https://doi.org/10.3390/healthcare10101940

Kazanskiy, N. L., S. N. Khonina, and M. A. Butt. "Smart Contact Lenses—A Step towards Non-Invasive Continuous Eye Health Monitoring." Biosensors (Basel) 13, no. 10 (2023): 933.

Khalid, N., A. Qayyum, M. Bilal, A. Al-Fuqaha, and J. Qadir. "Privacy-Preserving Artificial Intelligence in Healthcare: Techniques and Applications." Computers in Biology and Medicine 158 (2023): 106848. https://doi.org/10.1016/j.compbiomed.2023.106848

Khalid, W. B., A. Anwar, and O. T. Waheed. "Contactless Vitals Measurement Robot." In 2022 8th International Conference on Automation, Robotics and Applications (ICARA), IEEE, 2022.

Klein, A., A. S. De Freitas, C. D. Pedron, and S. Elaluf-Calderwood. "Who Is Afraid of Google Glass? Mapping the Controversy about Wearable and Ubiquitous Computing." Academy of Management Proceedings, 2015, Vancouver, Canada.

Klupa, T., L. Czupryniak, G. Dzida, P. Fichna, P. Jarosz-Chobot, J. Gumprecht, M. Mysliwiec, A. Szadkowska, D. Bomba-Opon, K. Czajkowski, M. T. Malecki, & D. A. Zozulinska-Ziolkiewicz. "Expanding the Role of Continuous Glucose Monitoring in Modern Diabetes Care Beyond Type 1 Disease." Diabetes Therapy 14, no. 8 (2023): 1241–1266.

Ku, M., J. Kim, J. E. Won, W. Kang, Y. G. Park, J. Park, and J. U. Park. "Smart, Soft Contact Lens for Wireless Immunosensing of Cortisol." Science Advances 6, no. 28 (2020): eabb2891.

Kufel, J., K. Bargieł-Łączek, S. Kocot, M. Koźlik, W. Bartnikowska, M. Janik, and K. Gruszczyńska. "What Is Machine Learning, Artificial Neural Networks and Deep Learning?—Examples of Practical Applications in Medicine." Diagnostics 13, no. 15 (2023): 2582.

Lee, C. C., V. C. Koo, T. S. Lim, Y. P. Lee, and H. Abidin. "A Multi-Layer Perceptron-Based Approach for Early Detection of BSR Disease in Oil Palm Trees Using Hyperspectral Images." Heliyon 8, no. 4 (2022).

Lee, J., P. Ng, B. Hamandi, S. Husain, M. J. Lefebvre, and M. Battistella. "Effect of Therapeutic Drug Monitoring and Cytochrome P450 2C19 Genotyping on Clinical Outcomes of Voriconazole: A Systematic Review." Annals of Pharmacotherapy 55, no. 4 (2021): 509–529.

Maleki Varnosfaderani, S., and M. Forouzanfar. "The Role of AI in Hospitals and Clinics: Transforming Healthcare in the 21st Century." Bioengineering (Basel, Switzerland) 11, no. 4 (2024): 337. https://doi.org/10.3390/bioengineering11040337

Manickam, P., S. A. Mariappan, S. M. Murugesan, S. Hansda, A. Kaushik, R. Shinde, and S. P. Thipperudraswamy. "Artificial Intelligence (AI) and Internet of Medical Things (IoMT) Assisted Biomedical Systems for Intelligent Healthcare." Biosensors 12, no. 8 (2022): 562.

Mansour, R. F., A. El Amraoui, I. Nouaouri, V. G. Díaz, D. Gupta, and S. Kumar. "Artificial Intelligence and Internet of Things Enabled Disease Diagnosis Model for Smart Healthcare Systems." IEEE Access 9 (2021): 45137–45146.

Mayerhoefer, M. E., C. C. Riedl, A. Kumar, A. Dogan, P. Gibbs, M. Weber, and H. Schoeder. "[18F] FDG-PET/CT Radiomics for Prediction of Bone Marrow Involvement in Mantle Cell Lymphoma: A Retrospective Study in 97 Patients." Cancers 12, no. 5 (2020): 1138.

McGlynn, E. A., and S. M. Asch. "Developing a Framework for Evaluating the Patient Safety of AI Technologies." Journal of Patient Safety 16, no. 2 (2020): 105–112.

Mekhail, N. A., Levy, R. M., Deer, T. R., Kapural, L., Li, S., Amirdelfan, K., Pope, J. E., Hunter, C. W., Rosen, S. M., Costandi, S. J., Falowski, S. M., Burgher, A. H., Gilmore, C. A., Qureshi, F. A., Staats, P. S., Scowcroft, J., Carlson, J., Kim, C. K., Yang, M. I., Stauss, T., Poree, L., Karantonis, D. M., Khurram, A., Leitner, A., Soliday, N., Obradovic, M., Ouyang, Z., Parker, J., Single, P., Mugan, D., Gmel, G. E., Gorman, R., Gould, I., Hanson, E., Hagedorn, J. M., Rauck, R., Kallewaard, J. W., Baranidharan, G., Taylor, R. S., and Poree, L. "ECAP-Controlled Closed-Loop Versus Open-Loop SCS for the Treatment of Chronic Pain: 36-Month Results of the EVOKE Blinded Randomized Clinical Trial." Regional Anesthesia and Pain Medicine 49, no. 5 (2024): 346–354.

Mohankumar, V., and G. V. Jayaramaiah. "Recent Progress on Bio-Mechanical Energy Harvesting System from the Human Body: Comprehensive Review." International Journal of Advanced Computer Science and Applications 12, no. 7 (2021): 184–194.

Morgado Areia, C., M. Santos, S. Vollam, M. Pimentel, L. Young, C. Roman, and P. Watkinson. "A Chest Patch for Continuous Vital Sign Monitoring: Clinical Validation Study During Movement and Controlled Hypoxia." Journal of Medical Internet Research 23, no. 9 (2021): e27547.

Mudgal, S. K., R. Agarwal, J. Chaturvedi, R. Gaur, and N. Ranjan. "Real-World Application, Challenges and Implication of Artificial Intelligence in Healthcare: An Essay." The Pan African Medical Journal 43 (2022): 3.

Mutharasan, R. K., and J. Walradt. "Population Health and Artificial Intelligence." JACC: Advances 3, no. 8 (2024): 101092. https://doi.org/10.1016/j.jacadv.2024.101092.

Najjar, R. "Redefining Radiology: A Review of Artificial Intelligence Integration in Medical Imaging." Diagnostics 13, no. 17 (2023).

Nelson, C. A., L. M. Pérez-Chada, A. Creadore, S. J. Li, K. Lo, P. Manjaly, and A. Mostaghimi. "Patient Perspectives on the Use of Artificial Intelligence for Skin Cancer Screening: A Qualitative Study." JAMA Dermatology 156, no. 5 (2020): 501–512.

Noorbakhsh-Sabet, N., R. Zand, Y. Zhang, and V. Abedi. "Artificial Intelligence Transforms the Future of Health Care." The American Journal of Medicine 132, no. 7 (2019): 795–801.

Nooruddin, S., M. M. Islam, and F. A. Sharna. "An IoT Based Device-Type Invariant Fall Detection System." Internet of Things 9 (2020): 100130.

Obuchowicz, R., J. Kruszyńska, and M. Strzelecki. "Classifying Median Nerves in Carpal Tunnel Syndrome: Ultrasound Image Analysis." Biocybernetics and Biomedical Engineering 41, no. 2 (2021): 335–351.

Olawade, D. B., O. J. Wada, A. C. David-Olawade, E. Kunonga, O. Abaire, and J. Ling. "Using Artificial Intelligence to Improve Public Health: A Narrative Review." Frontiers in Public Health 11 (2023): 1196397.

Pan, M., Z. Zhang, and L. Shang. "Smart Contact Lenses: Disease Monitoring and Treatment." Research (2025).

Park, W., H. Seo, J. Kim, Y.-M. Hong, H. Song, B. J. Joo, S. Kim, E. Kim, C.-G. Yae, J. Kim, J. Jin, J. Kim, Y.-H. Lee, J. Kim, H. K. Kim, and J.-U. Park. "In-Depth Correlation Analysis Between Tear Glucose and Blood Glucose Using a Wireless Smart Contact Lens." Nature Communications 15 (2024): 2828.

Pinto-Coelho, L. "How Artificial Intelligence Is Shaping Medical Imaging Technology: A Survey of Innovations and Applications." Bioengineering (Basel, Switzerland) 10, no. 12 (2023): 1435. https://doi.org/10.3390/bioengineering10121435

Pun, F. W., I. V. Ozerov, and A. Zhavoronkov. "AI-Powered Therapeutic Target Discovery." Trends in Pharmacological Sciences 44, no. 9 (2023): 561–572.

Quazi, S., R. P. Saha, and M. K. Singh. "Applications of Artificial Intelligence in Healthcare." Journal of Experimental Biology and Agricultural Sciences (2022): 211–222.

Rahim, A., R. Khatoon, T. A. Khan, K. Syed, I. Khan, T. Khalid, and B. Khalid. "Artificial Intelligence-Powered Dentistry: Probing the Potential, Challenges, and Ethicality of Artificial Intelligence in Dentistry." Digital Health 10 (2024): 20552076241291345.

Rivella, C., C. Ruffini, C. Bombonato, A. Capodieci, A. Frascari, G. M. Marzocchi, A. Mingozzi, C. Pecini, L. Traverso, M. C. Usai, and P. Viterbori. "TeleFE: A New Tool for the Tele-Assessment of Executive Functions in Children." Applied Sciences 13 (2023): 1728.

Samajdar, S. S., R. Chatterjee, S. Mukherjee, A. Dey, B. Saboo, J. Pal, and N. Chatterjee. "Artificial Intelligence in Healthcare: Current Trends and Future Directions." Current Medical Issues 23, no. 1 (2025): 53–60. https://doi.org/10.4103/cmi.cmi_93_24

Seo, H., W. G. Chung, Y. W. Kwon, S. Kim, Y. M. Hong, W. Park, E. Kim, J. Lee, S. Lee, M. Kim, K. Lim, I. Jeong, H. Song, and J.-U. Park. "Smart Contact Lenses as Wearable Ophthalmic Devices for Disease Monitoring and Health Management." Chemical Reviews 123, no. 19 (2023): 11488–11558.

Shajari, S., K. Kuruvinashetti, A. Komeili, and U. Sundararaj. "The Emergence of AI-Based Wearable Sensors for Digital Health Technology: A Review." Sensors 23, no. 23 (2023).

Sharma, A., M. Badea, S. Tiwari, and J. L. Marty. "Wearable Biosensors: An Alternative and Practical Approach in Healthcare and Disease Monitoring." Molecules 26, no. 3 (2021): 748.

Sheu, Y. H., C. Magdamo, M. Miller, S. Das, D. Blacker, and J. W. Smoller. "AI-Assisted Prediction of Differential Response to Antidepressant Classes Using Electronic Health Records." NPJ Digital Medicine 6, no. 1 (2023): 73.

Shiwlani, A., S. Kumar, and H. A. Qureshi. "Transforming Pediatric Leukemia Care: The Role of Artificial Intelligence in Diagnosing, Treating, and Optimizing Outcomes in Acute Lymphoblastic Leukemia." Journal of Development and Social Sciences 6, no. 1 (2025): 348–363.

Srivastav, S., R. Chandrakar, S. Gupta, V. Babhulkar, S. Agrawal, A. Jaiswal, and M. Wanjari. "ChatGPT in Radiology: The Advantages and Limitations of Artificial Intelligence for Medical Imaging Diagnosis." Cureus 15, no. 7 (2023).

Taiwo, O., and A. E. Ezugwu. "Smart Healthcare Support for Remote Patient Monitoring During COVID-19 Quarantine." Informatics in Medicine Unlocked 20 (2020): 1–12.

Tang, X. "The Role of Artificial Intelligence in Medical Imaging Research." BJR Open 2, no. 1 (2019): 20190031. https://doi.org/10.1259/bjro.20190031

Tao, X., T. B. Shaik, N. Higgins, R. Gururajan, and X. Zhou. "Remote Patient Monitoring Using Radio Frequency Identification (RFID) Technology and Machine Learning for Early Detection of Suicidal Behaviour in Mental Health Facilities." Sensors 21, no. 3 (2021): 776.

Thanoon, M. A., M. A. Zulkifley, M. A. A. Mohd Zainuri, and S. R. Abdani. "A Review of Deep Learning Techniques for Lung Cancer Screening and Diagnosis Based on CT Images." Diagnostics (Basel, Switzerland) 13, no. 16 (2023): 2617. https://doi.org/10.3390/diagnostics13162617

Topol, E. Deep Medicine: How Artificial Intelligence Can Make Healthcare Human Again. New York: Basic Books, 2019.

Tran, E. L., and L. K. Crawford. "Revisiting PNS Plasticity: How Uninjured Sensory Afferents Promote Neuropathic Pain." Frontiers in Cellular Neuroscience 14 (2020): 612982.

Tsvetanov, F. "Integrating AI Technologies into Remote Monitoring Patient Systems." Engineering Proceedings 70 (2024): 54.

Uhl, S., A. Choure, B. Rouse, A. Loblack, and P. Reaven. "Effectiveness of Continuous Glucose Monitoring on Metrics of Glycemic Control in Type 2 Diabetes Mellitus: A Systematic Review and Meta-Analysis of Randomized Controlled Trials." Journal of Clinical Endocrinology and Metabolism 109, no. 4 (2024): 1119–1131.

Wrzesińska, N. "The Use of Smart Glasses in Healthcare–Review." MEDtube Science 31 (2015): 31–34.

Wu, J., M. Dong, K. Ota, J. Li, and W. Yang. "Application-Aware Consensus Management for Software-Defined Intelligent Blockchain in IoT." IEEE Network 34, no. 1 (2020): 69–75.

Wu, K. Y., A. Dave, M. Carbonneau, and S. D. Tran. "Smart Contact Lenses in Ophthalmology: Innovations, Applications, and Future Prospects." Micromachines 15 (2024): 856.

Wüller, H., J. Behrens, M. Garthaus, S. Marquard, and H. Remmers. "A Scoping Review of Augmented Reality in Nursing." BMC Nursing 18, no. 1 (2019): 11.

Yang, Q., Y. Liu, Y. Cheng, Y. Kang, T. Chen, and H. Yu. "Federated Learning." Synthesis Lectures on Artificial Intelligence and Machine Learning 13, no. 3 (2019): 1–207.

Yen, C., C. L. Lin, and M. C. Chiang. "Exploring the Frontiers of Neuroimaging: A Review of Recent Advances in Understanding Brain Functioning and Disorders." Life (Basel, Switzerland) 13, no. 7 (2023): 1472. https://doi.org/10.3390/life13071472

Zafar, F., L. F. Alam, R. R. Vivas, J. Wang, S. J. Whei, S. Mehmood, and Z. Nazir. "The Role of Artificial Intelligence in Identifying Depression and Anxiety: A Comprehensive Literature Review." Cureus 16, no. 3 (2024).

Zhang, Y., Y. Hu, N. Han, A. Yang, X. Liu, and H. Cai. "A Survey of Drug-Target Interaction and Affinity Prediction Methods via Graph Neural Networks." Computers in Biology and Medicine 163 (2023): 107136. https://doi.org/10.1016/j.compbiomed.2023.107136

Zhao, Y., T. Heida, E. E. van Wegen, B. R. Bloem, and R. J. van Wezel. "E-Health Support in People with Parkinson's Disease with Smart Glasses: A Survey of User Requirements and Expectations in the Netherlands." Journal of Parkinson's Disease 5, no. 2 (2015): 369–378.

Zuidhof, N., S. Ben Allouch, O. Peters, and P. P. Verbeek. "Perspectives on the Acceptance and Social Implications of Smart Glasses: A Qualitative Focus Group Study in Healthcare." International Journal of Human–Computer Interaction 40, no. 2 (2024): 149–159.

Chapter 8

Behavioral drivers of mHealth adoption and their role in the smart healthcare ecosystem in India

Anushka Goel and Udita Taneja

8.1 INTRODUCTION

Mobile health (mHealth) applications have emerged as a powerful healthcare technology that can enhance the quality of care, provide better access to health services, and can also reduce the cost burden (Jacob et al., 2022). These technologies can improve the communication between healthcare practitioners and their patients (Addotey-Delove, Scott and Mars, 2022a; Jacob et al., 2022). mHealth enables the delivery of healthcare services through mobile devices such as mobile phones, tablets, smartwatches, personal digital assistants, and other wearable devices (Addotey-Delove, Scott and Mars, 2022b). mHealth can support healthcare delivery in the collection of clinical data, can help in information sharing, support research, and training, and can help in remote and real-time monitoring of patients (Addotey-Delove, Scott and Mars, 2022a). mHealth applications can also offer customized and confidential means of accessing answers to health-related questions (Deng, 2013a).

Though there have been enormous benefits to the adoption of mHealth applications, there is a constant inflow and outflow of users (Grundy, 2022). This acts as a potential barrier to the adoption of mHealth users. Thus, there is a need to attract and keep users to increase the mHealth adoption rate of users. To develop these mHealth applications and to sustain this technology in the healthcare ecosystem, there is a need to understand the factors that influence users' adoption process.

This emerging technology has gained a lot of scholarly interest in recent years, but more empirical work is being done to understand the technology acceptance variables and health-related aspects that are still neglected. Most researchers focus on the pre-adoption behavior and it is important to understand the pre-adoption behavior of users to build a proper framework for large-scale mHealth development.

We believe that health-related behavior can influence the adoption of healthcare technology and thus, it is vital to understand the antecedents for health technology acceptance behavior.

We aim to theoretically understand the determinants of mHealth applications by combining technology acceptance and health behavior

 DOI: 10.1201/9781003586951-8

theories. We also provide implications for health technology developers and policymakers.

This chapter is organized as follows. Section 8.2 includes the literature review and hypotheses development, while Section 8.3 presents the methodology. Section 8.4 presents the findings, followed by Section 8.5 that includes the discussion. Section 8.6 summarizes the implications of our study for research. Finally, we present the conclusions of this research.

8.2 LITERATURE REVIEW AND HYPOTHESES

This study explores the determinants of mHealth technology, and users' intention to adopt mHealth applications. Post the COVID-19 pandemic, people across the globe engaged in preventive health measures which involved social distancing, using technology for daily routine activities, and actively accepting technological interventions to access healthcare services.

In the field of information systems (IS), technology acceptance theories determine the factors which impact the user's intention to accept new technology. However, during the pandemic, preventive health behavior accelerated the adoption of technology in the healthcare ecosystem, and thus, understanding the behavioral factors which can facilitate or inhibit the adoption of mHealth technology is very important. The Health Belief Model (HBM) explains the health-related behavior of users which emerges due to disease threats.

Figure 8.1 shows the integration of Technology Acceptance Model (TAM) and HBM. These were, combined to understand both the technological and health-related behavioral aspects of users.

8.2.1 Health Belief Model

The HBM is a widely used model to predict health-related behaviors (Rosenstock, 1974). Health-related behavior can refer to activities undertaken by an individual to understand, protect, and maintain one's health (Zhao, Ni and Zhou, 2018). Thus, one can consider mHealth adoption as a health-related behavior as well.

HBM suggests that a person's willingness to take any health-related action is determined by evaluating both the net positive benefits of taking an action as well as the negative threats of not taking one (Zhao, Ni and Zhou, 2018). HBM focuses on two aspects of health behavior, perceived disease threat and behavioral evaluation to control or avoid it. Perceived disease threat comprises Perceived Susceptibility to Disease (subjective risks of contracting a disease/condition) and Perceived Severity (concerns regarding the seriousness of contracting a disease/condition or not treating it) (Huang, 2010; Rosenstock, 1974). Behavioral evaluations incorporate Perceived Benefits (reduction in the risk of contracting disease/condition) and Perceived Barriers (barriers related to negative aspects of action related to costs, family risk, etc.).

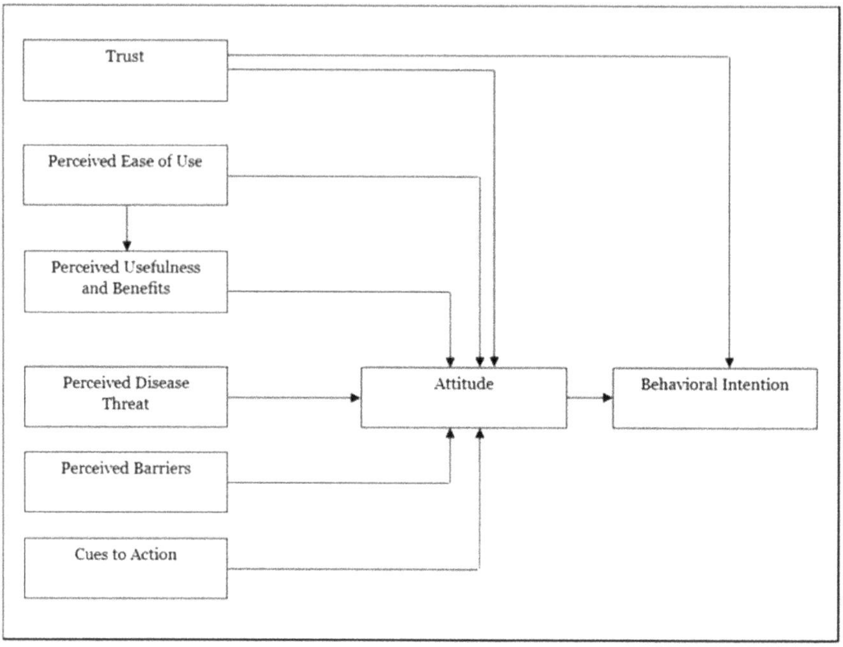

Figure 8.1 Conceptual model (authors' own creation).

Along with these, Cues To Action (trigger to appropriate action) are also a relevant variable of HBM (Huang, 2010; Rosenstock, 1974). The HBM is a widely used model in technology adoption literature. Many researchers have studied technology-based health behavior by comprising constructs of perceived disease threat along with technology acceptance theories. Wei et al. (2021) used HBM along with UTAUT for examining the factors affecting the adoption of mobile fitness applications. Sreelakshmi and Prathap (2020) proposed a model by combining HBM and Expectation Confirmation Model (ECM) to study the adoption of mobile-based payments. Sun, Wang, and Guo (2013) also used HBM constructs for understanding the acceptance of mHealth services.

8.2.1.1 Technology acceptance model

The TAM is considered the most influential model for tracing the influences on the attitudes of the user (Zhao, Ni and Zhou, 2018). TAM explains the users' acceptance of technology and Perceived Usefulness as the main variable of the model (Pai and Alathur, 2019). This model determines the Attitude and Behavioral Intention of users based on the key factors Perceived Usefulness reflects a person's belief that utilizing a particular system will enhance their task performance. In contrast, Perceived Ease of Use represents the belief that the system can be used with little or no effort (Davis, 1985; Zhao, Ni and Zhou, 2018). Behavioral Intention is defined as the probability that an individual will perform a specific behavior or not and is determined through Attitude (degree of influence toward the targeted behavior) (Davis, 1985). TAM is considered the most robust and powerful model for predicting users' acceptance of a technology. Hoque, Bao, and Sorwar (2017) and Zhao, Ni, and Zhou (2018) extended TAM to investigate the adoption of eHealth in Bangladesh. Manuel et al. (2011) adopted the TAM model to study technology adoption in the healthcare context.

8.2.1.2 Integration of HBM and TAM

The combination of HBM and TAM in our framework will explain whether health-related behavior (HBM) influences the adoption intention (TAM) of mHealth technology or not. This combined framework has gained scholarly interest in recent times and thus, it is widely used to study users' intentions toward healthcare technology. Schröder, Gewald, and Weeger (2021) combined HBM and TAM to understand the adoption of eHealth in a speech therapy context. Ahadzadeh et al. (2015) also used the HBM and TAM framework to investigate health-related internet use. Zhao, Ni, and Zhou (2018) discussed the relevance of combining these models to understand the adoption of mHealth services. Huang (2010) and Deng (2013b) also explored the combination of HBM and TAM to study adoption behavior in remote health monitoring.

We have added the variable Trust (degree to which individuals perceive that a technology is secure to use) in our model to facilitate the adoption of technology in healthcare. Trust can play a significant role in technology adoption (Zhao, Ni and Zhou, 2018). Thus, we propose the following hypotheses for our framework:

- H1: Attitude of users influences the Behavioral Intention to adopt mHealth applications.
- H2: Cues to Action influences the Attitude of users to adopt mHealth applications.
- H3: Perceived Barriers influences the Attitude of users to adopt mHealth applications.
- H4: Perceived Disease Threat influences the Attitude of users to adopt mHealth applications.
- H5: Perceived Ease of Use influences the Attitude of users to adopt mHealth technology.
- H6: Perceived Ease of Use has a direct impact of Perceived Usefulness and Benefits.
- H7: Perceived Usefulness and Benefits influences the Attitude of users to adopt mHealth technology.
- H8: Trust influences the Attitude of users to adopt mHealth technology.
- H9: Trust influences the Behavioral Intention to adopt mHealth applications.

Therefore, based on the proposed hypotheses, the following model has been developed and is studied in our study.

8.3 RESEARCH METHOD

8.3.1 Sample and procedure

Purposive sampling was used in this study, with the sample drawn from a population of the National Capital Territory of Delhi. A total of 250 respondents were approached, and 225 complete responses were received with a response rate of 90%. The majority of participants were aged 25–34 (64.8%), followed by less than 25 years (21.6%), and above 35 years (16.6%). Of the respondents, 51.2% were females, and 48.8% were males. Around 63.6% of respondents had used mHealth applications at least once before answering our questionnaire.

8.3.2 Data analysis

Data were analyzed using structural equation modeling (SEM) with partial least squares (Smart PLS 3.0).

8.3.3 Measures

A structured questionnaire was developed to measure the dependent and independent variables. The survey questionnaire was divided into two sections: Section A asked about the demographic profile of the respondents, and Section B consisted of statements based on our model. A 26-item questionnaire was made with a 5-point Likert scale ranging from one to five, with one for strongly disagree and five for strongly agree. PLS-SEM was used to test the measurement and structural models.

8.4 FINDINGS

8.4.1 Measurement model assessment

The measurement model was tested using the following criteria to ensure the robustness of the model. Table 8.1 shows the factor loadings, Cronbach's alpha (α), and Average Variance Extracted (AVE). Cronbach's alpha of each construct should be between 0.70 and 0.90 for it to be within the acceptable range for internal consistency or reliability (Hair et al., 2019). The factor loadings should be above 0.70 to indicate that the construct explains more than 50% of the indicator's variance and hence, to be accepted (Hair et al., 2019). The third measure of robustness is to assess the convergent validity of each measured construct, AVE is used to measure the convergent validity and it should be 0.50 or higher to be accepted (Hair et al., 2019). Table 8.2 shows the fourth measure, the discriminant validity, to understand to what extent a

Table 8.1 Convergent validity results for constructs

Construct	*Items*	*Factor loading*	*Cronbach's alpha*	*Average variance*
Perceived Usefulness and Benefits	PUB1	0.881	0.899	0.767
	PUB2	0.880		
	PUB3	0.864		
	PUB4	0.877		
Perceived Ease of Use	PEOU1	0.896	0.879	0.805
	PEOU2	0.914		
	PEOU3	0.881		
Perceived Disease Threat	PDT1	0.904	0.874	0.665
	PDT2	0.786		
	PDT3	0.739		
	PDT4	0.823		
Perceived Barriers	PB1	0.931	0.871	0.791
	PB2	0.953		
	PB3	0.773		
Trust	T1	0.908	0.772	0.814
	T2	0.896		
Cues To Action	CTA1	0.719	0.824	0.659
	CTA2	0.857		
	CTA3	0.883		
	CTA4	0.778		
Attitude	ATT1	0.911	0.872	0.796
	ATT2	0.904		
	ATT3	0.861		
Behavioral Intention	BI1	0.904	0.872	0.796
	BI2	0.887		
	BI3	0.884		

Source: Authors' own work.

Table 8.2 Discriminant validity results for constructs (HTMT)

	ATT	*BI*	*CTA*	*PB*	*PDT*	*PEOU*	*PSD*	*PUB*	*T*
ATT									
BI	0.884								
CTA	0.850	0.756							
PB	0.162	0.129	0.128						
PDT	0.074	0.087	0.195	0.316					
PEOU	0.584	0.631	0.583	0.079	0.129				
PUB	0.598	0.626	0.634	0.172	0.163	0.602	0.118		
T	0.702	0.718	0.760	0.251	0.126	0.587	0.059	0.515	

Source: Authors' own work.

construct is empirically distinct from others in the structural model. This can be measured using the heterotrait-monotrait (HTMT) ratio of correlations, HTMT value should be less than 0.90 to be conceptually distinct (Hair et al., 2019). In our study, all the factor loadings are above the required threshold values. Cronbach's alpha, AVE values, and HTMT ratio are also within the acceptable ranges and thus, the measurement model is robust and reliable.

8.4.2 Hypotheses testing

We further analyzed our results by bootstrapping using 5000 samples and tabulated the research in Table 8.3. The t-statistics and p-value were determined to test the hypotheses (Hair, Ringle, and Sarstedt, 2011). Table 8.4 displays the results of hypotheses testing.

Table 8.3 Bootstrapping results

	Original sample (O)	*Sample mean (M)*	*Standard deviation*	*t-Statistics*	*p-Value*
ATT → BI	0.697	0.696	0.060	11.573	0.000
CTA → ATT	0.504	0.500	0.084	5.982	0.000
PB → ATT	–0.015	–0.029	0.068	0.213	0.972
PDT → ATT	–0.066	–0.049	0.065	1.026	0.140
PEOU → ATT	0.115	0.108	0.069	1.706	0.048
PEOU → PUB	0.549	0.551	0.064	8.542	0.000
PUB → ATT	0.158	0.152	0.086	1.844	0.045
T → ATT	0.161	0.157	0.081	1.990	0.047
T → BI	0.223	0.223	0.064	3.463	0.001

Source: Authors' own work.
^Significant at 5% level.

Table 8.4 Hypotheses testing results

Hypothesis	*Relationship*	*t-Statistics*	*p-Value*	*Remarks*
H1	Attitude → Behavioral Intention	11.573	0.000	Supported
H2	Cues To Action → Attitude	5.982	0.000	Supported
H3	Perceived Barriers → Attitude	0.213	0.972	Not Supported
H4	Perceived Disease Threat → Attitude	1.026	0.140	Not Supported
H5	Perceived Ease of Use → Attitude	1.706	0.048	Supported
H6	Perceived Ease of Use → Perceived Usefulness and Benefits	8.542	0.000	Supported
H7	Perceived Usefulness and Benefits → Attitude	1.844	0.045	Supported
H8	Trust → Behavioral Intention	3.463	0.001	Supported
H9	Trust → Attitude	1.990	0.047	Supported

Source: Authors' own work.
^Significant at 5% level.

Table 8.5 Structure model results

	R square (R^2)	*R square adjusted (R^2)*	*SSO*	*SSE*	*Q^2*	*f^2*
ATT	0.584	0.568	486.000	268.680	0.443	1.137
BI	0.715	0.712	648.000	300.998	0.560	
PUB	0.302	0.297	648.000	507.213	0.217	0.034
CTA			648.000	648.000		0.316
PB			486.000	486.000		0.020
PDT			648.000	648.000		0.025
PEOU			486.000	486.000		0.020
T			324.000	324.000		0.036

Source: Authors' own work.

8.4.3 Structural model assessment

8.4.3.1 Evaluation for explanation power

To determine how well the model explains the variance in the dependent variables, R^2 values were examined. Based on commonly accepted thresholds, values around 0.25, 0.50, and 0.70 are interpreted as low, moderate, and high explanatory power, respectively. As indicated in Table 8.5, the results suggest that the model demonstrates a high level of explanatory strength among the constructs.

8.4.4 Evaluation of predictive relevance (Q^2) and effect size (f^2)

Predictive relevance was measured using Q^2 values. As per the guidelines by Hair et al. (2019), Q^2 values above 0.00, 0.15, and 0.50 suggest small, moderate, and high predictive relevance. The analysis shows that the model holds both moderate and high predictive capability for several constructs.

In terms of effect size, f^2 values were also reviewed. Thresholds of 0.02, 0.15, and 0.35 represent small, medium, and large effects, respectively. The variable Attitude was found to have a substantial (large) effect size in this context.

8.4.5 Goodness of Fit Index

Standardized root means square value (SRMR) is used to check the Goodness of Fit of the model (Hair et al., 2019). A value less than 0.08 shows a good fit, and our model shows a good fit with an SRMR value of 0.068.

8.5 DISCUSSION

Our study aimed to determine the factors influencing mHealth adoption and by doing so to improve the healthcare delivery process. Based on the theoretical foundation discussed, the antecedents of Attitude are Perceived Ease of Use, Trust, Perceived Usefulness and Benefits, Cues to Action, Perceived Barriers, and Perceived Disease Threat. These variables may reinforce Attitude that would further impact the Behavioral Intention of users to adopt mHealth applications. This study also validates the integrated framework of HBM and TAM to determine the factors influencing mHealth adoption.

The results confirm that users' attitudes significantly influences their behavioral intention to adopt mHealth applications. Overall, our study provides empirical support for the hypotheses proposed in the research. All the associations related to TAM constructs, H1, H5, H6, and H7 were supported, these findings are in line with prior literature (El-Wajeeh, Galal-Edeen and Mokhtar, 2014; Hoque, Bao and Sorwar, 2017). The associations for HBM constructs, H3 and H4 were not supported, whereas, H2 was supported. The findings from the study are discussed below.

As hypothesized, Perceived Ease of Use and Perceived Usefulness and Benefits significantly impact the Attitude of users to adopt mHealth applications. This implies that if users believe that using mHealth applications would be easier and can improve their healthcare process, then they will develop a positive attitude toward adopting mHealth applications. Significantly, Perceived Ease of Use also had an impact on Perceived Usefulness and Benefits with a path coefficient of 0.549 which implies that the mHealth applications would be perceived as more beneficial if they are easier to use by individuals.

Cues to Action significantly influenced the Attitude toward mHealth applications, whereas, Perceived Barriers and Perceived Disease Threat showed an insignificant impact on Attitude toward the mHealth applications. Cues to Action had a significant impact on the Attitude of users, this can be due to the recent pandemic which triggered the use of the mHealth application by users and acted as an effective tool to manage healthcare needs. The other two factors were insignificant in our study, as most of our study population were young adults under the age of 34 years who are tech-savvy and will face few barriers in using new technology. Perceived Barriers can play a significant role for the older generation who are more concerned about technology use than younger ones. Similarly, the younger generation perceived few threats from disease and thus, the impact of Perceived Disease Threat was also insignificant.

Our additional variable Trust also had a significant impact on both Attitude and Behavioral Intention of users to adopt mHealth applications. This implies that users' trust in mHealth applications can facilitate a positive attitude and increase the intention to adopt mHealth applications.

Our results show the high-explanatory power of Behavioral Intention with $R^2 = 0.715$. This means that Trust and Attitude have a significant impact on the Behavioral Intention to adopt mHealth applications, and the magnitude of Attitude is more significant than that of Trust. Building trust and influencing positive attitudes among users can significantly impact the adoption process of mHealth applications.

8.6 IMPLICATIONS

8.6.1 Theoretical implications

Our study developed an integrated framework to identify the factors determining mHealth adoption. The use of HBM and TAM constructs helped in understanding the impact of health-related and technology-related factors on users' intention to adopt mHealth applications. During the pandemic, health-related constraints accelerated the growth of technological interventions in different domains and thus, it was necessary to determine if health-related factors impact mHealth adoption post-pandemic. Hence, the conceptual model developed was tested to study adoption intention in the Indian context.

8.6.2 Implications for practice

The adoption of mHealth applications plays a significant role in streamlining the healthcare process between doctors and patients. mHealth applications can help in enhancing the coverage of healthcare facilities in India and can accelerate the process of attaining the Universal Health Coverage target of Sustainable Development Goal-3. By understanding the factors which impact the adoption of mHealth and other technological interventions in healthcare, health professionals and other stakeholders can better analyze healthcare needs and facilitate a more robust healthcare ecosystem for future crises. Users' positive attitude toward mHealth applications can increase their intention to adopt this technology for their healthcare needs.

Information and communication technology can connect people across the globe for building a more sustainable healthcare ecosystem. These technological interventions need to be properly implemented by governments to reach a wider audience and thus, ensure universal health coverage for all. The next factor which significantly impacted the intention of users was trust, the healthcare organizations and practitioners can play a significant role in developing trust among users for adopting mHealth technology. The potential adopters of mHealth applications might trust and adopt an application when it is recommended by their physicians. Practitioners and healthcare organizations can slowly integrate mHealth applications in their healthcare delivery process for follow-up messages, remote monitoring, storing prescriptions, and also for scheduling appointments. With proper facilitation and promotion of mHealth applications, users' cues can also act as an enabler for adopting this technology.

Additionally, technology entrepreneurs can also use this opportunity to create platforms that are easier to use, can help in real-time monitoring, and also involve the latest technology to create a robust healthcare ecosystem.

8.7 CONCLUSION

This study explored the factors of mHealth applications. From our analysis, we found Attitude, Cues to Action, and Trust as the important variables which impacted users' intention to adopt mHealth applications. Furthermore, Perceived Ease of Use and Perceived Usefulness and Benefits also influenced the attitude of users. This implies the importance of users' attitudes toward the adoption of any technology. We also integrated two theoretical models that would be beneficial for other researchers to study the technological and healthcare-related behavior of users.

8.8 LIMITATIONS AND FUTURE RESEARCH DIRECTIONS

This research has certain limitations. The study had a small sample size and needs to be carried out on a larger sample. There are other relevant theories and variables like awareness, social influence, privacy, and facilitating conditions that can also be studied in future studies to explore the adoption behavior of users.

REFERENCES

Addotey-Delove, M., Scott, R. E., & Mars, M. (2022a). A healthcare workers' mHealth adoption instrument for the developing world. *BMC Health Services Research*, *22*(1), 1–10. https://doi.org/10.1186/s12913-022-08592-0

Addotey-Delove, M., Scott, R. E., & Mars, M. (2022b). The development of an instrument to predict patients' adoption of mHealth in the developing world. *Informatics in Medicine Unlocked*, *29*(March), 100898. https://doi.org/10.1016/j.imu.2022.100898

Ahadzadeh, A. S., De, M. P., Sharif, S. P., De, M. P., & Ong, F. S. (2015). Integrating health belief model and technology acceptance model : An investigation of health-related internet use corresponding author. *Journal of Medical Internet Research*, *17*(2), 1–17. https://doi.org/10.2196/jmir.3564

Davis, F. D. (1985). Tissue culture studies of the human lymphocyte. *Massachusetts Institute of Technology*, *146*(3652), 1648–1655. https://doi.org/10.1126/science.146.3652.1648

Deng, Z. (2013a). *International Journal of Services Technology and Management and International Journal of Information Technology and Management*, Vol. 11, pp. 351–373. Electronic Markets.

Deng, Z. (2013b). Understanding public users' adoption of mobile health service. *International Journal of Mobile Communications*, *11*(4), 351–373. https://doi.org/10.1504/IJMC.2013.055748

El-Wajeeh, M., Galal-Edeen, G. H., & Mokhtar, H. (2014). Technology acceptance model for mobile health systems. *ISOR Journal of Mobile Computing & Application*, *1*(1), 21–33.

Grundy, Q. (2022). A review of the quality and impact of Mobile health apps. *Annual Review of Public Health*, *43*, 117–134. https://doi.org/10.1146/annurev-publhealth-052020-103738

Hair, J. F., Ringle, C. M., & Sarstedt, M. (2011). PLS-SEM: Indeed a silver bullet. *Journal of Marketing Theory and Practice*, *19*(2), 139–152. https://doi.org/10.2753/MTP1069-6679190202

Hair, J. F., Risher, J. J., Sarstedt, M., & Ringle, C. M. (2019). When to use and how to report the results of PLS-SEM. *European Business Review*, *31*(1), 2–24. https://doi.org/10.1108/EBR-11-2018-0203

Hoque, M. R., Bao, Y., & Sorwar, G. (2017). Investigating factors influencing the adoption of e-Health in developing countries: A patient's perspective. *Informatics for Health and Social Care*, *42*(1), 1–17. https://doi.org/10.3109/17538157.2015.1075541

Huang, J. C. (2010). Remote health monitoring adoption model based on artificial neural networks. *Expert Systems with Applications*, *37*(1), 307–314. https://doi.org/10.1016/j.eswa.2009.05.063

Jacob, C., Sezgin, E., Sanchez-Vazquez, A., & Ivory, C. (2022). Sociotechnical factors affecting patients' adoption of mobile health tools: Systematic literature review and narrative synthesis. *JMIR MHealth and UHealth*, *10*(5), 1–27. https://doi.org/10.2196/36284

Manuel, J., Egea, O., Victoria, M., & González, R. (2011). Explaining physicians' acceptance of EHCR systems: An extension of TAM with trust and risk factors. *Computers in Human Behavior, 27*, 319–332. https://doi.org/10.1016/j.chb.2010.08.010

Pai, R. R., & Alathur, S. (2019). Determinants of individuals' intention to use mobile health: Insights from India. *Transforming Government: People, Process and Policy*, *13*(3–4), 306–326. https://doi.org/10.1108/TG-04-2019-0027

Rosenstock, I. M. (1974). Historical origins of the health belief model. *Health Education & Behavior*, *2*(4), 328–335. https://doi.org/10.1177/109019817400200403

Schröder, T., Gewald, H., & Weeger, A. (2021). A research model to explain acceptance of speech therapy devices by senior aphasia patients. *Procedia Computer Science*, *198*(2021), 287–294. https://doi.org/10.1016/j.procs.2021.12.242

Sreelakshmi, C. C., & Prathap, S. K. (2020). Continuance adoption of mobile-based payments in Covid-19 context: An integrated framework of health belief model and expectation confirmation model. *International Journal of Pervasive Computing and Communications*, *16*(4), 351–369. https://doi.org/10.1108/IJPCC-06-2020-0069

Sun, Y., Wang, N., & Guo, X. (2013). Understanding the acceptance of mobile health services: A comparison and integration of alternative models. *Journal of Electronic Commerce Research*, *14*(2), 183–190.

Wei, J., Vinnikova, A., Lu, L., & Xu, J. (2021). Understanding and predicting the adoption of fitness mobile apps: Evidence from China. *Health Communication*, *36*(8), 950–961. https://doi.org/10.1080/10410236.2020.1724637

Zhao, Y., Ni, Q., & Zhou, R. (2018). What factors influence the mobile health service adoption? A meta-analysis and the moderating role of age. *International Journal of Information Management*, *43*(August 2017), 342–350. https://doi.org/10.1016/j.ijinfomgt.2017.08.006

Chapter 9

Implementation of projection-based AR application for learning enhancement

Anuj Kumar Goel

9.1 INTRODUCTION

Recent educational research has demonstrated that the utilization of augmented reality (AR) can improve students' ability to retain information, compared to conventional 2D diagrams typically found in textbooks (Ajit et al., 2021). This is advantageous in the study of anatomy, where the intricacy of diagrams and the 2D format of printed materials can pose difficulties in visualization (Almaadawy, 2019). The incorporation of AR technology, which is powered by AR core and cloud computing, offers medical students a more interactive and intuitive means of exploring the human body, thereby improving their comprehension and knowledge retention. An example of AR in education is the use of AR manuals (Bacca et al., 2014). AR textbooks use a smartphone or tablet to overlay digital content onto the pages of a physical textbook. These can include 3D models, animations, and videos, provide additional information, and help students visualize complex concepts (Bork et al., 2021; Cannizzaro et al., 2022).

AR textbooks have been used in subjects such as science and math, where visualizing abstract concepts can be difficult (Schrepp et al., 2017; Sheik-Ali et al., 2019) (Figure 9.1).

Numerous studies have been conducted worldwide to identify and propose innovative technological strategies that can significantly impact teaching practices and academic research (Shirazi & Behzadan, 2015). The continual advancements within computer simulation technology have led to the emergence of immersive technologies like virtual, augmented, and mixed realities (VR, AR, and MR) (Lee et al., 2021; Theodoropoulos & Lepouras, 2021). With insight from the "NMC/CoSN Horizon 2017 K-12 Edition" report, VR, AR, and MR are considered highly suitable for mid-range educational applications (2–3 years) with the potential for future improvements. Implementation of above-said technologies offer users a novel mode of multi-sensory, multi-dimensional experiences that integrate digital information with the real world, changing the way users view and interact with the digital world (Chytas et al., 2020; Lin et al., 2019). These techniques progressively implemented in various fields, including education, medicine, aviation, city planning, and gaming (Chang et al., 2020; Chatzopoulos et al., 2017; Eastman et al., 2009).

DOI: 10.1201/9781003586951-9

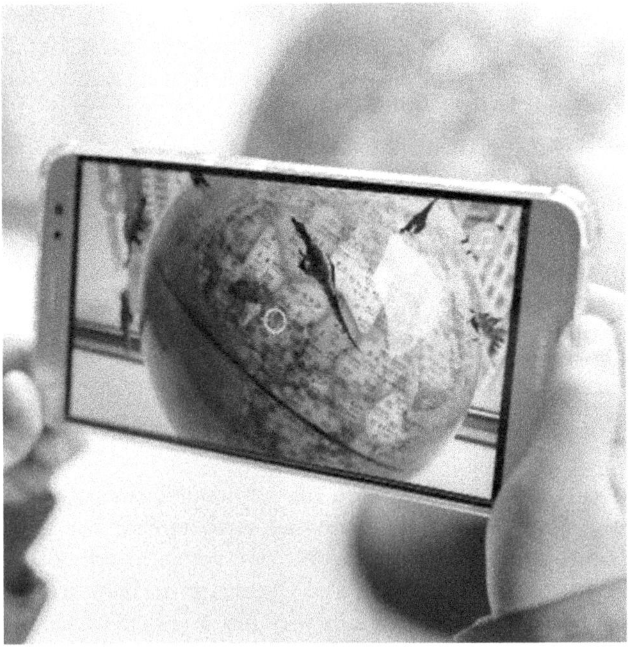

Figure 9.1 Visualization of the globe using AR tool. (From Cetin & Turkan, 2022.)

From a pedagogical standpoint, the progress of VR, AR, and MR not only provides beginners with novel erudition tools, platforms, and involvements, but also enables educators to enhance full human development (Alhassan, 2024; Boc et al., 2021; Buentello-Montoya et al., 2021). The ever-evolving worlds of VR, AR, and MR provide new-fangled prospects for creating an intellect of psychological immersion in a realistic environment that can be skilled, discovered, and operated (Alkhattabi, 2017; Barteit et al., 2021; Kharvari & Kaiser, 2022). Above said developments are developing quickly, allowing for a true overlay of additional content generated by this technology. While progressive submissions of VR, AR, and MR have been adopted in the entertainment business for a long time, educational researchers have realized their advantages and the benefits they can bring to students, making up for the lack of educational activities and addressing challenges in the field of education technology (Albishri & Blackmore, 2024; Ardiny & Khanmirza, 2018). For informal learning environments the adoption of VR, AR, and MR can also improve student learning (Campisi et al., 2020; Cho et al., 2020).

Jin j, Lui T, Wang Y, Wang C in 2019, proposed that ISEA (Interactive Science Education through Augmented Reality) is an innovative application of chemistry education that leverages AR technology to generate an immersive and collaborative learning experience for students. The ISEA platform allows students to visualize and interact with chemical structures in 3D, which helps them to better understand complex concepts and enhance their overall learning experience. By using AR technology, ISEA also allows for remote learning, making it a valuable tool in today's increasing digital world (Goel et al., 2022). Overall, ISEA represents an exciting development in chemistry education and has the potential to revolutionize the way students learn about chemistry.

Tang W, Ge Q, Zhou L, Zhu Q in 2018, uses AR interactive service modules, mobile user terminal module, web communication service module, and information recording module to construct a convenient user interaction system based on AR technology.

The proposed system achieves higher working efficiency, and the application of the proposed system can meet the needs of users' daily travel.

9.2 REVIEW OF LITERATURE

Lai J, Cheong K in 2022, projected the implementation of an AR in physics education. The use of AR creates an environment for next-generation AR-aided notes, virtual laboratory, and interactive problem-based learning with real-time automated generation of application-centric scenarios.

AR had already started to make its way into the field of school education, with numerous research studies exploring its potential impact on teaching and learning. Some of the key research areas and findings up to that point are as follows:

Enhancing Learning Experiences: Researchers investigated how AR can enhance traditional learning experiences by adding interactive and immersive elements to educational content. For example, AR can bring historical artifacts to life, enable 3D visualization of complex concepts in science, and provide interactive simulations for better understanding of abstract topics (Petrovich et al., 2018).

Improved Engagement: Studies examined the impact of AR on student engagement and motivation in the classroom. The interactive and visually stimulating nature of AR content was believed to capture students' attention and make learning more enjoyable (Rambli et al., 2013).

Personalized Learning: AR technology has the potential to cater to individual learning styles and pace. Research explored adaptive AR applications that could tailor content based on students' needs, helping them grasp concepts at their own pace.

Collaborative Learning: AR was also investigated for its potential to facilitate collaborative learning experiences. It enables students to work together on shared AR content, promoting teamwork and communication skills (Radu, 2014).

Teacher Training and Professional Development: Research explored how AR can be used in teacher training and professional development. AR simulations and practice scenarios were developed to improve teaching techniques and classroom management skills (Samad et al., 2021).

Accessibility and Inclusivity: AR offered new possibilities to make learning accessible to students with disabilities. Researchers explored how AR technology could be used to accommodate diverse learning needs (Gonzalez et al., 2020).

Assessment and Feedback: AR technology was examined as a means of assessing student learning and providing real-time feedback. Interactive quizzes and activities in AR format could offer insights into students' comprehension and progress (Sırakaya & Sırakaya, 2022).

Challenges and Implementation: Researchers also addressed the challenges of integrating AR into educational settings, such as the cost of hardware and software, the need for teacher training, and potential distractions in the classroom (Smith et al., 2011).

It's important to note that the field of AR is rapidly evolving, and there might be numerous new research studies and developments beyond my knowledge cutoff date.

9.3 METHODOLOGY

To create the AR applications the user must follow the mentioned steps. Additionally, the methods ought to be followed in the same sequence are:

a. Defining the Learning Space: The learning space includes the settings that contribute to learners' development and growth.
b. Discovering AR/VR: Explore progressions in VR technology and its probable utilization to enhance the erudition environment.
c. Impact on Perception and Control: Evaluate the projected reimbursements of AR technology in refining learners' perception, monitoring, and control of the physical environment.
d. Examining the Belongings of Virtual Education Settings: Conduct a systematic review of numerous studies that feature the positive effects of virtual learning environments on learners.
e. Identifying the Significance of 3D ability: Incorporates the crucial role of beginners' spatial ability in different learning environments. Further, investigate how VR technologies can leverage and enhance spatial ability.
f. Associating Learning Outcomes: Proposal of a controlled study comparing students who receive instruction using AR with a control group. Measure and examine the learning results, considering issues such as sympathetic, retaining, and academic performance.
g. Measuring Rendezvous in the Erudition Procedure: Appraise the level of pupil assignation during instruction with AR compared to traditional methods.

9.4 DESIGNED MODEL

Finally, the AR application is designed with the above-mentioned stages. Figures 9.2 and 9.3 demonstrate the configuration settings of the Vuforia platform to start working with AR applications (Williams et al., 2024). All validated results are presented, to have a better understanding of the concept.

- Step 1: Creation of an image token from Vuforia engine.
- Step 2: Then use its access keys in PlugXR to get target images. Users must have to create the keys to fully utilize the features of application software. Different packages are available for the development of AR and VR applications.
- Step 3: Then we made a 3D model of human heart using Blender. This model is used to create AR models that ought to be viewed by mobile applications (Figure 9.4).
- Step 4: Now implementing this 3D object in PlugXR as shown in Figure 9.5.

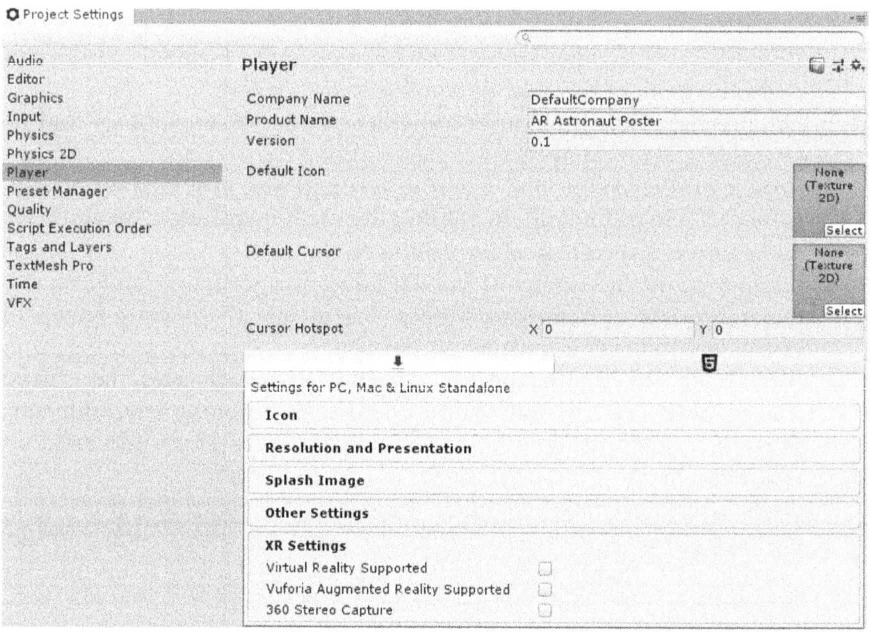

Figure 9.2 Project settings for development of AR model using Vuforia engine. (Created by the author.)

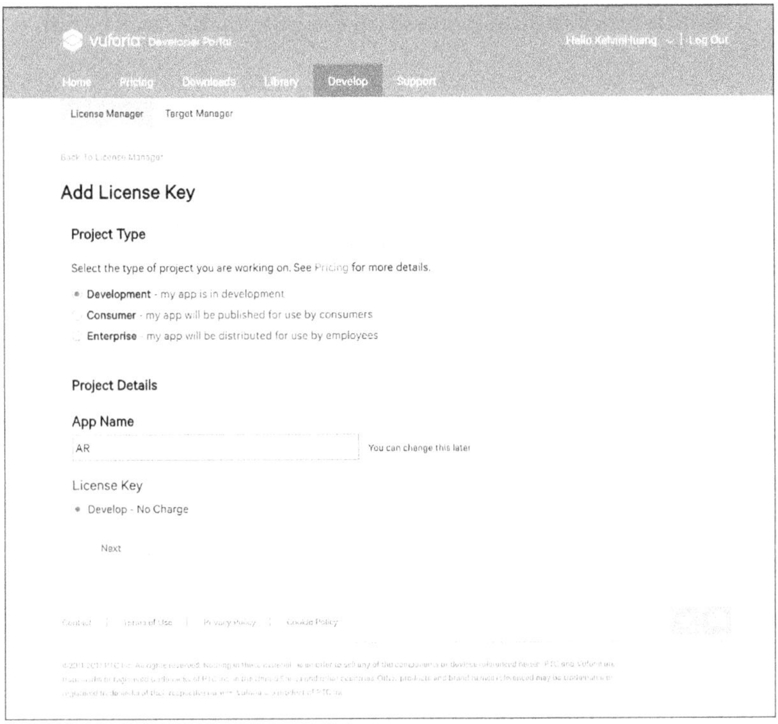

Figure 9.3 Licensing on Vuforia application platform using Vuforia engine. (Created by the author.)

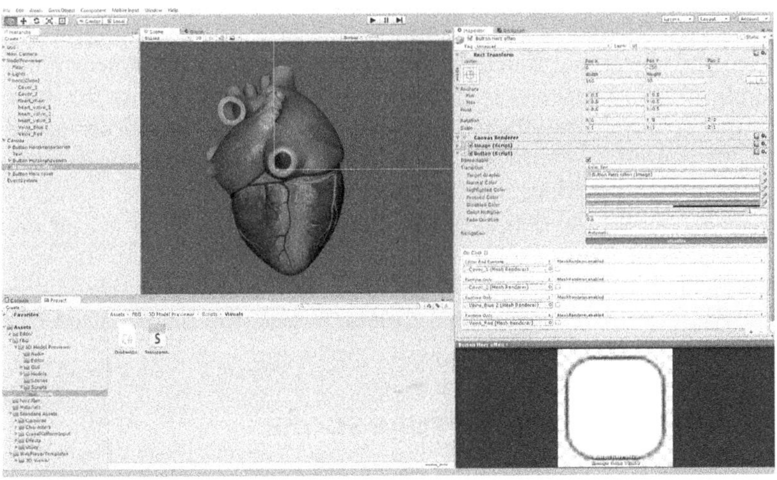

Figure 9.4 3D model of human heart using Blender using Vuforia engine. (Created by the author.)

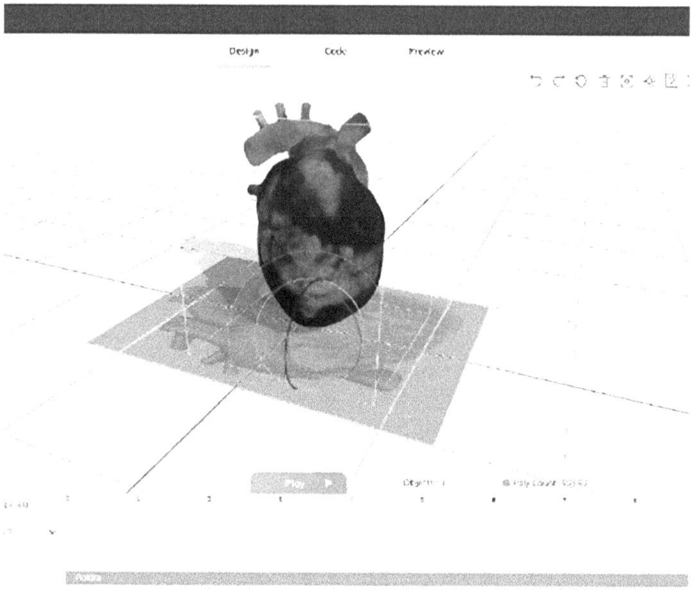

Figure 9.5 Implementation of 3D objects in PlugXR. (From Park & Yun, 2024.)

9.5 RESULTS

As discussed above in different sections, about design and implementation of AR applications, the results are presented in this section. Figure 9.6 shows the model that is imported in the AR tool. After publishing the marker-based AR, Figure 9.7 is used to augment a 3D heart model on our mobile device using the given link and marker.

Figure 9.7 shows the output of the finally implemented model for medical educational purposes.

Figure 9.8 shows the output in AR tool. In this model marker is used to present the final scene. Furthermore, these models can be enhanced with the incorporation of numerous medical descriptions.

9.6 CONCLUSION AND FUTURE SCORE

AR is poised to revolutionize medical education and healthcare. By providing immersive, interactive, and personalized experiences, AR technology offers unprecedented opportunities for students to understand complex concepts, surgeons to plan and perform surgeries with greater precision, and patients

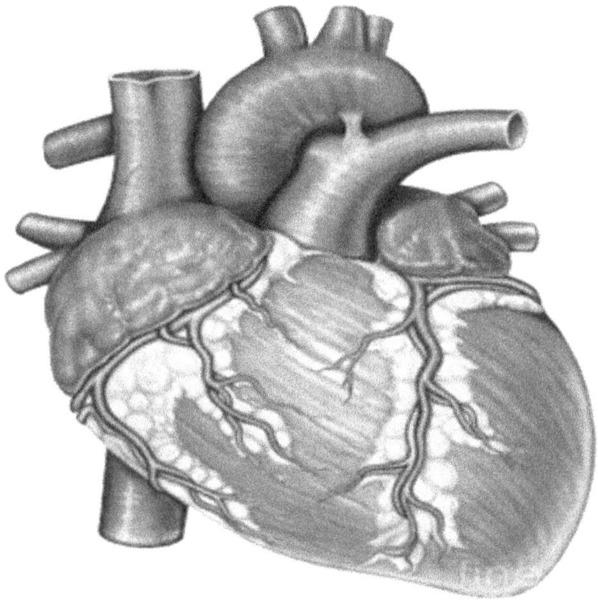

Figure 9.6 Added 3D model for AR. (Author's own work.)

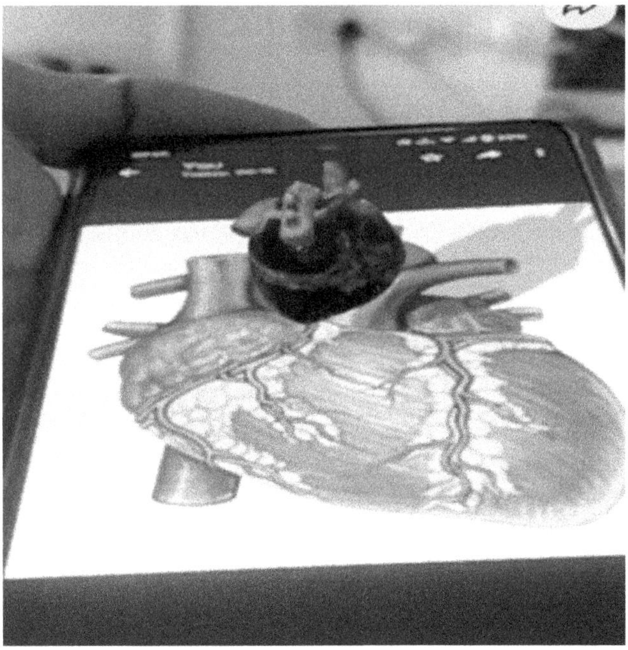

Figure 9.7 Implementation in mobile application. (Author's own work.)

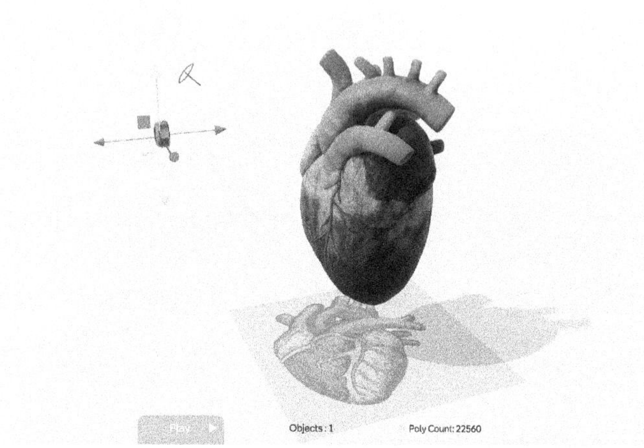

Figure 9.8 Implementation of AR for human heart using Vuforia engine. (Author's own work.)

to comprehend their medical conditions and treatment plans. As AR continues to evolve, its widespread adoption in medical education and healthcare is anticipated, leading to improved outcomes for patients, enhanced training for healthcare practitioners, and a transformative shift in the way medicine is taught and practiced.

REFERENCES

Ahmad, N., and S. Junaini. "Augmented Reality for Learning Mathematics: A Systematic Literature Review." *International Journal of Emerging Technologies in Learning (iJET)* 15, no. 16 (2020): 106–122.

Ajit, G., T. Lucas, and R. Kanyan. "A Systematic Review of Augmented Reality in STEM Education." *Studies of Applied Economics* 39, no. 1 (2021). https://doi.org/10.25115/eea.v39i1.4280.

Albishri, B., and K. L. Blackmore. "Duality in Barriers and Enablers of Augmented Reality Adoption in Education: A Systematic Review of Reviews." *Interactive Technology and Smart Education* (2024). https://doi.org/10.1108/ITSE-10-2023-0194.

Alhassan, A. I. "Analyzing the Application of Mixed Method Methodology in Medical Education: A Qualitative Study." *BMC Medical Education* 24, no. 1 (2024): 225. https://doi.org/10.1186/s12909-024-05242-3.

Alkhattabi, M. "Augmented Reality as E-Learning Tool in Primary Schools' Education: Barriers to Teachers' Adoption." *International Journal of Emerging Technologies in Learning (iJET)* 12, no. 2 (2017). https://doi.org/10.3991/ijet.v12i02.6158.

Almaadawy, M. A. N. "The Impact of the Difference in the Employment of Augmented Reality in Discovery-Based Learning Directed Versus Free on the Cognitive Load and the Enhancement of Scientific Curiosity in Science Among Sixth Graders." *Journal of Scientific Research in Education* 20, no. 5 (2019): 257–325. https://doi.org/10.21608/jsre.2019.59075.

Ardiny, H., and E. Khanmirza. "The Role of AR and VR Technologies in Education Developments: Opportunities and Challenges." In *2018 6th RSI International Conference on Robotics and Mechatronics (IcRoM)*, 482–487. IEEE, 2018.

Bacca, J., S. Baldiris, R. Fabregat, S. Graf, and Kinshuk. "Augmented Reality Trends in Education: A Systematic Review of Research and Applications." *Educational Technology & Society* 17, no. 4 (2014): 133–149. https://www.jstor.org/stable/jeductechsoci.17.4.133.

Barteit, S., L. Lanfermann, T. Bärnighausen, F. Neuhann, and C. Beiersmann. "Augmented, Mixed, and Virtual Reality-Based Head-Mounted Devices for Medical Education: Systematic Review." *JMIR Serious Games* 9, no. 3 (2021): e29080. https://doi.org/10.2196/29080.

Boc, R. G., R. L. Chiriac, and C. Antonya. "How Augmented Reality Could Improve the Student's Attraction to Learn Mechanisms." *Electronics* 10, no. 2 (2021): 175. https://doi.org/10.3390/electronics10020175.

Bork, F., A. Lehner, U. Eck, N. Navab, J. Waschke, and D. Kugelmann. "The Effectiveness of Collaborative Augmented Reality in Gross Anatomy Teaching: A Quantitative and Qualitative Pilot Study." *Anatomical Sciences Education* 14, no. 5 (2021): 590–604. https://doi.org/10.1002/ase.2016.

Buentello-Montoya, D. A., M. Lomelí-Plascencia, and L. Medina-Herrera. "The Role of Reality Enhancing Technologies in Teaching and Learning of Mathematics." *Computers and Electrical Engineering* 94 (2021): 107287. https://doi.org/10.1016/j.compeleceng.2021.107287.

Campisi, C. A., E. H. Li, D. E. Jimenez, and R. L. Milanaik. "Augmented Reality in Medical Education and Training: From Physicians to Patients." *Augmented Reality in Education: A New Technology for Teaching and Learning* (2020): 111–138. https://doi.org/10.1007/978-3-030-42156-4_7.

Cannizzaro, D., I. Zaed, A. Safa, A. J. Jelmoni, A. Composto, A. Bisoglio, K. Schmeizer, A. C. Becker, A. Pizzi, and A. Cardia. "Augmented Reality in Neurosurgery, State of Art and Future Projections: A Systematic Review." *Frontiers in Surgery* 9 (2022): 227. https://doi.org/10.3389/fsurg.2022.864792.

Cetin, H., and A. Turkan. "The Effect of Augmented Reality Based Applications on Achievement and Attitude towards Science Course in Distance Education Process." *Education and Information Technologies* 27 (2022): 1397–1415.

Chang, K. E., J. Zhang, Y. S. Huang, T. C. Liu, and Y. T. Sung. "Applying Augmented Reality in Physical Education on Motor Skills Learning." *Interactive Learning Environments* 28, no. 6 (2020): 685–697. https://doi.org/10.1080/10494820.2019.1636073.

Chatzopoulos, D., C. Bermejo, Z. Huang, and P. Hui. "Mobile Augmented Reality Survey: From Where We Are to Where We Go." *IEEE Access* 5 (2017): 6917–50.

Cho, J., S. Rahimpour, A. Cutler, C. Goodwin, S. P. Lad, and P. Codd. "Enhancing Reality: A Systematic Review of Augmented Reality in Neuronavigation and Education." *World Neurosurgery* 139 (2020): 186–195. https://doi.org/10.1016/j.wneu.2020.04.043.

Chytas, D., E. O. Johnson, M. Piagkou, A. Mazarakis, G. C. Babis, E. Chronopoulos, V. S. Nikolaou, N. Lazaridis, and K. Natsis. "The Role of Augmented Reality in Anatomical Education: An Overview." *Annals of Anatomy* 229 (2020): 151463. https://doi.org/10.1016/j.aanat.2020.151463.

Eastman, J. K., R. Iyer, and K. L. Eastman. "Interactive Technology in the Classroom: An Exploratory Look at Its Use and Effectiveness." *Contemporary Issues in Education Research* 2 (2009): 31–38.

Goel, A. K., A. Jain, C. Saini, R. Ashutosh, A. Das, and Deep. "Implementation of AI/ML for Human Emotion Detection Using Facial Recognition." In *Proceedings of the IEEE 4th International Conference on Cybernetics, Cognition and Machine Learning Applications (ICCCMLA)*, 511–515. 2022.

Gonzalez, A. A., P. A. Lizana, S. Pino, B. G. Miller, and C. Merino. "Augmented Reality-Based Learning for the Comprehension of Cardiac Physiology in Undergraduate Biomedical Students." *Advances in Physiology Education* 44, no. 3 (2020): 314–322. https://doi.org/10.1152/advan.00137.2019.

Kharvari, F., and L. E. Kaiser. "Impact of Extended Reality on Architectural Education and the Design Process." *Automation in Construction* 141 (2022): 104393. https://doi.org/10.1016/j.autcon.2022.104393.

Lai, J. W. and Cheong, K. H. "Educational Opportunities and Challenges in Augmented Reality: Featuring Implementations in Physics Education." IEEE Access, 10, 43143–43158. DOI: 10.1109/ACCESS.2022.3166478

Lee, G., S. Moshrefi, V. Fuertes, L. Veeravagu, R. Nazerali, and S. J. Lin. "What Is Your Reality? Virtual, Augmented, and Mixed Reality in Plastic Surgery Training, Education, and Practice." *Plastic and Reconstructive Surgery* 147, no. 2 (2021): 505–511. https://doi.org/10.1097/PRS.0000000000007595.

Lin, C., H. Chai, J. Wang, C. Chen, Y. Liu, and Y. Huang. "Augmented Reality in Educational Activities for Children with Disabilities." *Displays* 42 (2019): 51–54.

Park, S., and H. Yun. "Relationships between Students' Affective Experiences and Technology Acceptance in Augmented Reality Design Training in Higher Education." *Educational Technology Research and Development* 72, no. 2 (2024): 479–501.

Petrovich, M., M. Shah, and A. Foster. "Augmented Reality Experiences in Informal Education." In *2018 IEEE International Conference on Teaching, Assessment, and Learning for Engineering (TALE)*, 815–819. 2018. https://doi.org/10.1109/TALE.2018.8615396.

Radu, I. "Augmented Reality in Education: A Meta-Review and Cross-Media Analysis." *Personal and Ubiquitous Computing* 18 (2014): 1533–1543. https://doi.org/10.1007/s00779-013-0747-y.

Rambli, D., W. Matcha, and S. Sulaiman. "Fun Learning with AR Alphabet Book for Preschool Children." *Procedia Computer Science* 25 (2013): 211–219.

Samad, S., M. Nilashi, R. A. Abumalloh, F. Ghabban, E. Supriyanto, and O. Ibrahim. "Associated Advantages and Challenges of Augmented Reality in Educational Settings: A Systematic Review." *Journal of Soft Computing and Decision Support Systems* 8, no. 1 (2021).

Schrepp, M., A. Hinderks, and J. Thomaschewski. "Design and Evaluation of a Short Version of the User Experience Questionnaire (UEQ-S)." *International Journal of Interactive Multimedia and Artificial Intelligence* 4, no. 6 (2017).

Sheik-Ali, S., H. Edgcombe, and C. Paton. "Next-Generation Virtual and Augmented Reality in Surgical Education: A Narrative Review." *Surgical Technology International* 33 (2019). https://ora.ox.ac.uk/objects/uuid:c116032a-907f-4ac1-93ef-112901d159c1

Shirazi, A., and A. H. Behzadan. "Design and Assessment of a Mobile Augmented Reality-Based Information Delivery Tool for Construction and Civil Engineering Curriculum." *Journal of Professional Issues in Engineering Education and Practice* 141, no. 3 (2015): 4014012. https://doi.org/10.1061/(ASCE)EI.1943-5541.0000229.

Sırakaya, M., and D. A. Sırakaya. "Augmented Reality in STEM Education: A Systematic Review." *Interactive Learning Environments* 30, no. 8 (2022): 1556–1569. https://doi.org/10.1080/10494820.2020.1722713.

Smith, V., D. Devane, M. Begley, and M. Clarke. "Methodology in Conducting a Systematic Review of Systematic Reviews of Healthcare Interventions." *BMC Medical Research Methodology* 11, no. 1 (2011): 1–6. https://doi.org/10.1186/1471-2288-11-15.

Theodoropoulos, A., and G. Lepouras. "Augmented Reality and Programming Education: A Systematic Review." *International Journal of Child-Computer Interaction* 30 (2021): 100335. https://doi.org/10.1016/j.ijcci.2021.100335.
Wenjun Tang, Qihang Ge, Lei Zhou, and Quanyin Zhu, "Design of Spot Introduction and User Interaction System Based on AR Augmented Reality Technology," in Proceedings of the 2018 17th International Symposium on Distributed Computing and Applications for Business Engineering and Science (DCABES 2018) (Wuxi, China, October 2018), 76–79. DOI: 10.1109/DCABES.2018.00029
Williams, A., Z. Sun, and M. Vaccarezza. "Comparison of Augmented Reality with Other Teaching Methods in Learning Anatomy: A Systematic Review." *Clinical Anatomy* 38, no. 2 (2024): 168–185. https://doi.org/10.1002/ca.24234.

Chapter 10

Smart technologies for sustainable agriculture

Innovations and impacts within innovation ecosystems

Manju Priya R., Raji Rajan, N. Sathyanarayana, and Madireddi S. S. V. Srikumar

10.1 INTRODUCTION

Precision farming, often called smart agriculture, combines new technologies with traditional farm techniques and is at the forefront of agricultural development. In recent decades, the internet, mobile devices, and cloud computing have revolutionized numerous industries, including agriculture. Smart agriculture uses big data, Internet of Things (IoT), autonomous robotics, and satellite monitoring to automate this labor-intensive operation. Such technologies would reduce farmers' workloads and boost productivity by accomplishing more in less time. Therefore, smart agriculture has become a vital tool for economic growth, especially in third-world nations (Zhang et al., 2002).

Higher heat, erratic rain, and harsh weather are primary climatic changes that affect crop productivity. We need more sustainable and effective food production than ever, as the world population will hit 9 billion by 2050. In 2020, 690 million people were hungry, and the number is climbing. Smart agriculture addresses these issues, helping farmers adapt to climate change and increase food security (Wheeler & Von Braun 2013).

Smart farming uses hydroponics to grow crops without soil. It can also develop drought- and disease-resistant crops. Integrated Crop Management (ICM) reduces synthetic fertilizers and fossil fuel-powered machinery, enhancing crop productivity while protecting the environment.

10.1.1 Smart agriculture—Key benefits

Smart agriculture integrates cutting-edge technology to revolutionize farming practices, delivering numerous benefits:

1. **Improved Accuracy and Efficiency:** Sensor data, drones, and GPS have allowed farmers to refine resource management further. Efficiency in irrigation and applying nutrients or pesticides at proper levels reduces waste and strengthens crop health.
2. **Increased Efficiency:** Automation and advanced machinery, such as self-steering tractors, ensure that farming activities are done expeditiously.

 DOI: 10.1201/9781003586951-10

With such tools, farmers can automate activities such as planting and harvesting, therefore having time to attend to other important activities.

3. **Cost Reduction and Sustainability:** Smart farming technologies reduce resource wastage, lowering input costs. For instance, remote monitoring systems and Section Control technology economize on fuel and avoid extra consumption of fertilizers and pesticides for sustainability.
4. **Manual Labor:** Autonomous robots and smart devices do the most laborious jobs, such as planting, weeding, and harvesting. This lowers the amount of human labor needed and saves much time.
5. **Higher-yield and healthier crops:** This would be possible because the farmer could take action as real-time data on soil conditions, weather patterns, and even plant health becomes available.
6. **Livestock Management:** Wearable sensors and tracking devices provide unparalleled visibility into livestock health and behavior, enabling timely interventions and optimizing animal welfare.
7. **Ease of Use and Reporting:** Tools such as ISOBUS connectivity make it even easier because operations are plug-and-play compatible. Systems like FMIS perform seamless data recording, farm activity management, or financial forecasting.

10.1.2 Looking to the future of farming

The recent need to overcome global challenges in food security and environmental sustainability has made smart agriculture the cornerstone of modern farming. It is remodeling the future of farming through its capability to deliver accurate, efficient, and sustainable solutions.

a. **Increasing Resilience to Climate Change:** Technologies for smart farming are key to mitigating climate impacts, as they could lead to better management of water, land, and other resources.
b. **Feeding a Growing Population:** With growing populations worldwide, these technologies will be instrumental in ensuring an increase in agricultural productivity while reducing environmental strain to the barest minimum.

In the future, farming will adapt to smart technologies for agriculture to be adaptive and sustainable, meeting the world's food demand.

10.1.3 The objective of this chapter

The aim of this chapter, using advanced technologies, is to find how agriculture is getting reshaped and becoming more productive, with the transformative potential of smart technologies, to meet worldwide challenges in agriculture related to food security and climate change. This study sets out to analyze the innovation ecosystems supporting smart agriculture. Based on the approach developed for the study, this chapter examines how key stakeholders—governments, businesses, research institutions, and farmers—interact in agricultural innovation ecosystems to support the adoption and further development of smart farming technologies.

10.1.4 Scope

The chapter explains in detail the inclusion of smart technologies in farming: What the innovations are, how they are applied, and what benefits might be foreseen for farmers and the environment. It will involve studying precision farming techniques, predictive analytics, and digital platforms for resource optimization and better crop and livestock management. The coverage will also extend to the challenges that prevent widespread adoption, such as interoperability, high costs, and infrastructure limitations, by analyzing the policy frameworks in regions such as India, Japan, and Africa that support smart agriculture initiatives. The chapter concludes by assessing smart agriculture's sustainability and long-term environmental impacts, giving insights into future directions.

10.2 DEFINING INNOVATION ECOSYSTEMS IN AGRICULTURE

Much like other economic sectors, agriculture is rapidly altering due to technological and structural changes. Innovation ecosystems drive this dynamism, characterized by an interconnected web of entities with increasing importance to fuel innovation and progress. This ecosystem is important in agriculture to develop, test, and implement new technologies, approaches, and solutions that can meet the challenges of modern farming, such as climate change, resource scarcity, and food security.

10.2.1 Past studies

The number of research papers on the innovation ecosystem has grown recently, with most of these studies concentrating on the innovation ecosystem's composition, network structure, and dynamic progression between innovative subjects and the creative environment (Cui & Qingmei, 2021).

According to researchers, the innovation ecosystem is a web of interlinked and interacting businesses, suppliers, customers, complementary innovators, and other regulated activities. After a process of coevolution, members must contend with collaboration as well as competition. The innovation ecosystem is influenced by two key factors: The innovation subject and the innovation environment. The innovation environment is altered and changed by the innovation subject, and the innovation subject's collaborative creativity is encouraged and restrained by the innovation environment (Zhang et al., 2002).

Agriculture has a low entry barrier and is a weak business overall. Numerous challenges confront regional agriculture, particularly the creative growth of county agriculture. It is essential to rely on a more robust innovation platform since agricultural innovation subjects lack a reliable mechanism. Interfaces can only help foster and achieve deeper and more comprehensive integration of innovative resources in this manner. Rich innovative-related resources, including technical, financial, and informational, are available on the innovation platform with first-rate marketing, distribution, regulatory, and other advisory services (Su et al., 2018).

Integrating the innovation platform forms the interface ecology into the county's agricultural innovation ecosystem. There will be more integration between the innovation environment and the innovation subject, making them the same. Innovation platforms can play the role of an innovation topic in the macroenvironment and offer micro-innovation individuals a variety of resources, including cash, technology, culture, and other surroundings, to foster innovation.

Resource endowment should be the foundation for creating and growing an innovative agricultural ecosystem in county regions. Counties lack educational institutions, scientific research facilities, creative talent, and independent creativity in agriculture-related businesses (Yan et al., 2018). An increasing number of people are using agricultural innovation systems (AIS) to investigate how combined technological, social, and institutional innovations are organized in agriculture (Kilelu et al., 2013; Turner et al., 2016).

In practice, innovation platforms tend to attract enterprising champions who stimulate connectivity and interaction, foster learning, and provide adequate resources (Kilelu et al., 2013; Klerkx et al., 2012).

10.2.2 Key components and stakeholders of innovation ecosystem

Figure 10.1 illustrates that an innovation ecosystem involves various stakeholders operating in a dynamic and connected network to develop, produce, and deploy a new idea, technology, or product. Surrounding this regime are several actors, starting from government agencies down to farmers, technological businesses, and research institutes; it is highly dependent on all these in its innovation process. Interaction between various stakeholders would mean the flow of information, resources, and support crucial in advancing technology and addressing societal challenges.

Institutions of Research: Research institutions like universities, labs, and public and private research centers support creative environments. These institutes conduct basic and applied research to produce new ideas and knowledge. They enable researchers, scientists, and engineers to collaborate on innovative methods and technologies. Research institutes also teach and educate skilled personnel, improving their position in the innovation ecosystem. These institutions generate a steady talent flow that promotes innovation in many areas

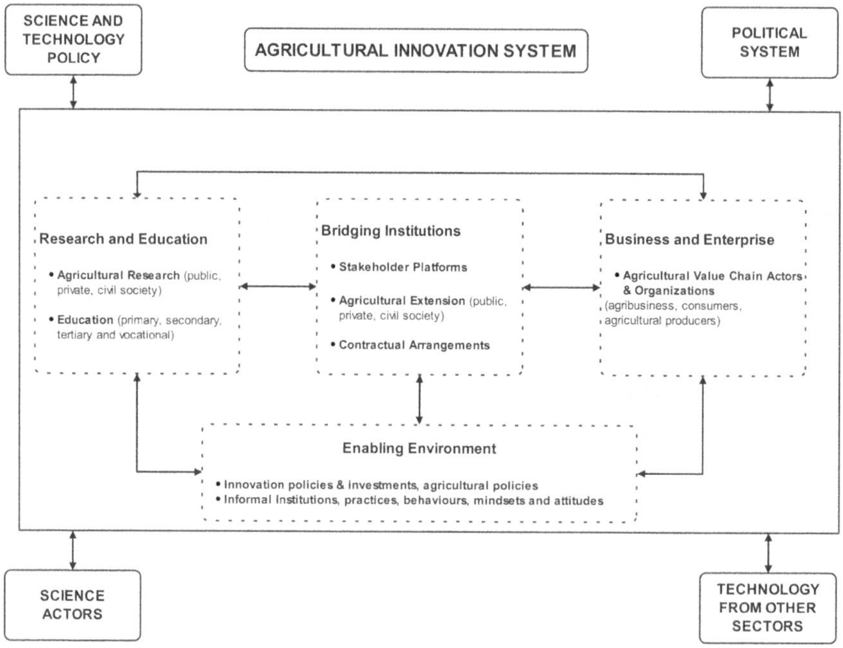

Figure 10.1 Conceptual diagram of an agricultural innovation system (Open source: https://tapipedia.org/framework/conceptual-diagram-agricultural-innovation-system)

by producing a bright workforce with advanced knowledge and technology skills. Research institutes collaborate with businesses to share information and commercialize research results. The partnership will connect theoretical research with practice, improving social welfare and economic development.

Business Enterprises: Technological Businesses or Tech companies—from startups to global conglomerates—are the center of an innovation ecosystem. They commercialize research and ideas. They have the infrastructure, skills, and funding to create, test, and implement new technologies. Tech businesses also drive innovation and progress by providing competition in this ecosystem. Most technology organizations conduct many research and product development projects simultaneously to stay competitive and tackle new challenges. These enterprises must collaborate with research institutes and other partners to fully commercialize contemporary scientific advances. Technology corporations also lend money and coach startups and entrepreneurs, adding to the innovation ecosystem. This assistance helps the ecosystem become inventive and enterprising (Autio et al., 2018).

Farmers or Cultivators: Farmers employ agricultural technology and inventions in agricultural and rural development. They are crucial to accepting and disseminating novel agricultural technologies that improve resilience, sustainability, and production (Altieri et al., 2015). Farmers can illuminate the agriculture industry's problems and help devise solutions. Farmers are involved in innovation chains to make them financially, socially, and technologically viable. Farmers help improve and scale technology for local circumstances through field demos, pilots, and feedback. Furthermore, farmer groups and cooperatives can facilitate grassroots innovation, capacity building, and information sharing between farmers and other stakeholders.

Government Authorities: The government will control, enable, and assist. The government fosters public and private sector innovation and entrepreneurship through R&D funding, tax benefits, and intellectual property protection regulations. It establishes national digital transformation, health, energy, and agricultural innovation targets. Aligning public expenditure with these priorities may help the government achieve sustainable development goals and address social challenges. In addition to direct funding, the government fosters stakeholder cooperation by constructing innovation centers, working with corporate society, and sharing platforms for academic institutions, technology businesses, and others. Dynamic and intricate, having a network of players who innovate separately. Farmers provide insights and discoveries, research institutions provide information and trained workers, technology businesses commercialize ideas, and the government promotes innovation. Stakeholders must collaborate to address social challenges, economic progress, and living standards. Teamwork, information communication, and enabling legislation are essential for innovation ecosystem development (Edler & Fagerberg, 2017).

10.3 SMART TECHNOLOGIES IN AGRICULTURE

The Food and Agriculture Organization (FAO) anticipates 70% food production growth by 2050 to feed a growing population. WHO estimates that meat output must rise 376 million tons by 2030 to meet demand. Resource constraints, diminishing arable land, and loss of carbon dioxide-induced nutritional value make these duties more challenging. Food security impacts survival, health, and social justice. Sustainable agriculture must replace obsolete methods to meet growing requirements and climate goals. By 2050, 70% of people will live in cities, making smart agricultural technology essential for sustainability (Food and Agriculture Organization, 2021).

Figure 10.2 illustrates that until recently, smart farming seemed utopian. Advanced AgTech breakthroughs advance farming, making it crucial to contemporary agriculture. Data aggregation and communication technologies that make equipment interact with each other and convey crucial information about crops, machinery, and activities are used in this agricultural management strategy. Real-time data helps farmers make better decisions and boost quality and quantity. Instead, smart farming generates actionable data to increase agricultural accuracy. Many small- and medium-sized farms, notably those cultivating berries, orchards, vineyards, or cattle, use smart farming tools. This comprises fundamental weather station applications, sensors,

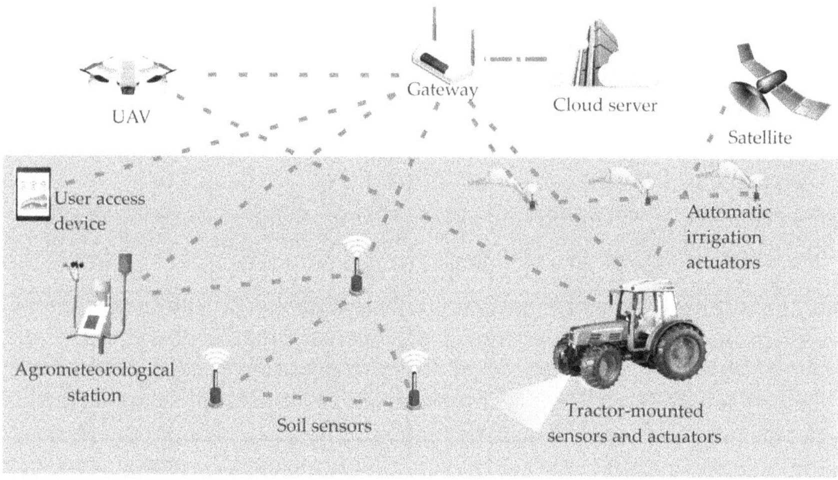

Figure 10.2 Precision agriculture remote monitoring of several crop characteristics. (Open source: https://www.mdpi.com/2078-2489/15/8/466)

robots, satellite navigation, GPS-enabled collars for livestock monitoring, and microclimate weather stations for orchard producers to obtain humidity or frost alerts. Monarch Tractor's MK-V technology enabled smart farming, electrification, and autonomy. The app-accessible Wingspan Ag Intelligence platform generates straightforward reports for sustainability and decision-making. Leads Connect, and other agri-tech firms push artificial intelligence (AI)-enabled irrigation, biotechnology, and automation.

Weather Surveillance: Climate impacts crops. Minor weather variations can cause pests, illnesses, and crop loss. Industrialized nations utilize indoor farming to protect their crops, but poor nations like India cannot afford it. Farmers use weather monitoring systems to adapt their operations. Modern weather monitoring instruments let farmers check soil moisture, temperature, humidity, wind speed, direction, and rainfall. Popular smart agricultural gadgets include weather stations with smart farming sensors. They upload regional environmental data to the cloud. Measurements can map the environment, pick crops, and increase productivity. This category comprises LoRaWAN, Wi-Fi, and Bluetooth agricultural IoT devices.

Distant Senses: Crops require definite watering, as overwatering and underwatering are detrimental to the plant. Hand-watered crops risk being watered inconsistently; however, with remote sensors, farmers can know what is going on in every part of their field. They send information on soil moisture and mineral content wirelessly, which may enable intelligent watering systems. The crop management devices track the temperature, amount of precipitation, and crop health, similar to the most conventional weather stations, all to assist farmers in finding anomalies that may lead to diseases or pests. Various small, low-power sensor nodes are designed to handle multiple tasks, such as sensing and computation, to ensure long-lasting and efficient monitoring in precision farming. Wireless sensor networks have many uses, including applications in event detection, smart cities, healthcare, and underwater monitoring (Kumar & Sharma, 2020).

Cloud-Based Software: Every smart technology employed in agriculture cooperates. Similarly, cloud computing is a crucial component of smart farming since it facilitates data integration from distant sensors. It supports farmers in forecasting weather, managing crop quality, and increasing production output. Various algorithms can be used to predict the effects of changes on crop development, such as the application of pesticides or herbicides. Farmers will receive detailed analyses and recommendations about water requirements, climate change adaptations, fertilizer requirements, etc., based on real-time data (Alex et al., 2023). Accurate resource allocation and precise planning will support sustainable agriculture and reduce waste.

Computational Intelligence: Technologies like AI are employed in practically every field. They can also help India's agriculture by offering real-time data on crop growth, harvesting, processing, and selling. Additionally, they help identify and treat illnesses that affect crops at different times of the year.

Predictive Analytics in Intelligent Agriculture: Predictive data analytics and precision agriculture go hand in hand. Farmers leverage real-time data from IoT and smart sensors to make precise forecasts regarding crop harvesting, disease risks, yield volume, and weather impacts about everything required to make farming more predictable. Artificial Neural Network (ANN) plays an important role in smart agriculture, especially yield prediction, by utilizing crop variety, soil, climate, and management techniques. While SoilScout helps farmers spend up to 50% less on irrigation water, such tools reduce fertilizer waste; platforms like Crop Performance help forecast crop quality and volume-benchmarking with weather resilience around the year (Irmak et al., 2006; Muangprathub et al., 2019).

GIS Technology: This technology enables visualization of complex field data by creating interactive, layered maps allowing remote crop monitoring. Farm automation is taken to the next level through GIS technology using drones, robotics, and satellite surveillance. Probably the most exciting agritech development is agricultural drones, or UAVs, which are far more efficient than satellites or aircraft in gathering data. The drones also carry out tasks related to planting crops, pest control, spraying, and monitoring of crops. For instance, drones owned by DroneSeed replant deforested areas and give human labor a six times higher productivity ratio. The economical drone called eBee SQ bases its use on multispectral imaging for assessing the health of crops (Friha et al., 2021).

Creative Greenhouses: Advanced greenhouse technologies will further enhance controlled-environment agriculture by implementing IoT sensors that can monitor real-time data on light, temperature, soil, and humidity. Automated systems, such as weather stations, change those conditions based on a pre-set setting (Jahnavi & Ahamed, 2015).

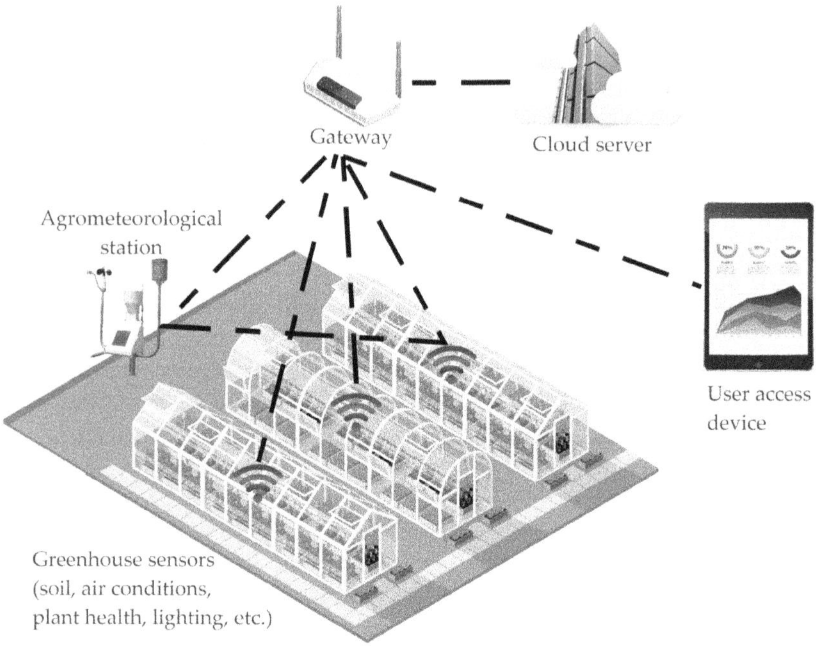

Figure 10.3 Visualization of remote greenhouse monitoring utilizing a wireless sensor network for data collection and processing. (Open source: https://www.mdpi.com/2078-2489/15/8/466)

Figure 10.3 illustrates that one example is the use of dye-sensitized solar cells (DSSC) for the greenhouse's glass cover. Temperature, humidity, and light are just a few of the elements that automation and AI-driven systems help regulate to maximize output and reduce resource use. Among other IoT agriculture products, Farmapp and Growlink provide these kinds of features.

Transportation and Logistics: Keeping track of who is where and what is being done has always been difficult, even on a small farm with a large land area. However, GPS technology and automation are transforming farm logistics, from autonomous harvesters to precision planting.

Agricultural Automation: Several types of agricultural automation have existed for many years. Systems for robotic dairy milking are one example. However, integrating automation into a moving vehicle like a tractor was quite a different matter. An electric, intelligent, and fully automated tractor best exemplifies the move toward comprehensive farm automation. Monarch's MK-V connects the dots between traditional farming methods and modern technology by offering the option of manual or autonomous operation. "Farm productivity management systems" are a more sophisticated way of looking at IoT goods in agriculture. They often come with various on-site IoT devices and sensors for agriculture, along with a robust dashboard with built-in accounting and reporting functions and analytical tools (Shamshiri et al., 2018).

Livestock Tech and Management: Figure 10.4 illustrates that health monitoring and tracking technologies are transforming animal husbandry. Sensors monitor an animal's health, location, and behavior to improve herd management further and provide personalized care in real time. Farmers can track animals with drones while staffing costs are minimal. Using smart sensors, devices like Allflex's SCR and Cowlar track herd health, activity, temperature, and nutrition both at the herd and individual levels (Anastasiou et al., 2023).

Accurate Farming: Precision agriculture, in other words, is the name given to productivity and data-driven decision-making wherein IoT sensors collect comprehensive field data on temperature, humidity, light, soil quality, CO_2 level, and infestation of pests. All this data helps farmers reduce costs by growing healthier crops and using optimum water levels, fertilizers, and pesticides. For instance, CropX offers IoT soil sensors that measure temperature, conductivity, and moisture to meet the specific needs of each crop. Tied in with geographic data, this technology makes for precise soil mapping for each field. Similarly, Motive employs smart farming solutions to help farmers reduce waste and enhance yields and farm sustainability (Liakos et al., 2018).

Autonomous Machines and Robots: Harvests, planting, and weeding make agribusiness change face with autonomous robots. Autonomous tractors from Bear Flag Robotics are examples of modern agrobots that follow designated routes with savings in labor costs and high degrees of efficiency. Robots like those from Eco Robotics use AI combined with computer vision to find weed growth and plant seeds at high levels of precision, lessening the impact further on the environment and fostering sustainability in farming. Indeed, this technological advancement promises a bright future for agriculture (Shamshiri et al., 2018).

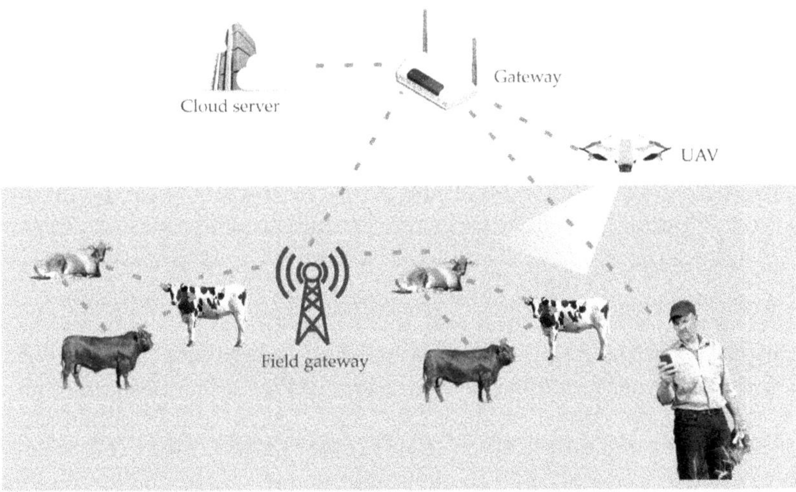

Figure 10.4 Diagram of IoT-based cattle health, behavior, and location monitoring. (Open source: https://www.mdpi.com/2078-2489/15/8/466)

10.4 IMPACT OF SMART TECHNOLOGIES ON AGRICULTURAL PRACTICES

Discussion on how smart farming technology can offer enormous cost savings and the following long-term advantages, revolving around:

1. The practice of smart farming can be very helpful in managing resources optimally. Farmers can cut down on waste and minimize their environmental impact through the precise application of water, fertilizers, and pesticides because of data analysis about crop health, weather patterns, and soil conditions.
2. Farmers may implement data-driven decisions in integrated pest management practices that reduce the demand for chemical pesticides and promote natural methods of controlling pests.
3. Technology can replace the determination of the best planting dates, the application of the correct type of crops, and the closely monitored weather conditions to improve agricultural yields.
4. Smart irrigation systems reduce water waste and stress on local water supplies by making sure plants/crops receive appropriate volumes of water based on real-time data.
5. Farmers can avoid erosion and degradation and promote the building and improvement of soil health over time by continuously monitoring their soils.
6. Smart farming helps lower greenhouse gas emissions by optimizing field activities and machinery use, reducing overall energy usage.
7. Farmers benefit by having more valuable information to answer challenges, adapt to constantly changing circumstances, and make informed decisions.
8. Smart farming combined with sustainable approaches can increase farmers' profitability, making it a financially attractive option.

Table 10.1 categorizes countries based on their geographical regions and development levels—high, middle, and low—with respect to smart agriculture adoption. It highlights disparities in agricultural advancement, infrastructure, and technological readiness. Higher-development countries like the United States, Japan, and Australia exhibit robust smart farming ecosystems driven by innovation, strong policies, and financial support. Middle-development nations such as Malaysia and South Africa show growing interest but face moderate challenges in infrastructure and policy execution. In contrast, lower-development countries, including India, Kenya, and Ethiopia, struggle with basic infrastructure, digital divides, and limited resources, hindering large-scale implementation. The table underscores the global inequality in adopting agricultural innovations and emphasizes the need for region-specific strategies to promote smart farming across varying levels of socio-economic development.

Table 10.1 Country classification table based on geographical location and development level

	Level of development		
Continent	*Higher-development countries*	*Middle-development countries*	*Lower-development countries*
Africa	Seychelles	South Africa	Ethiopia, Kenya, Uganda
Asia	Israel, Japan	Malaysia, Thailand	India, China, Pakistan
America	United States, Canada	Mexico, Colombia	*No reports found as of this review*
Europe	Denmark, Netherlands, Sweden	*No reports found as of this review. Moldova, Ukraine, Albania, Bosnia and Herzegovina, North Macedonia, Kosovo*	
Oceania	Australia, New Zealand	*No reports found as of this review. Fiji, Samoa, Tonga, Papua New Guinea, Solomon Islands, Vanuatu*	

Source: Open source https://www.mdpi.com/2078-2489/15/8/466

Precision Agriculture: Advanced technologies optimize resources and save expenses to boost crop output. Yield monitoring, GPS-guided machinery, and variable-rate input applications can evaluate every field or orchard at the right moment and with qualification. Improved soil monitoring and 70% lower water use with IoT systems like Agri-Talk boost operational efficiency (Chen et al., 2019). WSNs monitor water use to offer an appropriate environment for lemon and bell pepper crops, which allows good irrigation activity and conserves water. Variable-rate technology (VRT) ensures the right land application of fertilizers and pesticides to reduce waste and improve eco-friendly farming (Ncube et al., 2018). Satellite imagery and digital agricultural platforms allow farmers to monitor farms and handle environmental changes, improving decision-making remotely. These advances increase yields, save resources, and promote sustainable agriculture.

Advanced Greenhouse Technologies: Automatic fertilization, irrigation, and climate control systems in advanced greenhouses boost food yields and environmental sustainability. IoT-enabled remote monitoring and controlling of lighting, humidity, and temperature in high-tech greenhouses have produced high-quality crops while decreasing labor costs (Henriques et al., 2013). These improvements cut fertilizer consumption by 60% and pesticide use by 80%, lowering operational expenses. Greenhouse technologies enable year-round production, assuring a stable supply of fresh fruit for local and worldwide markets (Muñiz & Sánchez 2018). They employ precision agriculture, water management, and integrated pest control to conserve resources. This method boosts smallholder farmers' inclusive growth and local economic development. Government, business, and academics collaborate to investigate and improve greenhouse operations constantly. Due to their regulated environment, greenhouse projects promote climate-smart agriculture and minimize climate change, which is vital to sustainability. Many types of education and training help operators adapt new technology and best practices that promote a diversified and resilient greenhouse farming environment (Panchasara et al., 2021).

Livestock Care: Using behavioral analysis, remote health monitoring, and automated feeding, precision livestock farming has improved animal welfare and production. Technologies like robotic milking and sensor-based monitoring optimize milk production and herd management. Wearables on animal collars measure location, movement, and pasture feeding, allowing farmers to watch critical signals, including grazing patterns and well-being. Real-time data from IoT-enabled systems may improve animal health and help you make better care decisions. Broad immunizations and veterinary services are essential to keep animals healthy. Much of this research is done with international organizations to improve cattle nutrition, breeding, and disease control, making it more challenging. In rural regions with many jobs and earnings, livestock boosts GDP significantly (Dostain et al., 2021). Livestock development initiatives for agricultural resilience are needed during governance because rural livelihoods change. Modern technology, multinational partners, and government backing may assure development and sustainability in this sector.

Traceability of Food: Traceability measures in dairy, horticulture, and agriculture track product distribution, processing, and production against quality control requirements. Such technologies can help exporters and sellers verify a product's quality, boosting consumer confidence and food safety. IoT-based solutions enhance agricultural exports, especially fruit traceability, by preserving product integrity across the supply chain. Sweden studies blockchain and other digital methods to improve traceability, transparency, and efficiency. Traceability requires regulators and food producers to work together, comply with regulations, and handle food safety concerns quickly. Food origin-tracking initiatives support agri-food safety and legality (Ariga et al., 2019; Kungu et al., 2023). This approach is crucial for international trade and consumer protection in sectors that demand exact traceability. Agricultural modernization for traceability boosts productivity, export capability, and economic growth, especially in underdeveloped countries where agriculture accounts for 25% of GDP (Food and Agriculture Organization, 2021).

10.5 CHALLENGES AND BARRIERS

Apart from having so many advantages and advanced technologies, smart agriculture faces many challenges when adopted by farmers. Some of the key challenges are:

- **Interoperability among Different Standards:** Different devices and systems deployed in smart farming, using various technologies and often with proprietary standards, cause serious integration problems.

- **Complexity/Steep Learning Curve:** Many farmers find smart agriculture complex, mainly due to data acquisition, interpretation, analysis, and control encompassed in smart farming. Of course, the bottlenecks start with establishing IoT infrastructure or monitoring systems that may be difficult for the average farmer to handle.
- **Problems with Connectivity:** Poor network connectivity in rural areas hampers the adoption of digital technology for expanding smart farming.
- **Big Data in Agriculture:** Big data collection and analysis form the core of smart farming. However, sorting and understanding large chunks of data, which are contributed daily or weekly, may not be possible for a farmer.
- **Inability to Comprehend Farm Production Functions:** Most farming using modern intelligent systems requires understanding the operations and maintenance of such systems, which most farmers lack.
- **Management Zone Definition:** The number and size of the management zones depend on field variability, and their optimization often requires expert counseling and experimentation.
- **Barriers to Entry for New Firms:** Due to the advanced technologies required in smart farming, it is difficult for new agricultural businesses to access the necessary capital.
- **Software Configuration and Scalability:** Integrating various systems, equipment, and software packages increasingly used in modern farming will be challenging and often require specialized knowledge.
- **Energy Supply Challenges:** Smart farming technologies are energy-intensive, and most farms, especially in rural areas, cannot access reliable and sustainable energy sources.
- **Indoor Farming Challenges:** Indoor farming with artificially regulated environments has been quite advantageous, but the sector requires high initial investments. The cost of energy is extremely high, while the cost of technology acquisition is substantially higher. Hence, these act as big entry barriers for many farmers.

Table 10.2 outlines the major barriers and challenges hindering the adoption of smart agriculture across different global regions. In Africa, issues include limited access to localized datasets, privacy concerns related to drones, digital divides, high technology costs, and poor rural connectivity. The United States faces challenges such as labor shortages, climate change impacts, rising input costs, and land scarcity. Asia struggles with fragmented farming systems, small landholdings, lack of infrastructure, limited digital literacy, and inadequate governmental support. Meanwhile, Sweden contends with privacy issues and the high initial investment required for smart technologies like AI and IoT. These region-specific challenges highlight the need for targeted policy interventions, improved infrastructure, and capacity building to enable effective smart farming adoption and bridge technological gaps globally.

Table 10.2 Barriers and challenges in different countries

Country/Region	*Barriers and Challenges*
Africa	• Non-availability of diverse datasets on localized problems • Copyright and privacy issues with drones and satellite data • Digital divides among farmers • High costs are an entry barrier to new firms. • Bad connectivity in most rural areas • Lack of scalability and problems in configuration. • Energy depletion risks for operating smart farming technologies
USA	• Labor shortages • Climate change, such as abnormal weather conditions, may affect sanitation. • The increasing cost of inputs, including increasing prices of fertilizers, seeds, and machinery • Scarcity of land and water resources • High technology adoption costs and a lack of training among farmers.
Asia	• A fragmented agricultural landscape asks for tailored smart farming solutions. • Small-scale farmers may be unable to afford expensive technologies. • Poor infrastructure in rural areas, especially on issues touching on the internet. • Farmers are not well-educated and trained enough to use innovative farming technologies. • Climate change has impacts, such as unstable weather conditions, which affect crop yields. • The government has not implemented adequate policy and support; there is a lack of subsidies and incentives for adopting smart farming technologies.
Sweden	• These include copyright issues and privacy concerns, especially with monitoring by drones and satellite data. • High cost of initial investment for adopting AI and IoT

Source: Authors' contribution.

10.6 ROLE OF POLICY AND REGULATION

The technologies it indulges in might include AI, robotics, and the IoT, linked to increased agricultural productivity and sustainability. Smart farming encourages supportive policies, subsidies, and regulatory environments to help adaptive growth. Several nations have built their policy framework for supporting such innovations at an international level, and their effectiveness may vary due to regional challenges and resources. It considers policy and regulatory status concerning smart farming in India, Japan, and Africa through subheadings on the initiatives each has taken concerning its sustainable agricultural transformation.

10.6.1 Agricultural policy and regulations in India

India's economy relies on agriculture, contributing 17.8% to its GDP in 2019–20. The industry employed 41.5% of the workforce in 2020, demonstrating its significance. Yield plateaus, soil degradation, water stress, high oilseed imports, nutritional inadequacies, variable pricing, poor infrastructure, post-harvest losses, and decision-making information gaps plague Indian agriculture. Climate change is the most significant. Research claimed that natural catastrophes have devastated 85% of smallholders in India, destroying 5.04 million hectares of arable land by November 2021. The Indian government plans to improve agricultural efficiency and double farmers' income by 2022 compared to 2015–16 levels.

Smart Agriculture in India: Smart farming is growing in India. It offers higher efficiency than conservative approaches. It uses sensors, automatic watering, and temperature and soil moisture monitoring. It will let small farmers connect digital and physical infrastructure and remotely monitor crops. Agritech businesses offer farmers new and cost-effective solutions. Over 450 agritech firms launched in India in 2019 and have grown thanks to investment and funding. With climate-smart agriculture, these startups change farming methods (Dhanaraju et al., 2022).

Climate-Smart Agriculture: Growing populations and changing cuisines in India strain agricultural yields, degrade soil, and reduce biodiversity and water availability. Agriculture produces 14% of the nation's greenhouse gas emissions. Climate-smart agriculture integrates land, livestock, forest, and fisheries management to boost production, resilience, and the reduction of greenhouse gas emissions. Small and marginal farmers' food and income security are improved while embracing climatic threats (Aryal et al., 2020). Gujarat's Dhundi village farmers use solar irrigation. Farmers have incentives to sell power to the grid and minimize monsoon water use.

Government Partnerships: The Government of India has also undertaken some projects related to smart agriculture, which include the usage of AI in the prediction of crop yield, AI sensors for smart farming in collaboration with Microsoft, and drone-based monitoring of soil and the health of the crop. Projects such as sensor-based smart agriculture, aka SENSAGRI, and drone-powered funders' agencies include government funding, collecting real-time data, and providing it to farmers (Sishodia et al., 2020).

10.6.2 Japan Policy framework for smart farming enactment

Japan has established a solid policy framework to support the proactive adoption of smart farming technologies in consideration of and addressing critical agricultural challenges such as labor shortages and an aging farming population. The MAFF 2019 introduced a strategy for spreading innovative farming solutions through automation, robotics, and digital technologies, including drones. These are efforts directed toward farming modernization and productivity enhancement to place Japan at the frontier of agricultural innovation (MAFF, 2022a).

Reforms by Legislation-Enabling Technological Growth: The 2019 amendment to the Agricultural Land Law of Japan has been a strong driving force for smart farming. It relaxed the restrictions imposed on selling and renting farmlands to non-agricultural entities and opened up paths for companies other than farming. This encouraged incorporation, large-scale production of agricultural corporations, and the use of AI and automated machinery (Mi et al., 2023). This can be done through such expansion, enabling the scaling up of technology-driven agriculture in areas where this is needed in Japan.

Government Initiatives and Incentives: The Japanese government has initiated several programs to promote smart farming technologies. Among the important ones, WAGRI aims to be a data-sharing platform that consolidates vital data on weather patterns, soil data, and crop conditions into valuable insights for farmers. MAFF has also collaborated with research institutions and private companies in developing AI-based crop yield prediction models and technologies for controlling pests (MAFF, 2022b). The government also gives farmers financial support through subsidies and loans, especially for small-scale farmers farming in unfavorable geographic areas. These initiatives encourage the application of drone and robotic systems and others for precision farming (MAFF, 2022a). Agricultural cooperatives actively organize joint procurement initiatives for such technologies that reduce the cost for smaller farms (Washizu & Nakano, 2022).

Global Lessons from Japan's Smart Farming Journey: Japan's holistic approach to smart farming, which includes policy reforms, financial incentives, and technological collaboration, is instructive for all other nations. In her continuous quest to develop its digital agricultural transformation, Japan remained a model for regions facing similar challenges to labor shortages and climate-related agricultural issues (Nishi, 2022).

10.6.3 Smart agriculture policy initiatives in Africa

Smart farming is taking root in Africa, but the appropriate policy and regulatory frameworks support it. The African Union has gone the extra mile with initiatives like the Comprehensive Africa Agriculture Development Programme (CAADP), which encourages innovative and sustainable approaches to agriculture. CAADP focuses on agricultural transformation through integrating recent technologies like AI, robotics, and the IoT. These help address some of the critical issues affecting food security, climate change, and rural poverty in the entire continent (Newell et al., 2019).

Policy Initiatives for Support to Smart Farming: Various African countries have adopted smart farming technologies, supported by national and regional policies. For instance, Kenya has listed smart farming as one of the drivers of productivity and economic growth under its Vision 2030 strategy.

The government collaborates with the private sector and international organizations to promote access to precision agriculture technologies. It supports farmers with training and resources that will enable them to adopt such innovations. In response, the Government of Tanzania has integrated AI-based technologies in farming systems through its National Climate-Smart Agriculture Programme 2015–2025 to help smallholder farmers adapt to climate-related challenges (Newell et al., 2019).

Regulatory Frameworks and Investment Support: Regulatory frameworks also boost smart farming adoption. For example, Rwanda has provided subsidies or other fiscal incentives for drone-based monitoring, AI-driven soil analysis, and automated irrigation to scale up smart agriculture solutions through public-private partnerships. The World Bank and FAO have funded smart agricultural initiatives and helped African states create governance structures (Food and Agriculture Organization, 2021).

Overview of Challenges and Future Directions: Coupled with that, the access of all farmers, particularly smallholders, to smart farming technologies still has far to go. Inadequate internet connectivity, poor funding, and generally low levels of digital literacy are some factors that work against the increased usage of AI-based farming solutions. Many hindrances will require prolonged collaboration by African governments, technology providers, and international partners in infrastructural development, education, and financial support (Food and Agriculture Organization, 2021). Strategic policy interventions and regulatory frameworks lay the foundation for continued smart farming in Africa. The new African momentum goes a long way toward securing an appropriate situation where an agricultural economy can get a facelift toward improved food security and economic resilience.

10.7 SUSTAINABILITY AND ENVIRONMENTAL IMPACTS

Smart agriculture, therefore, enhances environmental sustainability through improved resource management, reduced use of chemicals, and efficient soil and water conservation care. In the present context of a global pursuit of sustainability in agriculture, adopting sophisticated digital tools and smart farming gives rise to several advantages that ensure the sustainable health of ecosystems (Wolfert et al., 2017).

Chemical Use Reduction: The major contribution of smart technology toward sustainability is reducing the application of harmful chemicals, such as pesticides and fertilizers. With precision agriculture involving GPS-guided systems and sensors, farmers can lower the chemical application rate. This reduces the overapplication of pesticides and fertilizers, lowering the cost and the environmental impact through its runoff into bodies of water. Drones and AI-powered sensors could detect pests and diseases early, hence allowing for precision spraying in the exact affected area in a field and not in a blanket mode in entire fields (Liakos et al., 2018). This reduced the usage of chemicals to a great extent at one go and avoided soil and water contamination to keep the ecosystem healthy.

Improved Management of Soil: Smart technologies improve soil management, a key farming variable. Soil health ensures production and a lengthy agricultural cycle. Farmers may better control soil erosion and deterioration by using soil-embedded sensors to detect moisture, nutrient levels, and pH. Crop rotation and cover cropping may be optimized with smart technology. Farmers will know what crops to plant at what time and how to rotate them so the soil can remain fertile and productive without synthetic inputs, which lowers its quality and productivity over time (Ghosh et al., 2021).

Water Conservation: However, water shortage is a major environmental issue. Agriculture is a major consumer of freshwater; therefore, real-time sensors and weather predictions give farmers irrigation choices to give plants the proper amount of water at the right time. Reduces water waste, eliminates over-irrigation, and ensures crops have enough water. In water-scarce areas, drip irrigation and soil moisture sensors preserve water (Kim et al., 2008). Runoff and evaporation are minimized in such systems because water is delivered at the root in regulated amounts. This allows agricultural development in drought-prone locations without diverting much water from the local environment.

Long-term Impacts on Sustainable Agricultural Practices: Their use affects agriculture's long-term viability. These technologies help farmers produce more with fewer inputs, which is crucial given the rising food demand. Smart agriculture helps farmers maximize yields, crop quality, and land utilization with little environmental effect. Smart technologies reduce farming's carbon impact and increase land use efficiency. Farming using automated machinery, drones, and precise tools reduces fossil fuel consumption and greenhouse gas emissions. Conservation tillage, which can be enabled by smart technology, sequesters carbon in the soil and helps mitigate climate change (Burney et al., 2010).

10.8 THE FUTURE OF SMART AGRICULTURE WITHIN INNOVATION ECOSYSTEMS

Smart farming uses cutting-edge technology to maximize agricultural productivity, food safety, and input costs. It is also transforming precision agriculture. Sensors and mapping technology help farmers optimize irrigation, fertilization, and other inputs by collecting data on soil, weather, and plant development. This method raised yields by 30% and lowered input expenses.

Another interesting smart agricultural technology is robots. Today, agricultural robots monitor crops, manage weeds, and harvest. When these technologies grow cheaper, they can drastically reduce agricultural labor expenses regardless of size (Duckett et al., 2018).

Smart farming will boost yield, but ecosystem development is difficult. Agriculture has many actors, including equipment makers, seed suppliers, and software providers; therefore, the ecosystem is fragmented with poorly linked data systems.

Deere and Monsanto are positioning themselves to dominate the agricultural data environment, but many solutions are still isolated. Large-scale data collaboration alone can spur farming innovation. Participation is encouraged via open data networks, user-friendly technologies, and economic incentives. Even if precision agriculture and robots are promising, conventional farming and corporate distrust will hinder their implementation. Farmers' independence and resistance to change make them skeptical of new technology (Klerkx et al., 2019).

Smart farming will be needed to increase production and address the continent's climate change and food insecurity issues. Other frontline projects like CAADP seek to expand access to agricultural technology. These actions will boost farmers' production and earnings. Laws, regulations, and private-sector collaborations must be improved to scale up biotechnology and climate-smart solutions to boost productivity and food security across the continent.

Advanced technology has transformed land and resource management, making smart farming the future of agriculture. Smart farming improves efficiency and reduces environmental impact. It will create fresher, healthier, and pesticide-free food for customers. It lowers food waste, extending shelf life. Due to these benefits, smart farming is profitable for farmers and businesses that will succeed in the future.

10.9 CONCLUSION

Smart technologies make agriculture more efficient, sustainable, and adaptable to future problems. Most case studies concluded that precision farming, AI, IoT, and data-driven decision-making have enhanced production and reduced resource waste and environmental impact compared to traditional agricultural practices (Wolfert et al., 2017). Real-time data on soil health, weather, crop development, and animal management helps farmers use inputs more efficiently, resulting in larger yields at lower costs.

These technologies might assist sustainable agriculture in reducing greenhouse gas emissions by saving water and soil resources, minimizing chemical usage, and conserving soil (Mulla, 2013). Smart irrigation, weather observation, and predictive analytics can help address climate change challenges, including irregular rainfall and rising temperatures. Technologies like those suggested will help farmers make better decisions, enhancing the farming system's resistance to external shocks.

However, intelligent technologies in agriculture have proved difficult to implement. The main obstacles are technological prices, farmer digital illiteracy, inadequate rural network access, and system complexity. Government funding, legislative improvements, and cooperation platforms are filling these gaps. Smart farming in Japan has advanced via supportive laws, public-private partnerships, and financial incentives for small- and medium-sized farmers. These gaps must be closed to scale up technology use and boost agricultural innovation. This change relies on an innovative agricultural ecosystem of research institutions, government agencies, technology businesses, and farmers. Research institutes may provide cutting-edge information and develop experts. Technology businesses may launch new goods and services, while farmers provide valuable input to refine and adapt technology to local settings. Financial aid, regulatory frameworks, and incentives from government policies assist innovation. Future technology developments will alter farming. As autonomous robots, AI analytics, and high-end IoT sensors become more affordable and widespread, farming will become more automated and data-driven (Shamshiri et al., 2018). The innovation ecosystem will evolve through better data sharing, collaboration, and a digital knowledge-sharing platform.

The full realization of these breakthroughs will require greater investment in infrastructure, education, and supporting legislation to make smart agriculture a reality at the farm level and considerably contribute to global sustainability goals (Food and Agriculture Organization, 2021).

REFERENCES

Alex, N., C. C. Sobin, and J. Ali. "A comprehensive study on smart agriculture applications in India." *Wireless Personal Communications* 129, no. 4 (2023): 2345–2385. http://dx.doi.org/10.1007/s11277-023-10234-5

Altieri, M. A., C. I. Nicholls, A. Henao, and M. A. Lana. "Agroecology and the design of climate change-resilient farming systems." *Agronomy for Sustainable Development* 35, no. 3 (2015): 869–890. https://doi.org/10.1007/s13593-015-0285-2

Anastasiou, E., A. T. Balafoutis, and S. Fountas. "Applications of extended reality (XR) in agriculture, livestock farming, and aquaculture: A review." *Smart Agricultural Technology* 3 (2023): 100105. https://doi.org/10.1016/j.atech.2022.100105

Ariga, J., E. Mabaya, M. Waithaka, and M. Wanzala-Mlobela. "Can improved agricultural technologies spur a green revolution in Africa? A multicountry analysis of seed and fertilizer delivery systems." *Agricultural Economics* 50 (2019): 63–74. https://doi.org/10.1111/agec.12533

Aryal, J. P., T. B. Sapkota, R. Khurana, A. Khatri-Chhetri, D. B. Rahut, and M. L. Jat. "Climate change and agriculture in South Asia: Adaptation options in smallholder production systems." *Environment, Development and Sustainability* 22, no. 6 (2020): 5045–5075. https://doi.org/10.1007/s10668-019-00414-4

Autio, E., S. Nambisan, L. D. W. Thomas, and M. Wright. "Digital affordances, spatial affordances, and the genesis of entrepreneurial ecosystems." *Strategic Entrepreneurship Journal* 12, no. 1 (2018): 72–95. https://doi.org/10.1002/sej.1266

Burney, J. A., S. J. Davis, and D. B. Lobell. "Greenhouse gas mitigation by agricultural intensification." *Proceedings of the National Academy of Sciences* 107, no. 26 (2010): 12052–12057. https://doi.org/10.1073/pnas.0914216107

Chen, W. L., Y. B. Lin, Y. W. Lin, R. Chen, J. K. Liao, F. L. Ng, and Y. Y. Chan et al. "AgriTalk: IoT for precision soil farming of turmeric cultivation." *IEEE Internet of Things Journal* 6, no. 3 (2019): 5209–5223. https://doi.org/10.1109/JIOT.2019.2899128

Cui, W., and T. Qingmei. "Research on the dynamic evolution of regional innovation ecosystem in the Yangtze River Delta: Based on the perspective of heterogeneity and symbiosis of innovation species." *Science & Technology Progress and Policy* 38, no. 5 (2021): 38–47. https://doi.org/10.6049/kjjbydc.2020060070

Dhanaraju, M., P. Chenniappan, K. Ramalingam, S. Pazhanivelan, and R. Kaliaperumal. "Smart farming: Internet of things (IoT)-based sustainable agriculture." *Agriculture* 12, no. 10 (2022): 1745. https://doi.org/10.3390/agriculture12101745

Dostain, A., I. A. Mengal, and S. H. Alizai. "Production trends in livestock and animal husbandry sector: A case study of Balochistan province." *Pakistan Journal of International Affairs* 4, no. 2 (2021). https://www.pjia.com.pk/index.php/pjia/article/view/343/252

Duckett, T., S. Pearson, S. Blackmore, B. Grieve, W. H. Chen, G. Cielniak, and J. Cleaversmith et al. "Agricultural robotics: The future of robotic agriculture." *arXiv Preprint Arxiv:1806.06762* (2018). https://doi.org/10.48550/arXiv.1806.06762

Edler, J., and J. Fagerberg. "Innovation policy: What, why, and how." *Oxford Review of Economic Policy* 33, no. 1 (2017): 2–23. https://doi.org/10.1093/oxrep/grx001

Food and Agriculture Organization (FAO). *Smart Farming in Africa: Opportunities and Challenges*. Rome: FAO, 2021, accessed July 4, 2025. https://www.fao.org.

Friha, O., M. A. Ferrag, L. Shu, L. Maglaras, and X. Wang. "Internet of things for the future of smart agriculture: A comprehensive survey of emerging technologies." *IEEE/CAA Journal of Automatica Sinica* 8, no. 4 (2021): 718–752. https://doi.org/10.1109/JAS.2021.1003925

Ghosh, D., A. Anand, S. S. Gautam, and A. Vidyarthi. "Soil fertility monitoring with internet of underground things: A survey." *IEEE Micro* 42, no. 1 (2021): 8–16. https://doi.org/10.1109/MM.2021.3121496

Henriques, I., B. W. Husted, and I. Montiel. "Spillover effects of voluntary environmental programs on greenhouse gas emissions: Lessons from Mexico." *Journal of Policy Analysis and Management* 32, no. 2 (2013): 296–322. https://doi.org/10.1002/pam.21675

Irmak, A., J. W. Jones, W. D. Batchelor, S. Irmak, K. J. Boote, and J. O. Paz. "Artificial neural network model as a data analysis tool in precision farming." *Transactions of the ASABE* 49, no. 6 (2006): 2027–2037. https://doi.org/10.13031/2013.22264

Jahnavi, V. S., and S. F. Ahamed. "Smart wireless sensor network for automated greenhouse." *IETE Journal of Research* 61, no. 2 (2015): 180–185. https://doi.org/10.1080/03772063.2014.999834

Kilelu, C. W., L. Klerkx, and C. Leeuwis. "Unravelling the role of innovation platforms in supporting co-evolution of innovation: Contributions and tensions in a smallholder dairy development programme." *Agricultural Systems* 118 (2013): 65–77. https://doi.org/10.1016/j.agsy.2013.03.003

Kim, Y., R. G. Evans, and W. M. Iversen. "Remote sensing and control of an irrigation system using a distributed wireless sensor network." *IEEE Transactions on Instrumentation and Measurement* 57, no. 7 (2008): 1379–1387. https://doi.org/10.1109/TIM.2008.917198

Klerkx, L., E. Jakku, and P. Labarthe. "A review of social science on digital agriculture, smart farming and agriculture 4.0: New contributions and a future research agenda." *NJAS-Wageningen Journal of Life Sciences* 90 (2019): 100315. https://doi.org/10.1016/j.njas.2019.100315

Klerkx, L., B. V. Mierlo, and C. Leeuwis. "Evolution of systems approaches to agricultural innovation: Concepts, analysis and interventions." *Farming Systems Research into the 21st Century: The New Dynamic* (2012): 457–483. https://doi.org/10.1007/978-94-007-4503-2_20

Kumar, N., and B. Sharma. "Opportunities and challenges with WSN's in smart technologies: A smart agriculture perspective." *Handbook of Wireless Sensor Networks: Issues and Challenges in Current Scenario's* (2020): 441–463. https://doi.org/10.1007/978-3-030-40305-8_22

Kungu, J. M., F. Ejobi, B. A. Okuyo, J. Namayanja, E. Namyalo, and R. Meeme. "Animal Source Foods Quality and Safety Standards Implementation in Uganda. A Review." (2023). https://doi.org/10.21203/rs.3.rs-2727321/v1

Liakos, K. G., P. Busato, D. Moshou, S. Pearson, and D. Bochtis. "Machine learning in agriculture: A review." *Sensors* 18, no. 8 (2018): 2674. https://doi.org/10.3390/s18082674

Mi, J., T. Nanseki, Y. Chomei, Y. Uenishi, and T. L. Nguyen. "Smart farming technology adoption and its determinants in Japan." In *Agricultural Innovation in Asia: Efficiency, Welfare, and Technology*, pp. 219–241. Singapore: Springer Nature Singapore, 2023.

Ministry of Agriculture, Forestry, and Fisheries (MAFF). *Subsidy Support for Farmland Utilization Efficiency (FY2022)* [in Japanese] Tokyo: MAFF, 2022a.

Ministry of Agriculture, Forestry, and Fisheries (MAFF). *Promotion of Smart Agriculture.* Tokyo: MAFF, 2022b.

Muangprathub, J., N. Boonnam, S. Kajornkasirat, N. Lekbangpong, A. Wanichsombat, and P. Nillaor. "IoT and agriculture data analysis for smart farm." *Computers and Electronics in Agriculture* 156 (2019): 467–474. https://doi.org/10.1016/j.compag.2018.12.011

Mulla, D. J. "Twenty five years of remote sensing in precision agriculture: Key advances and remaining knowledge gaps." *Biosystems Engineering* 114, no. 4 (2013): 358–371. https://doi.org/10.1016/j.biosystemseng.2012.08.009

Muñiz, I., and V. Sánchez. "Urban spatial form and structure and greenhouse-gas emissions from commuting in the metropolitan zone of Mexico Valley." *Ecological Economics* 147 (2018): 353–364. https://doi.org/10.1016/j.ecolecon.2018.01.035

Ncube, B., W. Mupangwa, and A. French. "Precision agriculture and food security in Africa." *Systems Analysis Approach for Complex Global Challenges* (2018): 159–178. https://doi.org/10.1007/978-3-319-71486-8_9

Newell, P., O. Taylor, L. O. Naess, J. Thompson, H. Mahmoud, P. Ndaki, R. Rurangwa, and A. Teshome. "Climate smart agriculture? Governing the sustainable development goals in Sub-Saharan Africa." *Frontiers in Sustainable Food Systems* 3 (2019): 55. https://doi.org/10.3389/fsufs.2019.00055

Nishi, M. "Japanese Land reform in the new era? Farmers' wellbeing and sustainable farmland management." *Environment and Planning C: Politics and Space* 41, no. 3 (2022): 429–447. https://doi.org/10.1177/23996544221138711 (Original work published 2023).

Panchasara, H., N. Hoque Samrat, and N. Islam. "Greenhouse gas emissions trends and mitigation measures in Australian agriculture sector—A review." *Agriculture* 11, no. 2 (2021): 85. https://doi.org/10.3390/agriculture11020085

Shamshiri, R., C. Weltzien, I. A. Hameed, I. J. Yule, T. E. Grift, S. K. Balasundram, L. Pitonakova, D. Ahmad, and G. Chowdhary. "Research and development in agricultural robotics: A perspective of digital farming." (2018). http://dx.doi.org/10.25165/j.ijabe.20181104.4278

Sishodia, R. P., R. L. Ray, and S. K. Singh. "Applications of remote sensing in precision agriculture: A review." *Remote Sensing* 12, no. 19 (2020): 3136. https://doi.org/10.3390/rs12193136

Su, Y. S., Z. X. Zheng, and J. Chen. "A multi-platform collaboration innovation ecosystem: The case of China." *Management Decision* 56, no. 1 (2018): 125–142. https://doi.org/10.1108/md-04-2017-0386

Turner, J. A., L. Klerkx, K. Rijswijk, T. Williams, and T. Barnard. "Systemic problems affecting co-innovation in the New Zealand Agricultural Innovation System: Identification of blocking mechanisms and underlying institutional logics." *NJAS-Wageningen Journal of Life Sciences* 76 (2016): 99–112. https://doi.org/10.1016/j.njas.2015.12.001

Washizu, A., and S. Nakano. "Exploring the characteristics of smart agricultural development in Japan: Analysis using a smart agricultural kaizen level technology map." *Computers and Electronics in Agriculture* 198 (2022): 107001. https://doi.org/10.1016/j.compag.2022.107001

Wheeler, T., and J. Von Braun. "Climate change impacts on global food security." *Science* 341, no. 6145 (2013): 508–513. https://doi.org/10.1126/science.1239402

Wolfert, S., L. Ge, C. Verdouw, and M. J. Bogaardt. "Big data in smart farming–a review." *Agricultural Systems* 153 (2017): 69–80. https://doi.org/10.1016/j.agsy.2017.01.023

Yan, M. R., K. M. Chien, L. Y. Hong, and T. N. Yang. "Evaluating the collaborative ecosystem for an innovation-driven economy: A systems analysis and case study of science parks." *Sustainability* 10, no. 3 (2018): 887. https://doi.org/10.3390/su10030887

Zhang, N., M. Wang, and N. Wang. "Precision agriculture—A worldwide overview." *Computers and Electronics in Agriculture* 36, no. 2–3 (2002): 113–132. https://doi.org/10.1016/S0168-1699(02)00096-0

Chapter 11

Innovative smart technologies for sustainability and development of smart cities

MaheshVasantrao Kulkarni, Ramesh Chandra Nayak, Sajid Hameed Siddiqui, Namrata Vasant Kumar Lotia, Biplab Sambit Kumar, Suryavanshi Bhagyeshkumar Vijaybhai, and Akshata Musale

11.1 INTRODUCTION

The establishment of smart cities brings forth a unique set of issues in terms of energy usage, environmental consequences, and public health. Key issues include the integration of separate energy systems to achieve a higher efficiency level while reducing the emissions of harmful gases (Ahad et al. 2020), smart cities resort to an integrated approach involving renewable sources of energy, energy-efficient infrastructure, and smart grid management. This transformation is, however full of challenges with regard to technology-advance energy storage solutions, improvement in energy distribution, real-time data analytics necessary to balance supply and demand, failure in these regions that lead to an increase in reliance on fossil fuels, thus perpetuating emission of greenhouse gases and jeopardizing the sustainability efforts. The roadside cook stoves have a low thermal efficiency; therefore, it increases the fuel consumption and also produces a high amount of smoke. Apart from the health impacts of the cook, WHO reported that household air pollution from solid fuels is one of the leading causes of premature death in the world? In the low- and middle-income countries, it occurs very frequently. In a smart city, the aim is to utilize data and technology maximally for the optimization of resources and infrastructure in an urban environment (Ajra, Majid, and Islam 2024). Under such a situation, the high health burden with these stoves provides an ironic contradiction. Since the frameworks of a smart city are meant to improve public health and environmental outcomes, this persistence with harmful cook stoves presents a critical hurdle toward attaining such outcomes. It is, therefore, of paramount importance to encourage the move toward cleaner cooking technologies to realize the smart city initiative in public health and environmental sustainability. Mitigating the adverse impacts of the traditional cook stoves within a smart city requires an intervention that integrates both health and public concerns with technological innovation in a manner engaging the community. Further to that, strategies such as house design and community kitchen design in the town are factors that will reduce health risk in addition to providing relationships in the social circle.

 DOI: 10.1201/9781003586951-11

Therefore, it is also a very important factor in turning toward a healthier urban ecosystem for cooking toward holistic objectives in the development of smart cities.

11.2 LITERATURE SURVEY

There are number of works related to green energy harvesting and environmental sustainability has been presented by various researchers' highlights the potential of industrial technologies to transform waste disposal in smart cities (Alajlan and Ibrahim 2025). By proposing the EBPO framework, the study shows how digital tools can enhance sustainability efforts in urban settings. However, there are still barriers to implementation, such as policy and technological limitations. The paper calls for improved integration of these technologies to overcome current challenges and to promote environmentally sustainable waste management. Alsyouf, Alsyouf, and Obaideen (2024) stress the significance of sustainability metrics in the creation of smart cities, showcasing their role in guiding urban policies. Their bibliometric review reveals growing global interest in sustainability and urban development, particularly in China. The study identifies gaps in resilience-focused research and emphasizes the need for more holistic indicators to balance technology with sustainability. Future studies ought to concentrate on integrating innovative solutions to address the complex challenges of urban development. It highlights the potential of Industry 4.0 technologies to transform the waste disposal in smart cities. Awasthi et al. (2024) examine the use of deep reinforcement learning (DRL) in the improvement of gust rejection performance in small unmanned aerial vehicles (UAVs). The authors provide evidence that shows that algorithms from DRL are effective at highly improving the UAVs' capacity to reduce environmental disturbances, that is, gusts from the wind, from a minimal suite of sensors. This methodology allows UAVs to respond adaptively to dynamic atmospheric conditions, which in turn minimizes reliance on extensive sensor arrays and resultant hardware complexity. The impact of this contribution is substantial for urban UAV flight in smart city environments, wherein environmental variability and safety concerns require effective and efficient mechanisms of flight stability. Bargavi et al. (2024) examine the tactical position of corporate social responsibility (CSR) as a driver toward sustainable urban development, especially within the context of public–private partnerships (PPPs). Their theoretical approach stresses the need to incorporate CSR activities into urban development strategies and assumes that multinational companies can contribute significantly in more than just making profits. Through socially responsible practices, companies can promote equitable urban development, ensure environmental sustainability, and promote overall well-being. The model promotes a shift in mindset to ensure corporate participation corresponds with wider urban development objectives, developing cooperative relationships among industry players and urban residents. Chen and Zhang (2024) establish a machine learning predictive model for residential occupancy rates that gives insights useful for urban energy consumption and infrastructure planning. Their study reveals the remarkable impact of urban spatial features that are of commercial center density and building height distributions on occupancy behavior. Chen, Deng, and Schneider (2024) carry out an extensive literature review that analyzes smart city projects' inclusivity, specifically with respect to their influence on disability populations. Their findings suggest that while smart city technology contains high potential for promoting urban innovation, it does not necessarily equate to value for marginalized communities. Chen and Cheng (2024) highlight the imperative of incorporating green energy systems in IoT-driven smart city models to promote urban sustainability goals. Their review extensively outlines the complex challenges of integrating renewable energy grids with smart city infrastructures, such as interoperability, scalability, and control problems. Da Silva Tomadon et al. (2024) propose an advanced Complex Event Processing (CEP) architecture specifically designed for real-time air quality monitoring under the backdrop of smart cities. The method combines data streams from a suite of pollutant sensors and takes advantage of CEP coupled with SPARQL queries to process and analyze big environmental data rapidly. De Oliveira Alves, de Oliveira, and Mühl (2024) suggest a holistic analytical framework for Home Energy administration system under the division of smart city infrastructures, focusing on energy spending reduction strategies. Their strategy uses Mixed-Integer Linear Programming (MILP) in order to regularize the scheduling of appliance operation and electric vehicle (EV) recharging, thus improving operational effectiveness. Dintén, López Martínez, and Zorrilla (2024) examine the potential of Digital Twins (DTs) to transform the creation of smart urban ecosystems. Employing the PRISMA systematic review structure, their study reviews studies to clarify the potential of DT technology for allowing dynamic modeling, simulation, and real-time data analysis in various urban fields ranging from infrastructure management to transportation systems and environmental surveillance. Despite their promise, the study acknowledges significant technological and real-world challenges, such as data integration, scalability, and stakeholder engagement issues must be resolved in order for DTs to completely support sustainable urban growth. With an emphasis on Industry 4.0, El Ghati et al. (2024) offer a model-based solution designed to streamline the design, setup, and implementation of data-driven programs for hybridized contexts. The tool addresses the challenges of integrating legacy systems with modern sensors in industrial settings by providing flexible, distributed architectures. A smart city use case involving environmental pollution data showcases how the tool can support real-time processing and storage across cloud, edge, and on-premise platforms. This tool promotes the efficient deployment of complex data applications, crucial for the smooth operation of smart cities. Eynolghozat, Ziyae, and Rezvani (2024) proposed a wireless integrated smart system that manages water and energy specifically designed for Indian smart cities.

The system incorporates the latest smart metering technologies intended to allow for real-time data capture and remote control of energy with water usage. This system takes advantage of LoRaWAN communication protocols to provide low-power, efficient connectivity suitable for deployment at large scales in urban infrastructure environments. The Telegram integration allows for timely alerts and operational monitoring improving utility management. Ferahtia et al. (2024) carried out an empirical study that looked at how the spatial processes of Durgapur, India—more specifically, its production-living-ecological space (PLES)—and the continuous transformation of urban morphology affect ecological services. The research indicates that the fast growth of cities, which saw an expansion of the urban core from 8.11% to 30.11%, has brought about significant environmental degradation. Fidvi et al. (2023) carry out an extensive survey to examine the social consequences of integrating smart home technology in the larger framework of decarbonization measures. Their study investigates the opinions of the public with respect to the implementation of intelligent home systems alongside renewable energy systems, outlining how these could provide savings in terms of energy consumption and enhance sustainability. Ghanavati, Matias, and Osório (2024) explore the operating models supporting scholarly smart city hubs, highlighting their capacity to spur urban innovation. From qualitative data gathered through surveys and interviews from seven different centers, the study outlines central difficulties in terms of resource distribution, organizational setup, and multi-stakeholder coordination within these interdisciplinary settings. Ghanbari et al. (2024) make an empirical examination of the impact of the Smart Cities Initiative (SCI) on child health outcomes in China, using an extended dataset that covers 2,638 counties between 2005 and 2018. They show that implementation of SCI is linked to sharp declines in the rates of child stunting and wasting. Ghosh and Sk (2024) suggest a novel technological solution to water shortage in arid and semi-arid areas through the installation of integrated hydro panel systems within building fabric. The hydro panels utilize solar thermal energy and atmospheric moisture to create potable water, thus promoting sustainable water resource management according to the United Nations Sustainable Development Goals (SDGs). Hameed Siddiqui et al. (2024) offer a thorough systematic review of smart waste thrash in urban waste disposal systems, highlighting their sophisticated functions including the separation of materials and automation. Although these smart bins are an essential part of optimizing waste collection efficiency as well as the enabling of resource recovery, some challenges remain, particularly in terms of realizing standardization and optimizing the conceptual models behind their deployment. Haughn, Harvey, and Inman (2024) discuss the integration of artificial intelligence (AI) and Visual Internet of Things (VIoT) technology in the building of smart city spaces, specifically in energy conservation. Their review highlights the promise of VIoT systems, which use cameras to collect data, in improving urban monitoring and control. The authors note major energy consumption issues, particularly relevant in outdoor battery-powered usage. The paper critically analyzes existing AI-based VIoT solutions, emphasizing the latest developments and suggesting power-efficient approaches to counter power requirements. This research makes important contributions to understanding the capabilities and constraints of VIoT systems for energy sustainability within smart city platforms. Huang et al. (2024) examine the impact of smart city construction (SCC) projects on destination tourism efficiency (TE) using prefecture-level cities in China as data. Their research clarifies the connection between urban development activities and tourism performance, providing empirical findings regarding the role of smart city measures in boosting tourism-related performance. The study emphasizes a general trend of increasing TE with time and underlines the spatial difference of SCC's effect, providing valuable insights for optimizing TE in smart city development programs. Jiang et al. (2024) analyze the economic contribution of smart city investments by the example of the Smart Columbus program. Applying a simple global balance model, the study analyzes the local economies' impact at the construction and operation phases of smart city initiatives. The analysis reveals varying economic impacts and stresses the importance of considering social equity during smart city planning. The findings show that while investments in smart cities can lead to economic growth, there is a need to concentrate on income distribution and social impacts for equitable development. Kumar (2024) conducts a bibliometric review of trends in research on smart urban governance across Asia. Drawing on 1,900 articles from the Scopus database, the study reveals prominent areas of research, top countries, and trends in urban governance. The analysis indicates growing interest in smart urban management, particularly from the United States and China. The landscape of research is also dominated by the dominance of social sciences, environmental science, and medicine. This review provides an insightful overview of ongoing trends in intelligent urban governance, pointing toward key directions of research due to current technological and societal developments. Compared to alternating current (AC) systems, DC micro grids are preferred due to their higher efficiency along with compatibility using renewable energy sources. The paper reviews current research on DC microgrid technology, highlighting concerns associated with safety, interactions, reliability of power with operational control. It discusses various energy management strategies that can improve operational performance, reduce costs, and enhance power system safety. The review underscores the need for further research to address these challenges and fully exploit the assurance of DC micro grids in energy from a renewable smart grid. Ma, Liu, and Zhan (2024) explore the concept of an IoT-based entrepreneurial city as a model for achieving economic sustainability. The paper investigates how IoT infrastructure can foster an entrepreneurial environment by enhancing urban management capabilities. Using a mixed-method research design, the study finds that developing IoT infrastructure accelerates the formation of an entrepreneurial city, leading to greater economic sustainability. However, the research also reveals that current conditions in some metropolitan areas are not conducive to this development. The study contributes to the entrepreneurship literature by defining the role of IoT in urban governance and providing insights into improving IoT-based entrepreneurial ecosystems. Makkonen and Inkinen (2024) examine the challenges and opportunities associated with ICT infrastructure in smart cities. The study identifies major issues such as information security risks, system incompatibilities, privacy concerns, and digital capability deficiencies. A survey conducted among Saudi nationals highlights public concerns about privacy breaches and information security, as well as issues related to system compatibility and adaptation costs. The paper emphasizes the need to address these challenges to enhance smart city infrastructure and improve public perception and engagement with smart city initiatives. Majhi et al. (2022) adopt China's "National Smart City" pilot project as an illustration to examine how the development of smart cities affects income inequality. The building of smart cities has considerably decreased regional income disparity, especially in areas with stronger financial growth, human resources buildup, and research and development expenditure, according to the study, which employs a difference-in-differences model to analyze data from 2008 to 2018. The report points out that improving public service delivery, closing the digital divide, and creating jobs through smart cities have the potential to lessen economic inequalities significantly. It stresses the importance of speeding up the development of smart cities to improve decent income distribution among the residents of cities. Nayak et al. (2023) suggest a revolutionary intelligent transportation system (ITS) based on machine learning algorithms for enhanced urban mobility and traffic management. Their study presents an optimal energy management system (EMS) that combines photovoltaic (PV) generation, energy storage systems, and EVs. Using the stochastic model-driven strategy in combination with the whale optimization algorithm, results reveal that the new EMS has the potential to reduce energy costs by as much as 45%. Nayak, Roul, and Sarangi (2018) investigate the intricacies of route planning in public transport in smart city environments. They propose a multi-objective optimization strategy taking into account factors including variability in traffic flows, operating expense, and environmental sustainability. Nayak, Roul, and Sarangi (2017a) provided a detailed view of sustainable infrastructure development through the use of the upcoming technologies to tackle the challenges of urbanization, climate change, and resource constraints are that are presently leading to major problems. In their study on energy efficiency and security in urban areas, Nayak, Roul, and Sarangi (2017b) introduce an original Intrusion Detection System (IDS) based on deep evolving Generative Adversarial Networks (GANs) to secure urban energy transactions. The research further enhances the GAN training optimization using a hybrid Bat Algorithm and Teaching-Learning-Based Optimization (TLBO) approach. With the integration of these innovations into real-world smart city environments, the authors illustrate the significant potential of their approach to improving urban energy management, security, and sustainability. Nayak, Roul, and Sarangi (2021a) investigated the integration of energy applications with transport in smart urban areas through the concept of a Subway Synergism Hub (SSH). This system, with advanced optimization algorithms and blockchain, demonstrates an efficient, secure, and cost-effective use of energy resources. The paper highlights that the system reduces the operational cost and enhances the overall efficiency of the infrastructure of smart cities, particularly in the transportation and energy fields. Nayak et al. (2021b) present a machine learning framework for urban mobility optimization of smart cities. This study utilizes a multi-level objective function along with a hybrid optimization technique to balance efficiency, energy use, and environmental sustainability of urban transport systems simultaneously. The approach complements ITS by providing a scalable and flexible solution that maximizes the sustainability value of urban mobility networks. By adopting this methodology, the research presents an inclusive mechanism for maximizing transportation operations, aligning operation performance with environmental factors to create more resilient and green urban transportation infrastructure. Nayak, Roul, and Roul (2022) propose a framework for integrating Active Mobility (AM) into urban planning, emphasizing its critical role in sustainable urban development. Despite the benefits of non-motorized transport, ITS have largely neglected AM. Thus, the paper demonstrates how urban planners can enhance accessibility and connectivity, leading to more inclusive, environmentally friendly and health-promoting mobility systems in smart cities. Nayak and Roul (2022) explore the potential of direct evaporative cooling as a sustainable solution for urban environments. Their experimental study on cellulose cooling pads demonstrates significant cooling effects, offering a partial solution for achieving sustainability goals. The authors propose that combining different cooling pad types could enhance performance, making this technology a viable option for smart city applications aimed at environmental conservation and energy efficiency. Nayak and Roul (2022) review the role of integrated energy technology in smart communities, highlighting key challenges such as energy loss and system uncertainties. De Oliveira (2022) compared the international smart techniques in order to lower the carbon emissions and improve the standard of living in smart urban cities. Nayak, Roul Sarangi (2021) investigate the role of commercial urban agriculture in promoting sustainability in cities. Through interviews with urban farmers in Porto Alegre, Brazil, the study reveals that urban agriculture contributes to food security, income generation, and environmental benefits. The authors suggest that incentivizing entrepreneurship and innovation in this sector could play a crucial role in achieving broader sustainable development goals in urban areas. Nayak et al. (2021c) focus on industrial energy efficiency as a critical component of smart city establishment with climate change prevention. Hence the review highlights the environmental impact of industrial processes and advocates for policies that promote energy efficiency, the circular use of materials, and sustainable economic practices. In order to combat climate change, the report emphasizes the necessity of international cooperation, especially through initiatives in green energy for sustainable production methods. Nayak et al. (2024) discuss the challenges and opportunities associated with solar energy adoption in developing countries, particularly within smart city frameworks. While financial and technical limitations deter widespread implementation, technological innovations in solar and beneficial policies are solutions to overcoming these limitations. The authors believe that smart cities are best positioned to lead the use of solar energy, facilitating sustainable urban development in regions facing energy deficits. Othengrafen, Sievers, and Reinecke (2025) propose a CSR framework to support the development of smart urban spaces. The authors argue that companies can play a major role in urban expansion via socially responsible practices, particularly in PPPs. The study presents ways in which CSR activities can aid smart cities by enhancing city sustainability, technological advancement, and the well-being of urban dwellers (Papageorgiou, Tsappi, and Wang 2024). Peldon et al. (2024) provide a comprehensive overview of technologies used to create sustainable smart cities. The paper discusses the role of AI, IoT, and wireless sensor networks in automating urban systems and improving decision-making processes. The authors identify key challenges, including technical, socio-economic, and environmental barriers, and offer mitigation strategies to assure that smart city technologies are implemented successfully (Nayak et al. 2024). Pervez et al. (2024) concluded about the advantages of implementation of innovative methods toward agriculture and environment sustainability.

11.2.1 Gap analysis

Cooking with smoky traditional chulhas in most of the rural village areas in the world happens to be a common exercise. Most of these cultural methods may be based more on the traditional ways or may just be an energy source; however, environmental and health issues are high. The burning of firewood also contributes to deforestation, but their fires emit particulate matter (PM) and other pollutants that degrade the air quality and augment climate change. More crucially, using smoke chulhas as a daily cooking practice is highly hazardous to one's health. Generally, people, women, and children being the more vulnerable in this regard are always highly exposed to these emissions due to their exposure during daily activities within an indoors area associated with respiratory, cardiovascular disorders, and more. Extended exposure to smoking is further linked to conditions of COPD and causes significant pulmonary infections like pneumonia (Roustaei et al. 2024). Apart from the foregoing, exposure to the above pollutants could also adversely influence the outcome of pregnancies and also generally affect child development. Hence, the cycle of poverty and health inequalities are sustained in these communities. Eliminating the dependency of these communities on smoke chulhas would, therefore be of paramount importance for achieving both environmental sustainability and well-being for such populations.

Conventional biomass-fueled chulhas continue to be widely employed in rural areas of developing nations (Safwat et al. 2024), where firewood is the major domestic energy source for household cooking. Although these stoves are culturally relevant and enable traditional cooking practices, their burning processes produce high levels of emissions of pollutants like PM, carbon monoxide (CO), polycyclic aromatic hydrocarbons (PAHs), and other toxicants. These emissions contaminate ambient and indoor air, leading to a heavy burden of respiratory and cardiovascular diseases in exposed populations, especially women and children who suffer from extended exposure. From an environmental point of view, unsustainable biomass harvesting practices lead to deforestation, land degradation, and enhanced greenhouse gas emissions, in particular CO_2 and methane (CH_4), thus accelerating climate change. The use of conventional chulhas therefore forms a nexus of health, environmental, and socio-economic issues (Saraswat, Kulhar, and Saraswat 2024). The need for transition is supported by the trends of urbanization and efforts toward sustainable and intelligent city development, which promote cleaner cooking technologies. Some possible interventions are the dissemination of upgraded cook stoves with higher combustion efficiency and lower emissions, and the use of renewable energy technologies like electric cooking with solar- or grid-based renewable electricity (Kumar et al. 2024). Scaling these technologies depends on overcoming barriers of affordability, cultural acceptability, and infrastructural preparedness. An effective transition requires a multi-disciplinary effort that integrates technological innovation with community acceptance, policy backing, and socio-economic incentives to make cleaner cooking options both affordable and acceptable. Such a strategic change is essential to enhance public health outcomes, protect environmental resources, and promote sustainable development. Energy harvesting is used to provide energy for smart cities in connection with connected devices and systems, though this is without its problems when integrating such a source. The four sources that are in use today and to be integrated into city infrastructure include solar, wind, thermal, and kinetic. All of them, depending on the location, have their problems regarding the availability and efficiency for (Sahoo et al. 2022). In the case of urban settings, placing solar panels in the open may be affected, while the output coming from kinetic energy differs, as it depends on changing traffic flow for (Sahoo et al. 2023). This level of complexity requires high-grade EMS to achieve optimization in energy storage while switching between sources for transportation and public safety services that remain stable (Lazaroiu et al. 2024). Energy harvesting systems in densely populated urban areas also present several significant challenges in terms of scale. As the urban populations grow, the demand for power to support IoT devices and smart infrastructures grows significantly, while the available space for large installations is usually limited, potentially causing an energy shortage during peak periods. Hence, effective solutions need to be efficient as well as compact Nayak et al (2021) (d). Continued research in scaling down energy converters and finding improvements in storage technologies should be complemented by effective innovative deployment strategies. To ensure the widespread adaptation of cost-effective technologies, the economic and regulatory barrier should be broken down effectively with collaborative action among government, industries, and academia in support of sustainable urban development in smart cities Nayak et al (2021) (e).

11.2.2 Overall presentation of the chapter

This study introduces a revolutionary technique that uses pressure energy, which is produced by the progress of people and cars to generate power in smart cities. The model uses pressure energy to generate electricity and comes with two attachments: simple gear train and epicyclic gear train setups. A concept that uses both main and secondary fuels is also shown in order to reduce air pollution caused by conventional open fires. Numerous studies have explored various methods of electricity generation, with pressure energy being one of the innovative approaches. The pressure exerted by foot traffic or vehicles on surfaces can be converted into electricity, and places with high footfall, such as railway stations, bus stops, temples, and highways, offer ideal opportunities for implementation (Singh et al. 2022). Various researchers have worked on generating electricity through such methods, and this paper aims to present practical models for smart city applications.

11.3 EXPERIMENTAL SETUP AND OBSERVATIONS

In order to improve the standard living for urban dwellers in smart communities, two models are designed and tested in this study: an energy harvesting model and a pollution control model. Both models are technology-driven and can be applied in various fields. The results from testing these models are presented here. Two types of energy harvesting models are presented here:

- A simple gear train-based energy harvesting model
- An epicyclic gear train-based energy harvesting model

11.3.1 Simple gear train-based energy harvesting model

This model's main objective is to use the pressure that walking over a track exerts to produce energy. The experimental setup includes a track equipped with a simple gear train mechanism consisting of a large and small gear, the bearings, a flywheel, a belt with a motor. This system is mounted on a track where high foot traffic and vehicle movement occur. Table 11.1 shows the equipment details along with their prerequisites and volume.

Figure 11.1 confirms the overall array of the arrangement. The mechanism functions for the designed model as follows: the track presses against a lever

Table 11.1 Prerequisites of equipment for the simple gear train model

Sl. No.	*Equipment details*	*Prerequisites*	*Volume*
1	Heavy wheel	35 kg	One
2	Driver gear	110 teeth	One
3	Driven gear	40 teeth	One
4	Transmission device (belt)	1,425 mm	Flat belt type 1 unit
5	Supporting parts	Roller support bearing	6 units
6	Electric generator	24 volt, 2 amp	One
7	Lane dimension	2,130 × 420 × 210 mm	One
8	Driver pulley	216 mm diameter	One
9	Driven pulley	85 mm diameter	One
10	Electric energy storage (battery)	24 volt, 2 amp	One

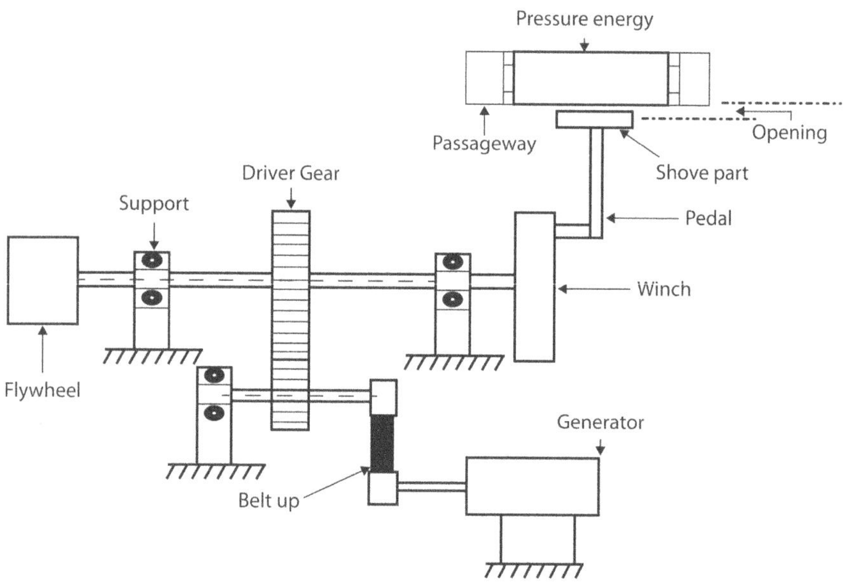

Figure 11.1 Arrangement details of simple gear train-based model.

arm when a weight is applied, which rotates the pulley. The pulley is attached to one end of a shaft, with a larger gear located in the middle and a flywheel at the other end. The rotation of the larger gear transfers motion to a smaller gear attached to another shaft. An electromotive force (emf) is produced as the generator's armature swings due to the belt that transfers the rotational power from the lower gear to the generator. The system also includes storage and transmission setup consisting of a 24-volt, 2-amp generator and a 24-volt, 2-amp battery, includes a transmission line intended to deliver the power produced. The model uses a plastic track made from ABS (acrylonitrile butadiene

Table 11.2 Specific outputs available from the simple gear train model

Sl. No.	*Equipment details*	*Quantities*
1	Cost per unit	Rs. 9,000
2	Total number of lane strikes per day	15,000 times
3	Per-strike energy	5 joule
4	Total lane energy over a 24-hour period	8,000 joule
5	Average weight (kg) per strike	40 kg
6	Produced voltage per one strike	3.6 volt
7	Typical current produced	2.4 m amp
8	Required strike to charge 12-volt battery	12,000 steps
9	Time required for a 12-volt battery to be completely accused	9 hours
10	Power output	60 watt

styrene) for its high stiffness. Its track dimension is 2,130 × 420 × 210 mm, placed within a concrete frame with a 10 mm gap. Spring attachments allow for slight displacement when a load is applied, enabling the track to return to its original position after the load is removed. The potential output from the planned setup is displayed in Table 11.2.

11.3.2 Observations from the simple gear train model

- A 5-watt LED light bulb may be lightening for eight hours using the system's 60-watt power output.
- The track can process 15,000 steps a day, and it takes about 12,000 steps to completely recharge a 12-volt battery.
- While the energy output may seem modest, this clean, renewable energy source has immense prospects for sustainable metropolitan applications.

11.4 EPICYCLIC GEAR TRAIN-BASED ENERGY HARVESTING MODEL

Three planet gears, one sun gear, and one fixed ring gear make up the epicyclic gear train unit included in this model. A shovel portion is positioned at a designated opening, and the track is fitted into the route. The shovel is fastened to a pedal that is eccentrically coupled to a winch. When load is applied to the track, the shovel moves the pedal, rotating the winch and its attached shaft. This rotation is transferred to the planet carrier, causing the planet gears to revolve around the sun gear, which then transmits rotational energy to the generator, producing electricity (Figure 11.2). Table 11.3 shows the equipment details along with their scope and measure.

The technical output from the epicyclic gear train energy harvesting model is presented in Table 11.4.

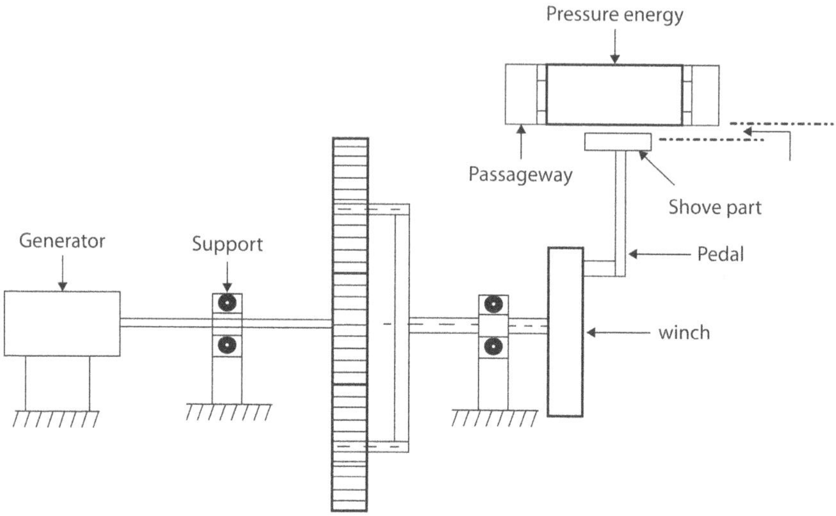

Figure 11.2 Arrangement details of epicyclic gear train-based energy harvesting model.

Table 11.3 Characteristics details of parts for the developed model

Sl. No.	*Equipments*	*Characteristics*	*Units*
1	Sun gear	95 teeth	One
2	Planet gear	35 teeth	One
3	Ring gear	155 teeth	One
4	Support unit	Roller type support	Four numbers
5	Winch diameter	25 inch	One
6	Electric generator	24 volt, 2 amp	One
7	Lane dimension	2,130 × 420 × 210 mm	One
8	Opening	165 mm Diameter	One
9	Electric energy storage (battery)	24 volt, 2 amp	One

Table 11.4 Specific outputs available from the designed model

Sl. No.	*Equipment details*	*Quantities*
1	Cost per unit	Rs. 11,000
2	Total number of lane strikes per day	15,000 times
3	Per-strike energy	9 joule
4	Total lane energy over a 24-hour period	8,000 joule
5	Average weight (kg) per strike	40 kg
6	Produced voltage per one strike	3.6 volt
7	Typical current produced	2.4 m amp
8	Required strike to charge 12-volt battery	7,000 steps
9	Time required for a 12-volt battery to be completely accused	5 hours
10	Power output	100 watt

11.4.1 Observations from the epicyclic gear train model

- The 100 watts of electricity produced by the epicyclic gear train system is enough for lightening a 10-watt LED light bulb for eight hours.
- This system requires fewer steps (7,000) and less time (5 hours) to fully charge a 12-volt battery compared to the simple gear train model.
- The increased efficiency is due to the higher rotation speed at the generator shaft, making this model more effective for urban energy harvesting applications.

11.5 COMMUNITY KITCHEN SMOKELESS CHULHA

Smart cities focus on enhancing residents' quality of life while minimizing environmental impact, yet challenges like air pollution persist. In particular, traditional cooking methods using wood and charcoal in mud stoves contribute significantly to air pollution. This project offers a smokeless chulha for a community kitchen that tackles fuel economy and air pollution. The stovetop appliance provides a more efficient and environmentally friendly cooking option by using wood as the main fuel and elevated steam as its supplementary fuel. The dimension details of the designed chulha are given in Figure 11.3.

Table 11.5 shows the dimension particulars of the designed chulha. The smokeless chulha setup includes components such as a water tube, super heater,

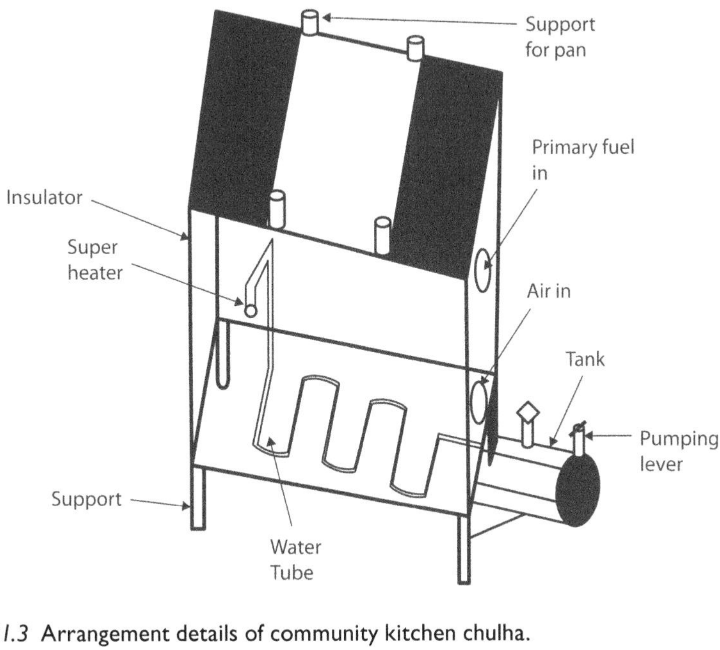

Figure 11.3 Arrangement details of community kitchen chulha.

Table 11.5 Dimension particulars of the designed chulha

Sl. No.	*Elements*	*Information*
1	Chulha height (L)	40 inch
2	Chulha outside diameter	30 inch
3	Chulha combustion unit diameter	16 inch
4	Secondary fuel tank	6 inch × 14 inch
5	Tube	1.5 inch internal dia
6	Super heater unit	4 inch height × 3 inch dia
7	Insulating material	Mineral wool
8	Draft hole diameter	2 inch
9	Fuel supply unit	3 inch dia
10	Wood (primary fuel) size	4 inch length × 2 inch dia

insulator, and support for cooking pans, a secondary fuel tank, and air intake. The process begins by filling the water tank and igniting a small amount of wood (around 200 grams) with kerosene to initiate the combustion process. After a few minutes, a blower is activated, and water in the tube is converted into superheated steam. This steam is atomized and injected into the stove, enhancing the combustion of the primary fuel and increasing the flame temperature.

The designed model is community kitchen specific, as it can cook for 50–60 people at one time. Starting the working procedure, a person first fills the tank with water, then connects the blower, and finally closes the valve of the tank. The user does 5–7 lever pumps to pressurize the water tank. To that, 200 grams of wood are put in the chulha and 20 milliliters of kerosene oil are put to light the ignition. Now the wood is lighted up and for a few 2–3 minutes time, an initial burning time is allowed so that fire gets set up, then the blower is put on for some short period, which will convert the water present in the tube into superheated steam. With time, the user pumps the lever again and a fine atomized spray of superheated steam mixes with the burning fuel in the initial process. The mixture enhances the flame temperature, hence efficiency in cooking.

11.5.1 Observations from the smokeless chulha

Four tests have been conducted; Table 11.6 presents the performance report of the chulha, Table 11.7 provides the parameters related to environmental factors during testing and Table 11.8 compares the conventional chulha with the designed chulha.

- The designed chulha achieves a maximum flame temperature of 700–800°C, significantly higher than the 300–350°C of traditional stoves.
- The stove is capable of cooking for 50–60 people at a time, making it suitable for community kitchens.

Table 11.6 Performance report of chulha

Sl. No.	*Flame velocity(m/s)*	*Main fuel requirement (kg/hr)*	*Calorific value of main fuel (kJ/kg)*	*Efficiency (%)*
Test(i)	7.12	7.2	20,050	41
Test (ii)	7.15	7.4	20,150	39
Test (iii)	7.21	7.1	20,252	38
Test (iv)	8.02	7.3	20,351	37

Table 11.7 Parameter related to environmental factor

Sl. No.	*Combustion temp. (°C)*	*Outside body temp. (°C)*	*Carbon monoxide (ppm)*	*Total particulate matter (mg/MJ)*	*Dilution ratio*
Test (i)	770	35	355	134.1	49.23
Test (ii)	730	32	356	128.6	48.31
Test (iii)	755	33	352	131.4	46.19
Test (iv)	801	37	365	129.2	47.30

Table 11.8 Comparative analysis between the conventional chulha with designed chulha

Sl. No.	*Parameters*	*Conventional chulha*	*Designed chulha*
1	Maximum combustion temperature	280–320°C	730–800°C
2	Possibility carbon monoxide (ppm)	590 ppm	323–386 ppm
3	Maximum calorific value	16,660 kJ/kg	20,350 kJ/kg
4	Method to use	Complex	Trouble-free
5	Combustion efficiency	19–24%	38–41%

- It demonstrates improved fuel efficiency and calorific value compared to traditional stoves, with thermal efficiency reaching 43%.
- The stove significantly reduces air pollution, emitting lower levels of CO and particulate matter compared to traditional stoves.

11.6 ENVIRONMENTAL AND PERFORMANCE RESULTS

- Flame velocities ranged from 7 to 8 m/s, with fuel consumption between 7.1 and 7.3 kg/hr, and calorific values ranging from 20,050 to 20,235 kJ/kg.
- Environmental monitoring revealed significant reductions in CO emissions and particulate matter, highlighting the stove's environmental benefits.
- The stove proved to be easily useable, making it deployable among non-trained community members. Its high thermal efficiency combined with its environmentally beneficial emission profiles indicate it's potential as an effective strategy for reducing urban air pollution.

11.7 CONCLUSIONS

The three different alternative models of energy generation were examined in this research are (1) energy extraction from a simple gear train, (2) energy extraction from an epicyclic gear train, and (3) a smokeless community kitchen burner, also known as a "chulha." After experimental testing of each configuration, their respective performance indicators and results were documented and compared systematically.

1. Simple Gear Train Arrangement for Energy harvesting
   A specifically designed path was utilized to measure more than 10,000 steps in a day for model testing. It was discovered that charging a 12-volt battery required 9,600 steps. This model generated approximately 60 watts of electricity, capable of charging a 5-watt LED light bulb for a complete eight hours. The model demonstrated the potential for capturing the human movement to generate clean, green energy, despite the fact that it only generates a small output.
2. Epicyclic Gear Train Arrangement for Energy Harvesting
   This configuration, which produced 100 watts of power, was more efficient and could illuminate a 10-watt LED light bulb for a complete eight hours. In comparison to the simple gear train, the epicyclic gear train increased the system's overall productivity by allowing the generator shaft to rotate more times every track step. Though these outputs may seem small, the significance lies in the use of clean and green energy. The practical application of this model ignites optimism and highlights the potential of small-scale energy harvesting. This innovation can inspire the next generation to explore sustainable energy solutions, aligning with the broader goal of achieving a greener, more sustainable Earth.
3. Smokeless Community Kitchen Stove (Chulha)
   The designed smokeless chulha has remarkable environmental advantages by significantly decreasing particulate and gaseous emissions when compared to conventional open-fire cooking practices and conventional solid-fuel stoves. Conventional biomass combustion methods are linked with elevated indoor air pollution levels and contribute immensely to greenhouse gas emission with health and environmental impacts. Eco-friendly stoves based on biofuel provide a sustainable option with enhanced ambient and indoor air quality and reduced carbon footprint and black carbon emissions. The constructed chulha had a higher temperature flame than the conventional models, and thus it was more appropriate for extensive use like community kitchens. Its easy operating design means that unskilled or semi-skilled labor can operate it without any difficulty, hence allowing its wider utilization. The innovation itself is a good step for reducing health risks and environmental damage caused by biomass combustion, fostering sustainable energy transitions in rural and peri-urban communities.

4. Therefore, these new models confirm the applicability and feasibility of incorporating sustainable technological solutions into the context of smart urban areas. The inclusion of energy-harvesting systems represents a forward-thinking measure for promoting cleaner, renewable energy sources, thus furthering environmental sustainability goals. In particular, the development of smokeless chulha presents a feasible and scalable solution to immensely lower indoor and outdoor air pollutants of conventional biomass burning, as well as related health impacts on exposed populations. With further development, such new technologies have tremendous potential to enable smarter, cleaner, and more sustainable urban environments in accordance with next-generation models of sustainability.

11.8 DISCUSSION AND FUTURE SCOPE

The findings in this research highlight the viability of integrating pressure energy harvesting with smokeless chulha technology in smart city infrastructure for increased sustainability. Models for simple and epicyclic gear train models, even with minimal energy output, reveal significant potential for decentralized micro-energy generation from pedestrian and vehicular movement. These technologies have the potential to make renewable energy devices more efficient and reduce reliance on conventional power sources (Saraswat, Kulhar, and Saraswat 2024). The smokeless chulha, with improved combustion efficiency and lower emissions, addresses vital health and environmental problems arising from conventional cooking (Singh et al. 2024). Future research should focus on scaling these models for large-scale applications, material improvement for increased durability, and integration of IoT-driven monitoring systems for real-time observation of energy generation and air quality impacts (Sushma et al. 2024; Wang et al. 2024). In addition, integrating these technologies into community-wide systems requires policies that drive affordability and user adoption (Xu, Yu, and Shao 2024; Xiao et al. 2024). Advances in compact energy storage and autonomous controls can further enhance efficiency and use. Interdisciplinary collaboration between engineers, policymakers, and communities will be instrumental to overcome implementation challenges and to realize the vision of smart cities as inclusive, clean, and sustainable urban environments (Zou et al. 2024; Zoumpoulis et al. 2024). Future innovations will improve energy resilience while improving the life quality of people in cities and peri-urban areas.

REFERENCES

Ahad, M. A., S. Paiva, G. Tripathi, and N. Feroz. 2020. "Enabling Technologies and Sustainable Smart Cities." *Sustainable Cities and Society* 61: Article 102301.

Ajra, H., M. A. Majid, and M. S. Islam. 2024. "IoT-Enabled Smart Cities towards Green Energy Systems: A Review." *International Journal of Reconfigurable and Embedded Systems* 13 (3): 708–723.

Alajlan, N. N., and D. M. Ibrahim. 2025. "Deep Smart Cities: A Review of Smart Cities Applications Based on Inference of Deep Learning upon IoT Devices." *Journal of Advanced Research in Applied Sciences and Engineering Technology* 47 (2): 94–120.

Alsyouf, L., I. Alsyouf, and K. Obaideen. 2024. "Advancing Water Security in Arid Regions: A Technological and Bibliometric Study on Integrating Hydro Panels in Buildings for Sustainable Development Goals." *Energy Nexus* 15: Article 100312.

Awasthi, A., M. Geeta Yadav, S. Baswaraju, G. Nijhawan, S. Ziara, and A. Kumar. 2024. "Detailed Analysis of Sustainable Infrastructure Design and Benefits for Urban Cities." *E3S Web of Conferences* 552: Article 01054.

Bargavi, M., B. Saraswat, P. Naval, and K. S. Kulhar. 2024. "Industrial Energy Efficiency and Climate Change Mitigation in Smart Cities: A Comprehensive Review." *E3S Web of Conferences* 540: Article 08001.

Chen, Z., and J. Cheng. 2024. "Economic Impact of Smart City Investment: Evidence from the Smart Columbus Projects." *Journal of Planning Education and Research* 44 (3): 1881–1897.

Chen, H., K. Deng, and F. Schneider. 2024. "Does City Smartness Improve Equality? Research on the Impact of Smart City Construction on Income Inequality." *Pacific Economic Review* 29 (3): 328–353.

Chen, G., and J. W. Zhang. 2024. "Intelligent Transportation Systems: Machine Learning Approaches for Urban Mobility in Smart Cities." *Sustainable Cities and Society* 107: Article 105369.

Da Silva Tomadon, L., E. V. do Couto, W. T. de Vries, and Y. Moretto. 2024. "Smart City and Sustainability Indicators: A Bibliometric Literature Review." *Discover Sustainability* 5 (1): Article 143.

De Oliveira Alves, D., L. de Oliveira, and D. D. Mühl. 2024. "Commercial Urban Agriculture for Sustainable Cities." *Cities* 150: Article 105017.

Dintén, R., P. López Martínez, and M. Zorrilla. 2024. "Model-Based Tool for the Design, Configuration and Deployment of Data-Intensive Applications in Hybrid Environments: An Industry 4.0 Case Study." *Journal of Industrial Information Integration* 41: Article 100668.

El Ghati, O., O. Alaoui-Fdili, O. Chahbouni, N. Alioua, and W. Bouarifi. 2024. "Artificial Intelligence-Powered Visual Internet of Things in Smart Cities: A Comprehensive Review." *Sustainable Computing: Informatics and Systems* 43: Article 101004.

Eynolghozat, M., B. Ziyae, and M. Rezvani. 2024. "IoT-Based Entrepreneurial City: A New Model of Urban Governance to Achieve Economic Sustainability." *Kybernetes* 53 (9): 2871–2888.

Ferahtia, S., A. Houari, T. Cioara, M. Bouznit, H. Rezk, A. Djerioui, et al. 2024. "Recent Advances on Energy Management and Control of Direct Current Microgrid for Smart Cities and Industry: A Survey." *Applied Energy* 368: Article 123501.

Fidvi, H., P. C. Ghutke, S. M. Gondane, M. V. Kulkarni, R. C. Nayak, and D. Padhi. 2023. "Advanced Smokeless Stove towards Green Environment and for Sustainable Development of Rural Women." *E3S Web of Conferences* 455: Article 02017.

Ghanavati, F., J. C. O. Matias, and G. J. Osório. 2024. "Towards Sustainable Smart Cities: Integration of Home Energy Management System for Efficient Energy Utilization." *Sustainable Cities and Society* 111: Article 105579.

Ghanbari, H., R. M. Soe, K. Toiskallio, and L. Mora. 2024. "How Do Academic Smart City Centres Operate in Complex Environments? A Business Model Perspective." *Cities* 152: Article 105211.

Ghosh, S., and J. Sk. 2024. "Impact of Changing Urban Form and Production-Living-Ecological Space on Changing Ecosystem Services of a Smart City of Eastern India, Durgapur." *Environmental Monitoring and Assessment* 196 (9): Article 804.

Hameed Siddiqui, S., R. C. Nayak, H. Fidvi, N. H. Khan, M. V. Kulkarni, P. C. Ghutke, and M. K. Roul. 2024. "Thermoplastic Polymer Track with Gearing Mechanism to Harvest Green Energy for Sustainable Development." *E3S Web of Conferences* 552: Article 01011.

Haughn, K. P. T., C. Harvey, and D. J. Inman. 2024. "Deep Learning Reduces Sensor Requirements for Gust Rejection on a Small Uncrewed Aerial Vehicle Morphing Wing." *Communications Engineering* 3 (1): Article 53.

Huang, J., W. Zonghui, D. D. Koroteev, and M. Rynkovskaya. 2024. "Energy Efficiency Security in Urban Areas: Challenges and Implementation." *Sustainable Cities and Society* 107: Article 105380.

Jiang, J., Y. Li, C. Li, L. Yu, and L. Li. 2024. "Smart Transportation Systems Using Learning Method for Urban Mobility and Management in Modern Cities." *Sustainable Cities and Society* 108: Article 105428.

Kumar, A. 2024. "A Novel Framework for Waste Management in Smart City Transformation with Industry 4.0 Technologies." *Research in Globalization* 9: Article 100234.

Kumar, S. S., R. Chandra, and S. Agarwal. 2024. "Rule-Based Complex Event Processing for an Air Quality Monitoring System in Smart City." *Su*stainab*le Cities and So*ciety 112: Article 105609.

Lazaroiu, A. C., M. Roscia, V. S. Dancu, and G. Balaban. 2024. "Social Impact of Decarbonization Objectives Through Smart Homes: Survey and Analysis." *Renewable Energy* 230: Article 120872.

Ma, L., C. Liu, and Y. Zhan. 2024. "How Smart City Construction Affects Destination Tourism Efficiency." *Tourism Economics* 30 (6): 1600–1623.

Majhi, A., A. Das, S. Panda, E. Pagal, M. Singh, R. C. Nayak, and S. K. Dash. 2022. "An Attachment with the Hand Pump to Lift Water without Any External Source." *Materials Today: Proceedings* 62 (P12): 6755–6758.

Makkonen, T., and T. Inkinen. 2024. "Inclusive Smart Cities? Technology-Driven Urban Development and Disabilities." *Cities* 154: Article 105334.

Nayak, R. C., M. K. Roul, and S. K. Sarangi. 2017a. "Experimental Investigation of Natural Convection Heat Transfer in Heated Vertical Tubes." *International Journal of Applied Engineering Research* 12 (10): 2538–2550.

Nayak, R. C., M. K. Roul, and S. K. Sarangi. 2017b. "Experimental Investigation of Natural Convection Heat Transfer in Heated Vertical Tubes with Discrete Rings." *Experimental Techniques* 41 (6): 585–603.

Nayak, R. C., M. K. Roul, and S. K. Sarangi. 2018. "Natural Convection Heat Transfer in Heated Vertical Tubes with Internal Rings." *Archives of Thermodynamics* 39 (4): 85–111.

Nayak, R. C., M. K. Roul, P. Padhi, and S. K. Sarangi. 2021 (a). "Design and Development of Improved Methods of Curing of Bricks during Manufacturing Process and Construction Work to Save Water, Minimize Pollution and Human Effort." In *Ecological and Health Effects of Building Materials*, 259–274. Springer International Publishing.

Nayak, R. C., M. K. Roul, and S. K. Sarangi. 2021 (b). "Forced Draft and Superheated Steam for Design and Development of Community Smokeless Chulha to Help Women in Rural Areas." In *Lecture Notes in Mechanical Engineering*, 93–102. Advanced Manufacturing Systems and Innovative Product Design: Select Proceedings of IPDIMS 2020. pp. 93–102. Singapore: Springer Singapore.

Nayak, R. C., M. K. Roul, and S. K. Sarangi. 2021 (c). "Enhancement of Natural Convection Inside Vertical Tubes Using Internal Obstacles as Rings with Different Arrangements." In Current Advances in Mechanical Engineering: Select Proceedings of ICRAMERD 2020. pp. 191–201. Singapore: Springer Singapore.

Nayak, R. C., M. K. Roul, and S. K. Sarangi. 2021 (d). "Innovative Methods to Enhance Irrigation in Rural Areas for Cultivation Purpose." *Journal of The Institution of Engineers (India): Series A* 102 (4): 1045–1051.

Nayak, R. C., M. K. Roul, A. Sarangi, A. Sarangi, and A. Sahoo. 2021 (e). "Mechanical Concept on Design and Development of Irrigation System to Help Rural Farmers for Their Agriculture Purpose During Unavailability of External Power." *IOP Conference Series: Materials Science and Engineering* 1059 (1): Article 012048.

Nayak, R. C., M. K. Roul, and P. D. Roul. 2022. "Design and Development of Smokeless Stove for a Sustainable Growth." *Archives of Thermodynamics* 43 (1): 109–125.

Nayak, R. C., M. K. Roul. 2022. "Technology to Develop a Smokeless Stove for Sustainable Future of Rural Women and Also to Develop a Green Environment." *Journal of the Institution of Engineers (India): Series A* 103 (1): 97–104.

Nayak, R. C., C. Samal, M. K. Roul, and P. Padhi. 2023. "A New Irrigation System Without Any External Sources." *Journal of The Institution of Engineers (India): Series A* 104 (2): 281–289.

Nayak, R. C., S. Majumdar, V. Kandpal, and L. G. Anthopoulos. 2024. "New Urban Technology Towards Energy Harvesting and Environmental Initiatives for a Smart City." In *Smart Cities. S.M.A.R.T. Environments*, edited by S. Majumdar, V. Kandpal, and L. G. Anthopoulos. Cham: Springer. https://doi.org/10.1007/978-3-031-59846-3_14.

Othengrafen, F., L. Sievers, and E. Reinecke. 2025. "From Vision to Reality: The Use of Artificial Intelligence in Different Urban Planning Phases." *Urban Planning* 10: Article 8576.

Papageorgiou, G., E. Tsappi, and T. Wang. 2024. "Smart Urban Systems Planning for Active Mobility and Sustainability." *IFAC-PapersOnLine* 58 (10): 261–266.

Peldon, D., S. Banihashemi, K. LeNguyen, and S. Derrible. 2024. "Navigating Urban Complexity: The Transformative Role of Digital Twins in Smart City Development." *Sustainable Cities and Society* 111: Article 105583.

Pervez, M. S., R. C. Nayak, H. Fidvi, A. Sahoo, M. V. Kulkarni, P. C. Ghutke, and R. N. Bhoi. 2024. "Mechanical Energy Harvester for Sustainable Development towards Green Environment." *E3S Web of Conferences* 529: Article 04018.

Roustaei, M., T. Niknam, J. Aghaei, M. Sheikh, H. Chabok, A. Kavousi-Fard, V. Vahidinasab, and J. M. Guerrero. 2024. "Enhancing Smart City Operation Management: Integrating Energy Systems with a Subway Synergism Hub." *Sustainable Cities and Society* 107: Article 105446.

Safwat, H., M. A. Elsayed, M. Alghrieb, and I. El-Mahallawi. 2024. "Direct Evaporative Cooling for Sustainability and Environmental Conservation." *Innovative Infrastructure Solutions* 9 (7): Article 256.

Sahoo, A., R. C. Nayak, A. K. Senapati, and M. K. Roul. 2022. "Validation of Experimental Results with Theoretical by Using ANSYS Workbench on Vertical Tube Subjected to Natural Convection Heat Transfer Without Internal Obstacles." *Materials Today: Proceedings* 52: 1348–1353.

Sahoo, A., R. C. Nayak, A. K. Senapati, M. K. Roul, and K. Khuntia. 2023. "Experimental Study of Fluid Flow Properties at Exit of the Heated Tube Subjected to Natural Convection Heat Transfer." *Materials Today: Proceedings.*

Samal, C., R. C. Nayak, M. K. Roul, D. Das, and R. Kumar. 2023a "A Comprehensive Review on Performance Evaluation Methods of Biomass Cook Stoves." In *Waste Recovery and Management: An Approach toward Sustainable Development Goals*, 95–106. CRC Press.

Samal, C., R. C. Nayak, M. K. Roul, D. Das, and R. Kumar. 2023b. "Performance Evaluation of an Improved Biomass-Fired Cook Stove: A Comparative Analysis." In *Waste Recovery and Management: An Approach toward Sustainable Development Goals*, 107–120. CRC Press.

Saraswat, B., K. S. Kulhar, and N. Saraswat. 2024. "Enabling Smart Cities Through Renewable Energy Integration." *E3S Web of Conferences* 540: Article 04003.

Saraswat, N., M. Pandeya, R. K. Pareek, and K. S. Kulhar. 2024. "Solar Energy in Developing Countries: Challenges and Opportunities for Smart Cities." *E3S Web of Conferences* 540: Article 04003.

Singh, K., P. Kolar, S. Nanduri, V. Seetha Ram, and D. Kumar. 2024. "CSR and Smart City Progress: A Proposed Model for Urban Development." *Letters in Spatial and Resource Sciences* 17 (1): Article 3.

Singh, M., A. Majhi, S. Panda, R. C. Nayak, S. K. Dash, and E. Pagal. 2022. "An Attachment with the Hand Pump to Lift Water Without Any External Source." *Materials Today: Proceedings* 62 (P12): 6755–6758.

Sushma, N., H. N. Suresh, L. J. Mohana, and K. B. Santhosh Kumar. 2024. "Experimental Investigation on Wireless Integrated Smart System for Energy and Water Resource Management in Indian Smart Cities." *Results in Engineering* 23: Article 102687.

Wang, Y., Y. Li, Y. Cao, W. Li, X. Yang, Y. Xu, H. Chen, and M. Shahidehpour. 2024. "Technical and Practical Application of Integrated Energy Technology in Smart Community." *Electric Power Systems Research* 232: Article 110422.

Xiao, M., L. Chen, H. Feng, Z. Peng, and Q. Long. 2024. "Smart City Public Transportation Route Planning Based on Multi-Objective Optimization: A Review." *Archives of Computational Methods in Engineering* 31 (6): 3351–3375.

Xu, L., H. Yu, and S. Shao. 2024. "The Smarter the City Develops, the Healthier the Child Grows? Evidence from China." *Cities* 152: Article 105195.

Zou, Y., W. Xie, S. Lou, Y. Huang, D. Xia, X. Yang, and C. Feng. 2024. "Prediction Model Establishment for Residential Community Occupancy Considering Urban Environment." *Journal of Building Engineering* 96: Article 110463.

Zoumpoulis, P., F. K. Konstantinidis, G. Tsimiklis, and A. Amditis. 2024. "Smart Bins for Enhanced Resource Recovery and Sustainable Urban Waste Practices in Smart Cities: A Systematic Literature Review." *Cities* 152: Article 105150.

Chapter 12

IoT, AI, Big Data, and Automation

Sankar Padmanabhan, Anand Joseph Daniel D., and Latha Ramamoorthy

12.1 INTRODUCTION

Digital skills are outlined by the globalization of many technology fields, like robotics, artificial intelligence (AI), data management, and the Internet of Things (IoT), increasingly affecting the world. In the last few years, Big Data and distributed information management strategies have rapidly expanded and made it considerably more difficult to create deep learning (DL) models more ecologically efficiently in terms of privacy, security, and distributed data. For example, worrying that modern centralized methods of training models often require that data be gathered in a single place therefore increases data privacy and security risks and increases the risk of non-compliance with legal standards (Ahmed and Shakoor, 2025). An IoT-based system is also suggested which can detect, trap, and classify the snakes. Effective mitigation techniques are necessary to prevent human-snake interactions and address the public health danger posed by snakebites, especially in hot and warm climatic countries. An inventive non-invasive method of preventing snakebites is offered with the creation of an IoT-based smart snake-trapping gadget that can autonomously capture and identify snakes. The system's connectivity lowers the risk of snakebite by enabling prompt action in the event of venomous species. Furthermore, the system contributes to the study of more general ecological issues by offering data on snake habits of motion and species distribution (Ahuja et al., 2025). AI, through better decision-making and predictions, has transformed how the agricultural sector has responded to weather forecasts in very significant ways. This shift is being accelerated by developments in big data, IoT, remote sensing, and AI. The system demonstrates that sensor fusion and the IoT can change small-scale farming by having higher sustainability, less human labor, and higher plant production and health (Abu et al., 2022). Transformer designs are becoming increasingly vital for boosting agricultural management techniques. A notable transition from countryside to urban areas is taking place in the current century, mirroring the global urbanization trend. Economic opportunities as well as a desire for better services and a higher standard of living are the driving

 DOI: 10.1201/9781003586951-12

forces behind this movement. Urban development provides great opportunities as well as threats (Kow et al., 2025; Mehanna and Morsy, 2022). Yes, growth provides benefits to living standards, however, it can bring problems such as traffic congestion, noise, and pollution. Creative solutions will be required to mitigate the issues related to urban development to promote sustainable development. Smart cities taking advantage of the IoT, a large amount of data and machine learning (ML), could provide some approaches to managing these challenges.

The central theme of this chapter will be analyzing how these technologies cooperate to influence processes and industries. With the IoT, the usable data provides real-time contextual information. When this data is integrated together with analytics and decision-making capabilities offered by AI, the data can trigger actionable insights in terms of predicted maintenance, automation of processes, and optimized processes. All of these capabilities involve Big Data, since they all provide the scale, speed, and diversity of data necessary to develop AI models for automation and efficiency. These technologies work together to provide a seamless integration of services across industries and sectors, such as manufacturing and logistics, healthcare, smart cities, and agriculture (Sankar, 2022). When paired with intelligent systems that increase productivity, improve efficacy, and ultimately allow for enhanced decision-making, this is where the industry innovation is emerging. This chapter provides key insights and framework for using these technologies to build smarter and smarter connected ecosystems to help solve very significant multi-layered challenges like interoperability, scalability, and data security.

12.2 BACKGROUND

This report provides an insight into emerging areas of agriculture dealing with the integration of Big Data, AI, and IoT with the aim of monitoring and reducing carbon emissions in agriculture. The authors stress how these approaches facilitate the adoption of Climate Smart Agriculture (CSA) and can help facilitate sustainable agriculture. The report provides a comprehensive account of how Big Data, AI, and IoT can work together to account for carbon footprints and aid sustainable agriculture in a broader context. The report discusses possibilities for the future, such as transparent blockchain carbon credit trading, remote sensing, and ML based precision agriculture and scalable IoT carbon markets. The report aims to offer action-oriented examples of how leading edge technologies, using Big Data, AI and IoT will assist the agricultural sector and contribute to an overall goal zero emission and sustainable development (Zakar, 2024). Figure 12.1 gives the bibliographic visualization of around two thousand papers connecting AI, Big Data, IoT, and automation for the last five years.

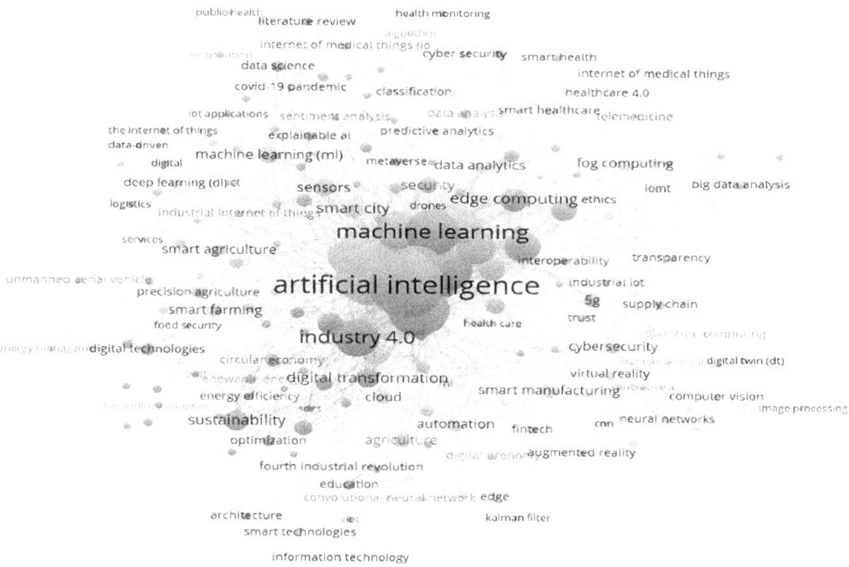

Figure 12.1 Bibliographic visualization of the AI, Big Data, IoT and automation literatures.

12.3 ARTIFICIAL INTELLIGENCE

Robots are now able to perform tasks that usually require human intelligence due to rapid advancements in AI. AI has the potential to revolutionize industries and enhance sustainability in line with the UN's Sustainable Development Goals (SDGs), particularly SDG 9 (Industry, Innovation, and Infrastructure), SDG 11 (Sustainable Cities and Communities), and SDG 13 (Climate Action).

12.3.1 Sustainable Development Goal 9

Sustainable Development Goal 9, or otherwise in short known as SDG 9, relies on building resilient infrastructure, promoting inclusivity and sustainable industrialization, thus fostering innovation. This goal is highly crucial, as it can be a pioneer in achieving sustainable development globally, as it addresses key challenges in infrastructure development, industrial growth, and technological advancement. Constructing a withholding, sustainable, and high-quality infrastructure is crucial for both human welfare and economic growth. This comprises transborder and regional infrastructure that prioritizes fair and inexpensive access. By 2030, the goal is to greatly increase the industry's proportion of GDP and employment, particularly in the least developed nations, by promoting inclusive and sustainable industrialization. Improving scientific research and the industrial sectors' technical capacities is essential. This entails raising funding for research and development as well as increasing the percentage of investigators per million people (Yao and Zhou, 2025; Zia et al., 2025).

12.3.1.1 Roles associated with IoT, automation, and AI in achieving SDG 9

Through the provision of real-time usage and efficiency statistics, predictive maintenance, and resource allocation optimization, IoT improves infrastructure management. Automation supports sustainable industrialization by making it less expensive, improving productivity, and streamlining processes. In addition, through the processing of large and complex data and data sets, AI helps deliver creativity to the table by optimizing industrial processes, improving planning for the built environment, and enabling innovations that will allow for technological development in the area of sustainable development.

12.3.2 Sustainable Development Goal 11

Sustainable Development Goal 11 (SDG 11) aims to achieve inclusive, safe, resilient, and sustainable cities and urban settlements. As urban areas increase in population, demographic and socio-economic implications for social equity, revenue, jobs, and environmental sustainability are large, necessitating goal 11. This brief overview of SDG 11 demonstrates automation, AI, and IoT contributions to achieve the goal. The pillars of SDG 11 are sustainable urbanization, the resilience of infrastructure, and the creation of inclusive and safe cities.

12.3.3 The impact of IoT, automation, and AI on SDG 11

The IoT has transformed urban management with improving data streams through real-time data regarding traffic situations, environmental conditions, and infrastructure utilization to make intended decisions. Through IoT sensors; energy consumption, garbage disposal and air quality are monitored in order to facilitate sustainability. Automated services; garbage collection and traffic management are more efficient and cost effective when applied. AI technologies help to explain the predicted situation through IoT data and develop measures for improved city planning, allowing for improvements in public safety and optimal planning for the future city to manage traffic. AI can allow for efficiencies in energy usage while managing what is needed for demand and estimating what is needed to create the buildings. Smart transportation planning, energy conservation, and waste management methods are significant technology milestones aligned with a mitigation against SDG 11 intentions of flexibility, sustainability, and equity.

12.3.4 Sustainable Development Goal 13

Sustainable Development Goal 13 emphasizes the need for urgent action to combat climate change and its impacts (SDG 13). This goal is critical for climate change mitigation and adaptation, key elements of a sustainable future for world society.

The three main pillars of SDG 13 include its purpose statement, climate change impacts, and global cooperation. The goal of SDG 13 is crucial for addressing climate change, a crisis affecting every aspect of life on Earth. Current efforts alone will not be adequate to keep global warming below 2°C, as outlined in the Paris Agreement. Challenges include global emissions reduction, cooperation, and adaptation and resilience.

12.3.4.1 Role of IoT, automation, and AI in achieving SDG 13

IoT is used to support climate monitoring through real-time information and tracking of weather issues, pollution and environmental issues. IoT sensors monitor all wildlife population changes, while monitoring for deforestation and pre-warning of disasters. Automation can also improve some processes in the renewable energy generation sector including enabling wind turbines movements and maximizing the amount of energy from a solar panel. Automation can also improve forms of energy efficiency for industrial processes and buildings. AI allows for better tracking of climate occurring events through analyzing huge amount of data, developing climate models and forecasting weather and climate. AI can also help with decision by understanding social and economic impact of climate change. An AI management system is shown in Figure 12.2.

Figure 12.2 AI-powered energy management system.

12.4 APPLICATIONS AND BENEFITS OF AI

AI-powered predictive maintenance may help minimize downtime and optimize resource use in sectors such as industry and energy. AI can decrease waste and maximize energy use in buildings and enterprises. AI can help achieve SDG 3, a good quality of life, by increasing the accuracy of diagnoses and customizing treatment regimens. AI can increase productivity and automate tedious chores, supporting SDG 8 for decent employment and economic development. AI supports SDG 16 justice, peace, and institutional strength by analyzing vast data sets and offering insights that guide decision-making. SDG 12 is sustainable consumption and manufacturing can benefit from AI's ability to minimize waste and maximize resource use.

12.4.1 How do automation, Big Data, and IoT play a major role in AI?

In today's world, AI has swiftly emerged as one of the most transformative technologies that empowers machines and systems to emulate human intelligence, understanding from experience, patterns, making informed decisions on each step, and even solving problems. The IoT acts as the eyes and ears of AI. Well, talking about IOT, it refers to an enlarged network of intelligent devices such as sensors, cameras, wearables, and machines that collect and share real-time data from their surroundings. It goes as a synonym with humans, like how they use their senses to gather information before making decisions. AI uses data collected through IoT to understand what's happening in the real world. For example, in agriculture, farmers are now using IoT sensors in their fields to monitor and measure the level of soil moisture, amount of sunlight, and temperature at the same time even noting the movement of pests. Drones can scan large crop areas to capture visual health indicators. This information is passed on to an AI system that analyzes it and recommends when to water the crops, apply fertilizers, or take preventive actions. The result is more accurate farming, higher crop yields, and reduced waste of resources. IoT allows AI to respond to live data rather than outdated or static information, significantly improving decision-making (Gupta and Kumar, 2024).

A similar concept is used in smart homes. Devices comprising of sophisticated thermostats, lights, appliances, etc. gather data on how people live, at what time they're home, what their preferences are, etc., because AI learns these patterns over time. As a result, it can adjust the home's temperature before you arrive or turn off lights in empty rooms, helping you save energy without even thinking about it (Sanchez et.al., 2025).

Once AI receives data from IoT devices, there is high need to process it. This is where the field of Big Data Biology pitches into play. Well, Big Data refers to large, raw, and complex set of data that the traditional computing systems cannot handle efficiently. It's not just about having a lot of information; it's also about how fast that information comes in, how varied it is in format, and whether it's trustworthy. These factors are often summarized as volume, velocity, variety, veracity, and value. Big Data technologies help clean, sort, and organize this information so AI systems can use it to learn. For instance, when an ML algorithm is trained to recognize fraud in banking transactions, it must look at millions of past transactions to know the difference between normal and suspicious. This is only possible through the tools and frameworks developed to process Big Data, such as Hadoop, Apache Spark, Kafka, Flink, and others.

E-commerce is another area where Big Data and AI work together. Every time a customer browses product, reads reviews, or makes a purchase, those behaviors are recorded. AI analyzes this mountain of data to recommend other products the customer might like. Platforms like Amazon and Netflix thrive on this approach. Without Big Data to supply these insights, AI would operate with limited knowledge and fail to deliver personalized user experiences.

The healthcare sector applies the potential in Big Data as memory and learning which runs AI when dealing with imaging data, lab results, electronic health records, and data generated from wearables. In demonstrating AI's ability to transform healthcare, it can process Big Data to analyze trends in population data to determine where disease outbreaks may occur, identify early symptomology for chronic disease management, and assess the best therapies based on Big Data.

The last component of the process is automation, known as the hands or the muscle of AI, and it is simply using technology to conduct tasks without human intervention. While automation takes care of specific tasks, AI provides insights and recommendations. Examples of this include robots in production lines or software bots that manage Outlook email. When combining automation with AI, we can create intelligent automated systems capable of thinking and acting independently.

These days, sensors are being inserted into manufacturing machinery to track what it is doing. AI will consider information such as temperature change and vibrating to determine if a machine is about to fail, it can trigger maintenance schedules, engineer notifications, or even halt the machine entirely to prevent damage. Production lines can operate without interruption, no downtime to worry about and avoid costly repairs.

Robotic Process Automation (RPA) brings together automation and AI in the financial services industry. In commercial banks, RPA bots are tasked to create reports, review compliance documents, and then process a loan application. AI can enhance these RPA processes, searching for outliers or compliance issues, leveraging predictive analytics for both enhanced speed and accuracy. In smart cities, for example, thousands of sensors on roads, traffic lights, and public utilities for all types of data on traffic flows, electrical use, air quality, etc., are collected, stored, and analyzed using Big Data platforms, all while AI predicts traffic jams and patterns of high use.

Automated systems then respond by redirecting public transport, altering routing using the appropriate signaling, and relaying to individual vehicles, all without human monitoring, improving urban life, energy consumption, and pollution levels.

Healthcare is evolving due to three interrelated aspects: wearable devices that gather continuous health data, Big Data platforms that organize health data, and AI that detects preliminary symptoms and health risks. Automation will alert stakeholders or stimulate emergency action. Patients with chronic illness or the elderly can take the most advantage from active and passive health monitoring. In this case, AI algorithms model trends to predict adverse situations. Automation acts by signaling public officials or starting safety contingencies.

In summary, AI acts as the intelligence layer that enables machines to think, but it depends heavily on IoT, Big Data, and Automation to function at full potential. IoT collects data from the real world, Big Data prepares and supplies that information to AI, and automation makes sure the insights are put into action quickly and accurately. Collectively, they are revolutionizing entire sectors, enhancing public services, and developing safer, more intelligent, and more sustainable lifestyles and workplaces. These are not merely complementary technologies. They are tightly interconnected, and their combined influence is what really drives the AI-driven world of the present and the future.

12.5 BIG DATA

Big Data consists of the vast amount of structured, semi-structured, or unstructured data that continuously inundates organizations and businesses. These data possess the characteristics of large volume and complexity at the same is also driven by the velocity of their occurrence. In each step to take the informed decisions, to identify correlations and associated trends, Big Data explores beyond the boundaries of collecting large quantities of information; it requires extracting valuable insights and patterns from that data. The phrase describes that data sets are too large or complex for a conventional way of data processing, for the software to handle effectively. Henceforth, it is has become a necessity for the use of Big Data technologies and analytics as they serve as a vital entity for eliciting insights from large data sets. Thus, it could be a boon for those businesses and organizations that are planning to improve decision-making, streamline operations, and gain a competitive edge through effective data utilization.

12.5.1 Characteristics of Big Data with example

The five different characteristics of Big Data are the volume, variety, veracity, variety, and value. These characteristics of Big Data are picturized in Figure 12.3.

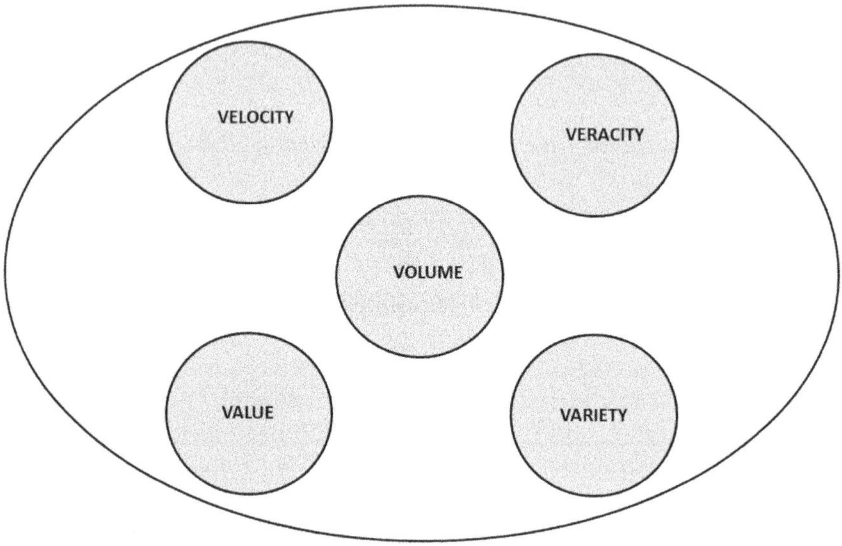

Figure 12.3 Characteristics of Big Data

Social media platforms like Facebook generate vast amounts of user-generated content daily; for instance, thousands of posts, images, and videos are shared every minute. This illustrates the concept of volume, which refers to the extensive amount of data produced. In contrast, velocity refers to the speed at which this data is created and processed. The banking industry, which generates real-time stock market data, is a prime example. Millions of transactions are made on stock exchanges every second, therefore, traders and investors must process and analyze these transactions quickly to make timely decisions. Variety is a reference to the various forms and kinds of data. Consider a healthcare system that collects patient information. It includes semi-structured data (such as lab reports), unstructured data (like doctor's notes or medical photos), and structured data (including patient demographics). Integration and analysis are difficult when dealing with such a variety of data formats. The veracity has to do with how accurate and dependable the data is. An example of IoT devices in a smart city is that sensors collect data regarding environmental conditions, traffic flows, and aspects of other activities. However, at times sensors may supply inaccurate readings and causes errors. Verifying and validating data accuracy are vitally important for sound decisions based on data. The quintessential intent of Big Data is to leverage it into smart and productive insights. Let's think of e-retailers like Amazon. They will consider purchase history, browsing patterns, and other purchasing behaviors when recommending products to customers. By utilizing targeted marketing techniques and examining large amounts of data, they simultaneously develop customer experiences and increase sales.

As the examples demonstrate, it is both promising and challenging to derive insight from larger, more unconventional, and diverse data sets, while also demonstrating how Big Data characteristics become relative across fields and industries.

12.5.2 Data storage and processing techniques

Organizations across many industries are now experiencing both new opportunities and challenges. Data is developing incredibly quickly, especially during the era of the digital economy, and in this time, Big Data has become a leading transformational feature of organizations that allows them to create their own position and democratizes innovation along the way as well as unlocking real insights to create a competitive advantage. Big Data refers to the ingestion, storing, processing, and analysis of immense amounts of data to provide patterns and trends. In this review, we have focused on the tools, frameworks, and methodologies that are influencing how Big data can be processed.

Organizations leverage distributed computing, parallel processing, and advanced analytical techniques to take advantage of Big Data, which facilitates organizations ability to more effectively handle large quantities of data and processing and analysis of that data. Big Data processing, which is commonly abbreviated as "Big Data," encompasses a variety of technologies, techniques, and approaches for managing the volume, velocity, and variety of data that are now created and available in our digital world. Technologies leveraged in Big Data processing include interactive querying, ML, batch processing, and real-time streaming, and all are intended to address the specific requirements for the appropriate use case or application. Figure 12.4 illustrates the significance of Big Data within the digital era cycle.

12.5.3 Commonly used tools and frameworks

The most popular frameworks and tools around are Hadoop, Apache Frameworks like Spark, Flink, Kafka, and NiFi.

Hadoop: Hadoop is a critical platform for dealing with Big Data, as Hadoop performs distributed processing and storage of large data sets. There are two core components of Hadoop: the programming model known as MapReduce for data processing in parallel; and the service, called Hadoop Distributed File System (HDFS) used to store data that will be clustered together. Enterprises can do more with Hadoop through the capabilities of tools like Apache Hive as data warehousing;

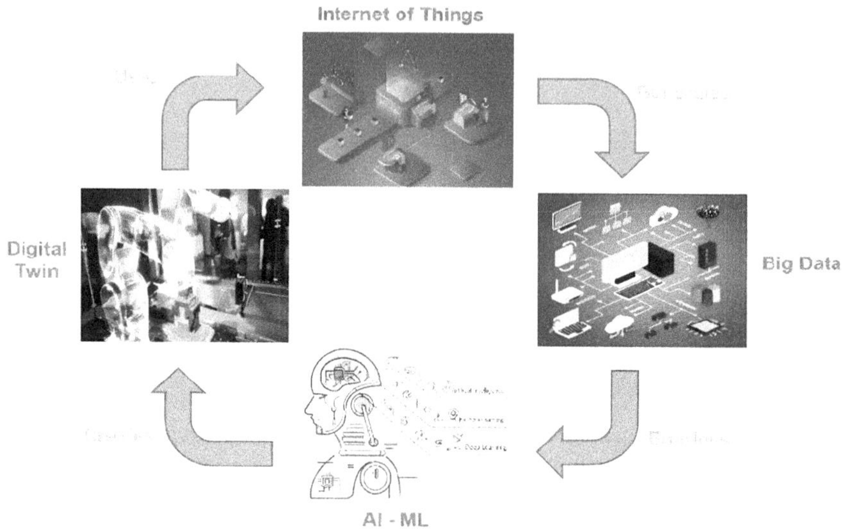

Figure 12.4 Big Data in the cycle of the digital era.

Pig as a batch processing residual language; or leveraging the capabilities of Apache Spark to manage process, analytics, and ML for larger more complex data sets in these two examples.

Apache Spark: It is a fast, flexible cluster computing framework for managing large amounts of data. It allows for many kinds of workloads to be done all at once through preparing for Batch, Streaming, SQL Queries, and ML through the same code base having the data processing run in fault tolerant in-memory. Developers can write code in a concise, expressive way in SQL, Scala, or Python and Java even, as Spark is a programming model of Specification or unified a framework that offers an abstraction of a resilient, distributed data set used to effectively and fault-tolerantly to manage Big Data and data processing in the context of a Spark Cluster.

Apache Flink: Apache Flink is a comprehensive stream processing framework for real-time data analytics and event-driven applications. Flink has various features that positively impact its effectiveness when applied to fraud detection, anomaly detection, or real-time monitoring applications. Stateful stream processing, event time processing, and exactly once semantics are just a few of the features of Apache Flink that can help organizations achieve high-throughput and low-latency processing facets of Flink to derive analytic value through streaming data, where value is expressed in milliseconds.

Apache Kafka: Apache Kafka is a distributed streaming platform for building event-driven architectures and real-time data pipelines. Kafka has proven itself to be resilient, scalable, and fault-tolerant, which allows organizations to ingest, publish, and process vast amounts of analytic event streams reliably and efficiently. Kafka makes it easy to integrate many data sources and systems into a seamless data processing operation utilizing distributed commit logs, message replication, and topic partitions.

Apache NiFi: Apache NiFi is a powerful data flow management tool for coordinating, directing, and converting data in many contexts. Its visual interface and intuitive flow-based programming model allow organizations to easily create and implement data pipelines. Its scalable architecture, along with features for data provenance, security, and governance, ensures comprehensive data visibility, control, and compliance.

12.5.4 Features and use cases

The features and the use cases of the Big Data tools are of various types.

Batch Processing: Performing data processing operations on massive amounts of data in distinct groups or chunks is known as batch processing. Usually, data is consumed, processed, and saved in batch mode according to the "collect-process-store" paradigm.

Use Case: Batch processing works well for offline data processing, historical analysis, and batch reporting. Data warehousing, batch analytics tasks, and ETL (Load, Extract, Transform) processes are a few examples.

Real-Time Stream Processing: By analyzing continuous streams of data almost instantly, real-time stream processing helps businesses respond quickly to trends, events, and variances as they arise.

Use Case: Applications for processing streams in real time include IoT, analytics, fraud detection, and real-time monitoring. Real-time dashboards, detection of anomalies systems, and pipelines for processing sensor data are a few examples.

Interactive Querying: Large data sets can be queried and analyzed interactively with low latency and response times thanks to interactive querying, which also makes random evaluation, exploratory analysis of data, and interactive data visualization easier.

Use Case: In situations like business intelligence, investigating data, and data discovery, interactive querying is helpful. Data exploration tools, interactive notebooks, and SQL-based query engines are a few examples.

ML and Advanced Analytics: Statistical, mathematical, and computational methods are applied in the areas of ML and advanced analytics to glean data-driven insights, trends, and projections.

Use Case: Applications for advanced analytics and ML include anomaly detection, recommendation systems, and predictive modeling. Predictive analytics platforms, training model frameworks, and ML libraries are a few examples.

12.5.5 How do IoT, automation, and AI play a major role in Big Data?

In the big data realm, the combination of IoT, automated processes, and AI is revolutionary because it makes it possible to gather, process, and analyze data effectively and produce insights that can be applied to a variety of industries. Through sensors built into actual objects, like wearables, IoT devices, such as smart city infrastructure, manufacturing equipment, and smart appliances, generate vast volumes of data. Volume (size), velocity (speed), variety (various formats), and veracity (accuracy) are the "4 Vs" of big data that define this data. Because IoT continuously streams real-time information about processes, environments, and user interactions, it serves as a main source for Big Data analytics. This data is then processed by big data frameworks to improve client experiences, forecast maintenance requirements, optimize operations, and create new products.

IoT sensors in smart cities enhance waste management systems, optimize distribution of electricity in smart grids, and control traffic systems in real time. In the medical field, wearable IoT devices gather patient health information for proactive care and individualized treatments. IoT sensors in manufacturing identify equipment problems early to reduce downtime and increase supply chain transparency.

By lowering the need for human involvement, automation streamlines the processing of big data. Automated systems can speed processes, find predictive patterns, and prepare data for examination. In predictive analytic processes, we are creating automated processes as a subset of automation, such as have been developed using the Data Science Machine, that allow volumes of large, complicated data to be processed in hours, rather than months (Data Science accepted, 2018). Unlike automated systems, automation supports self-service systems for non-technical experts (like data scientists) to scale better, have better operational efficiency, and can provide lower-cost systems for the companies that deploy them. Automation can help uncover relevant pieces of data that will assist decision-making, focusing on the banking sector and e-commerce through uncovering predictability to patterns in data. The following are aspects of automation that are the common understandings of what automation is used for: Hadoop automation to provide distributed data management and processes at a very large scale; self-service solutions to enable analytics job descriptions are useable for non-technical users; workflows integrating across databases and applications creating a coherent user experience while processing data.

AI enhances Big Data by offering sophisticated analytical tools that extract meaning from intricate data sets. Algorithms for ML examine unstructured as well as structured information to spot patterns, forecast results, and make decisions on their own. Businesses can use AI-driven analytics to: identify irregularities or corruption in financial transactions; estimate consumer behavior for tailored marketing efforts; and use predictive forecasting to optimize supply chains. AI effectively manages the intricacy of Big

Data by automating anomaly detection and pattern recognition. It processes fast streams of information from IoT devices to facilitate real-time decision-making. AI-powered solutions increase accuracy while lessening the cognitive strain on human analysts.

12.6 INTERNET OF THINGS (IoT): A FOUNDATION FOR SMARTER AND SUSTAINABLE SYSTEMS

The IoT is a vast, interconnected ecosystem of physical devices embedded with sensors, software, and network connectivity. This includes automobiles, home appliances, industrial machinery, wearable technology, and even city infrastructure. These embedded technologies enable the devices to collect, transmit, and sometimes respond to data without requiring human intervention. In simple terms, IoT transforms everyday objects into smart tools that can "talk" to each other, share real-time data, and help both humans and machines make informed decisions.

The power of IoT resides in its capacity to connect the digital and physical worlds. By collecting data from its environment and sharing it over networks, IoT helps unlock insights that drive efficiency, enhance safety, reduce waste, and automate tasks. This makes it a core enabler of digital transformation across industries. What makes IoT particularly impactful is that its applications directly contribute to sustainable development. When implemented thoughtfully, IoT technologies align closely with several SDGs, helping build a more resilient and equitable future.

12.6.1 Real-world applications of IoT and their role in supporting SDGs

12.6.1.1 Smarter cities (Aligning with SDG 11: Sustainable cities and communities)

One of the most promising applications of the IoT is in the development of smart cities. Urban areas are becoming increasingly complex and densely populated, which puts pressure on infrastructure, energy systems, waste management, transportation networks, and environmental health. IoT enables cities to become more responsive and efficient by embedding sensors across traffic lights, streetlights, water systems, and energy grids. Figure 12.5 shows an IoT-based smarter meter grid in a smart city.

For instance, intelligent traffic signals can change according to real-time congestion, reducing emissions and travel time. Waste bins fitted with sensors can notify city services when they're full, preventing overflow and improving cleanliness. Smart meters can track electricity or water usage and help conserve resources.

By enabling better planning, improved energy efficiency, and more sustainable transportation systems, IoT is essential for SDG 11, which aims to make cities inclusive, safe, and resilient sustainable.

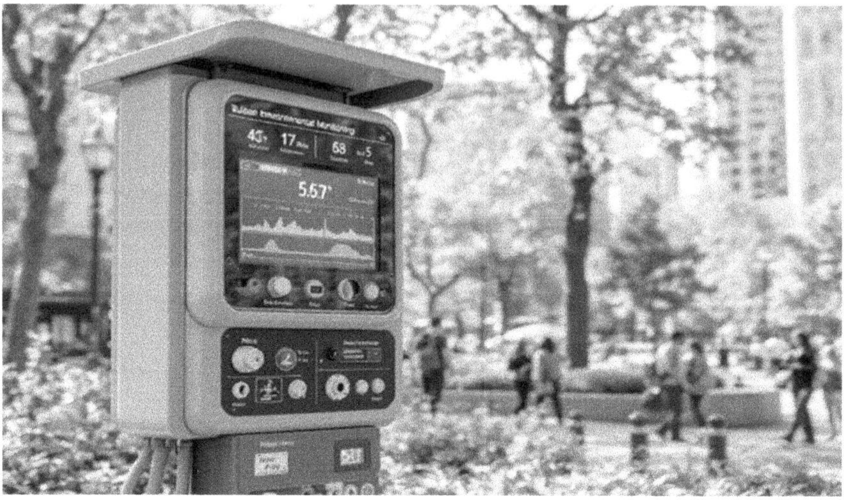

Figure 12.5 An IoT-based smarter meter grid in a smart city.

12.6.1.2 Industrial automation (Supporting SDG 9: Industry, innovation, and infrastructure)

In the industrial sector, IoT is reshaping how manufacturing and logistics operate. Through Industrial IoT (IIoT) applications, factories and supply chains are becoming smarter, more automated, and more productive. Machines on the factory floor can communicate their operating conditions in real time, allowing predictive maintenance that prevents breakdowns and minimizes downtime. Inventory levels can be tracked automatically, reducing waste and optimizing supply chain management. Even quality control is enhanced through real-time monitoring of production metrics. All these improvements contribute to higher efficiency, cost savings, and better product quality. More importantly, they support SDG 9 emphasizes resilient infrastructure, inclusive sustainable industrialization, and innovation with modern technologies.

12.6.1.3 Environmental monitoring (Advancing SDG 13: Climate action and SDG 14: Life below water)

IoT is another area where work on sustainability and environmental responsibility is needed. The harm from environmental degradation, pollution, and climate change represent critical challenges for the planet. IoT devices can help in the battle by providing a constant data feed on the environment with regard to air quality, water quality, noise pollution, and soil health. For example, air quality sensors could be placed around a city to notify authorities about pollutants in real time. In agriculture, sensors can monitor soil and weather data to prevent excessive water and chemical use. In a marine context, IoT systems are being developed to monitor submarine ecosystems, track marine biodiversity, identify oil spills, and monitor illegal fishing activity. There are clear links to the Sustainable Development Goals of climate action (SDG 13) and conserve and sustainably use oceans, seas and marine resources (SDG 14).

12.7 VISUALIZATION AND INFRASTRUCTURE

It explains how these ideas are also embodied through the IoT-enabled infrastructure, assuming a context of technology cheered on by urban settings. For example, a smart city infrastructure may involve a purpose-built network of sensors that coordinate traffic flow, measure pollution levels, conserve energy, and enhance emergency response. These sensors "talk" to the centralized platforms, enhanced by AI and BIG data, that enable service improvements for both the public and public safety as well as reducing resource consumption.

12.7.1 Challenges and benefits of IoT

IoT boils down to the process of connecting everyday items like your phone, your car, your refrigerator, or your lights to the web to allow them "communicate" and work together in a more intelligent way. In many respects, these gadgets already make life easier by gathering and sharing data. Imagine your home lights going on automatically at night or your smartwatch recording your steps. IoT in action is that, It is increasingly influencing the way we work, live, and even care for our environment and health.

There are numerous advantages. IoT simplifies life, to start. Your appliances can automatically adapt to your habits. It can be used by cities to save energy or enhance traffic flow. To save time and water, farmers employ smart sensors to water crops only when necessary. It is used by hospitals to keep an eye on patients in real time, even when they are far away. Companies use it to monitor deliveries or identify issues with machines before they become serious. All things considered, IoT makes our lives smarter, our work faster, and our safety safer. An article suggests a system that allows one to identify diseases on tomato plants with deep learning (CNNs). The accuracy of precision agriculture exhibited by their model is that it achieved high levels of accuracy in a classification of different types of diseases depending on leaf images. Another significant factor identified in the study is the need to diagnose early to reduce losses of crops and increase output with the help of diagnostic strategies that are automatic and image-based (Jelali, 2024).

It isn't flawless, though, like anything else. IoT devices gathers a lot of personal data, like where you travel, how you feel, and when you go home, so it's critical to protect privacy information. It might be abused if it ends up in the wrong hands. Security is another problem. Hackers might gain access to a smart system and interfere with traffic lights or unlock your smart door, among other things. The issue of compatibility is another. It might be annoying when brands don't complement one another. Additionally, some devices may completely stop functioning if your internet connection is lost if a sensor provides inaccurate data. So, while IoT is opening up a world of amazing possibilities, we also need to be thoughtful about how to use it. With strong security, good design, and better privacy controls, one can enjoy the best of what IoT has to offer while keeping ourselves safe and in control.

12.8 AUTOMATION

Automation refers to the use of technology to perform tasks with little to no human intervention. Across all industries, it improves efficiency, streamlines methods, and reduces errors. Automation changes workflows in a variety of industries, including manufacturing, healthcare, and customer service, by utilizing technologies like robotics, artificially intelligent, and software systems.

12.8.1 Types of automation

There are several types of automation. They are fixed, programmable, flexible, intelligent process, and robotic process automations. The chemical manufacturing processes, assembly lines, conveyor systems are examples of a fixed automation system where there is rigidity and designed for high-volume tasks which are repetitive in nature. For CNC machines and industrial robotics batch production is necessary. Different tasks require flexible programmable automation with reprogramming. With minimal changeover time, flexible manufacturing systems, computer-integrated manufacturing combines high production rates for handling diverse products with adaptability. In business processes, RPA helps and mimic human action for repetitive tasks whereas in healthcare RPA integrates intelligence for adaptive automation capabilities.

12.8.2 Benefits of automation

Machines perform much faster than human beings and reduces production cycles. It also does tasks much faster than humans and also gives better output. The benefits of automation are given in Figure 12.6. They are increased productivity, improved accuracy, cost saving, also enhances safety and scalability. Human errors are minimized and it ensure quality. When the work flows are streamlined it reduces labor costs and operational expenses. These systems can adjust the production levels to meet the necessary pre-requisites and also minimizes risks in hazardous environments.

12.8.2.1 Examples of automation across various industries

There are several examples of automation in various industries. Some of the examples are manufacturing, healthcare, customer care, and warehouse

Figure 12.6 The benefits of automation.

management. Automation optimizes logistics. Nike uses picking robots to speed up order fulfillment in Japan. Automated storage and retrieval systems reduce human error and enhance inventory accuracy. Automated ticketing systems effectively classify consumer inquiries. Interactive voice response (IVR) systems effectively manage large call volumes. Automation improves operational efficiency and patient care. Robotic-assisted surgery offers more precise, minimally invasive procedures. AI diagnostics enables faster diagnosis by analyzing medical images. Automated prescription dispensing systems enhance pharmacy processes and reduce medication errors. Manufacturing automation has transformed the industry by improving task accuracy and speed, boosting mass production efficiency, and enhancing warehouse management with barcode scanning, as seen at Amazon.

12.8.3 Drawbacks

Automation has benefits but also drawbacks, including high initial costs, the need for trained staff, and potential job displacement from reduced physical labor. To effectively harness the benefits of automation, it is crucial to prioritize workforce reskilling initiatives and address ethical considerations. Balancing these elements will ensure a responsible and empowering transition for all involved.

12.8.3.1 How do IoT, Big Data, and AI play a major role in automation?

Automation technology is being transformed through big data, AI, and the IoT. These technology advances yield substantial gains in productivity, precision, and agility across industries.

AI is a must to embody bona fide automation, because it lends automation systems with learning, adaptation, and the ability to make decisions. With AI, automation systems can apply natural language processing and ML to analyze data, identify patterns, and achieve identifiable objectives without human decision-making. AI-enabled automation systems are capable of analyzing big data and can make conclusions and navigate changes in the environment. Chatbots are examples of technology, that can utilize NLP to respond to customer service questions by themselves any time of day. AI is useful, for reducing human errors as well as completing highly repetitive and, detailed tasks such as data entry, and processing invoices. Intelligent bots can also enable appointment scheduling, or claims-related workflows and others, to improve customer service and performance. Big data analytics revolutionizes automation by allowing businesses to analyze vast amounts of both structured and unstructured data. With automated solutions based on big data, better decision-making, streamlined processes, and trend prediction are all possible. Businesses can predict consumer behavior and equipment failures through big data automation relying on predictive patterns. Automated big data systems scale operations and dramatically reduce operating expense. An example is MIT's "Data Science Machine" which outperformed human analysts in predictive modeling tasks. Companies also use big data tools to enhance the speed of decision-making. E-commerce companies study patterns in consumer behavior to improve product recommendations.

Through the connection of numerous sensors and devices, IoT makes real-time data collecting and monitoring possible. By integrating automation systems with the IoT, industries can achieve more efficient and safer operations. IoT sensors monitor equipment conditions to anticipate malfunctions and plan maintenance, which helps reduce downtime in sectors like manufacturing. By optimizing processes, IoT enhances production efficiency. For example, Toyota improved its productivity by 30% through the use of IoT for production line monitoring. In agriculture, farmers leverage IoT for various purposes, including monitoring insects, managing fertilization, and controlling irrigation, all of which contribute to increased agricultural yields (Singh and Zhang, 2025).

12.9 SYNERGY BETWEEN IoT, BIG DATA, AUTOMATION, AND AI

IoT supplies the raw data streams, and these technologies work together to form a potent ecosystem. This data is efficiently processed at scale by automation. From the analyzed data, AI derives insights that can be put to use.

Together, they benefit companies in a range of industries, including manufacturing, healthcare, smart cities, retail, and more, and achieve new heights of operational efficiency, innovation, and customer-centric strategies. The future of automation using AI in transforming urban transportation to include practical time-based traffic forecasts combined with intelligent parking system optimization. They exploit ML algorithms to promote their system to ameliorate traffic choking and enhance parking effectiveness, making transportation systems within cities more sustainable and responsive (Chen et al, 2025).

12.10 CASE STUDY – AN INDOOR PLANT CARE SYSTEM

Caring for indoor plants can be challenging because they require consistent light and moisture levels to thrive. Often, unhealthy plants result from irregular watering, inadequate lighting, and poor care practices. The system employs light-dependent resistors (LDRs) and soil moisture sensors to optimize plant growth indoors. A microcontroller monitors soil moisture and ambient light. When moisture levels fall below a set threshold, a water pump automatically irrigates the plants. If the LDR registers inadequate light, the system will adjust the artificial lighting for the plants. By providing these sensors, they can capture a real-time and processed, data providing details about what is required for the requirements of distinct plant species. The intended system interface indicated an easy to use application, for remote information and control management of indoor gardens. In addition, this system will continue to give the least impact on the environment through energy-efficient components, which allows for the need of sustainability for urban agriculture. The thesis presents an exploration into building and using a multi-sensor integrated smart mechanism for watering plants utilizing voice recognition technologies. This system has three components constitutes the smartphone with voice recognition module robot, a soil moisture sensor, and an M5 Stack Basic Core displays temperature and humidity. The water pump operates on a timer and automated water systems, such that it will irrigate plants properly without any human effort, while considering the issues of common plant care that associates with traditional systems, such as inconsistent watering and labor costs. A previous study has designed a simple dual-sensor soil moisture detection system that enhances irrigation management by measuring moisture levels to a depth of 40 cm deep within the effective root zone; therefore, the researcher can better determine irrigation depths. This involved the setup of two soil moisture sensors spaced 20 cm apart, using Arduino microcontrollers which provided for a better understanding of moisture distribution in the root zone, while allowing farmer/researchers to make more informed irrigation decisions. Rather than using a single sensor installation, which may or may not be measuring moisture at a depth of interest, it was demonstrated that, through using the additional sensor, improved irrigation management could occur. Although study was conducted using both shallow and deep moisture sensors, and in various soil types, the results demonstrated

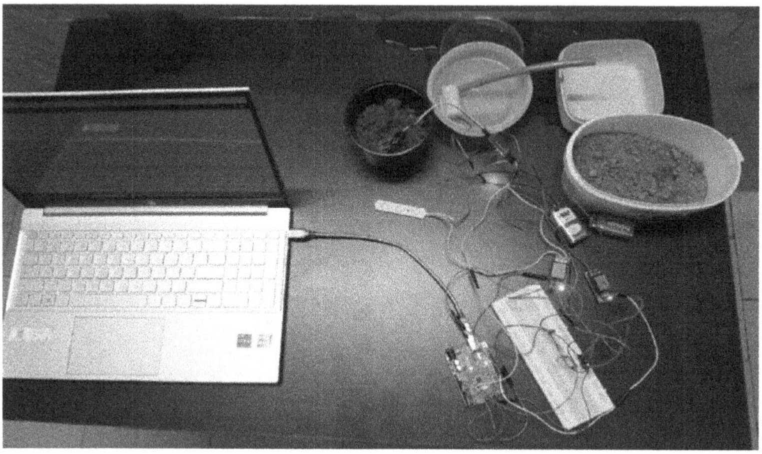

Figure 12.7 A simple prototype of an indoor plant care system.

that the dual-sensor was effective at measuring moisture in the root zone and could easily be a determining factor in maximizing water used and crop yield.

This research presents an adaptive indoor plant care system that utilizes principles of automation and IoT, and it was demonstrated using a functional prototype. The objective of the system was to provide a perfect environment for the growth of their plants, implementing smart lighting, an automated watering system, and a soil moisture sensor. The system implemented adaptive lighting, giving the plants their required amount of light through automated or smart lighting. The adaptive lighting used sensors to capture the amount of ambient light and automatically adjust the artificial light. Furthermore, a soil moisture sensor continuously captured soil moisture content to avoid over-watering and under-watering and turns on the irrigation system, if necessary. This plant care solution promotes healthy plant growth and improves water and energy efficiencies to make indoor gardening sustainable and easier. Figure 12.7 is a simple prototype of an Indoor plant care system. This concept exemplifies how technology combined with automation and real-time monitoring helps make indoor plant care simple and easy for everyone.

Digital twins enable farmers to simulate, monitor, and optimize the work in their farms with the implementation that has been previously unknown, and are virtual representations of real farming that gather data in real time (Kam 2024). This can be an insight into future research.

12.11 SUMMARY

The chapter primarily illustrates how rapidly the world is moving to the digital realm. Technologies like AI and connected devices are making this transition quicker. There is, however, an issue – having one big pool of data raises concerns around privacy and security and thus there is an ever-increasing movement toward more intelligent systems that allow us to utilize the data effectively but more safely. A great example is the application of AI, Big Data, and IoT with farmers when trying to assess the weather, crop management, and making better decisions.

One of the primary ideas in the chapter, is that IoT devices are collecting a large amount of data in real time. This data is just raw data until it is parsed and given meaning by Big Data tools. AI's job starts then and finally comes automation so a human does not have to stand between data and action. For example, sensors are detecting dry soil, AI is determining how much water is required, irrigation effluent is turned on by itself.

The five essential characteristics of big data – volume, velocity, diversity, truth, and value, are also explained in this chapter. It provides real-world examples like how social media platforms, banks, and hospitals manage massive amounts of data every day. Tools like Hadoop, Apache Spark, Flink, Kafka, and NiFi are introduced as the engines that help analyze and use this data efficiently. These tools allow for both large-batch analysis and real-time data streaming. Another primary focus is how these technologies work together. IoT collects the data, Big Data organizes it, AI understands it, and automation acts on it. Real-life examples include innovative city traffic systems, wearable health monitors, and factory machines that know when they are about to break down and alert the system in advance. The chapter showed how IoT is being used today, whether in smart homes, farming, or environmental monitoring. It highlights the benefits of saving time, cutting costs, and making better decisions, and discusses challenges, such as protecting data privacy, ensuring different devices can work together, dealing with internet outages, and keeping systems safe from hackers. Automation also plays a big role. It is used in many areas, such as robotic-assisted surgeries, automated customer service, and robots managing warehouses. It improves speed, reduces human error, and saves money, although it also brings challenges like high setup costs and the need to retrain workers.

Finally, the chapter ends with a simple but powerful example: an innovative indoor plant care system. It uses sensors to monitor the soil and light. With the help of AI and automation, it gives the plant exactly what it needs. This small use case shows how effective and practical these technologies can be when they work together and how they can improve daily life more innovatively and sustainably.

REFERENCES

Abu, N. S., W. M. Bukhari, M. A. Firdaus, and N. M. Sukri. 2022. "Smart Indoor Plantation Using Soil Moisture Sensor and Light Dependent Resistor Sensor." *International Journal of Robotics and Control Systems* 3 (1): 721–738.

Ahmed, N., and N. Shakoor. 2025. "Advancing Agriculture through IoT, Big Data, and AI: A Review of Smart Technologies Enabling Sustainability." *Smart Agricultural Technology* 10: 100848.

Ahuja, N. J., N. Pasi, H. Naz, and R. Chamola. 2025. "Smart IoT-Based Snake Trapping Device for Automated Snake Capture and Identification." *Environmental Monitoring and Assessment* 197 (3): 258.

Jelali, M. 2024. "Deep learning networks-based tomato disease and pest detection: a first review of research studies using real field datasets." *Frontiers in Plant Science* 15.

Kow, P. Y., Y. T. Wang, Y. W. Chang, L. C. Chang, and F. J. Chang. 2025. "AI-Driven Weather Downscaling for Smart Agriculture Using Autoencoders and Transformers." *Computers and Electronics in Agriculture* 232: 110129.

Gupta, G., Kumar Pal, S. 2024. "Applications of AI in precision agriculture." *Discover Agriculture* 3: 61.

Mehanna, H. M., and A. S. Morsy. 2022. "Double Soil Moisture Detector for Irrigation Management Using Arduino." *International Journal of Multidisciplinary Research and Growth Evaluation.*

Rim, Zakar. 2024. "Design and Implementation of an Indoor Smart Plant Watering System with Voice Control and Multi-Sensor Integration." Thesis. https://www.theseus.fi/handle/10024/861587.

Sanchez, T W., M. Brenman, and X. Ye 2025. "The Ethical Concerns of Artificial Intelligence in Urban Planning." Journal of the American Planning Association 91 (2): 294–307.

Sankar, P. 2022. "The Vehicle Parking System in Smart Cities." In T. Kavitha, V. A Devi, S. N. Pari and S. Ramanathan (eds.) *Internet of Everything: Smart Sensing Technologies*, Nova science publisher.

Singh, A., and L. Zhang. 2025. "Smart Irrigation Systems Using Wireless Sensor Networks: A Review." *Environmental Technology & Innovation* 20: 102345.

Kam W Ng. 2024. "Digital Twins in Smart Agriculture: Bridging the Physical and Virtual Worlds." *Computers in Industry* 150: 104567

Yao, Xu, and Wei-hong Zhou. 2025. "The 'New Engine' of University Digit-Intelligent Governance: Algorithmic Administration—Opportunities and Challenges." *SSRN Preprint.* https://ssrn.com/abstract=5137214.

Zia, Huma, Bara Fteiha, Nour Alrifaee, Jawad Yousaf, Mohammed Ghazal, and Nick Harris. 2025. "Advancing Smart Cities with Artificial Intelligence: A Systematic Literature Review of Challenges and Future Directions." *SSRN Preprint.* https://ssrn.com/abstract=5156689.

Chapter 13

Emerging trend in smart technology for sustainability

Surveillance of CCTVs using IoT devices

Yashaswini S., Jayanthi M. G., and Shreekanth M. Prabhu

13.1 INTRODUCTION

A growing alarm has been raised in recent years over the absence of proper surveillance solutions in different environments, resulting in catastrophic events that could have been averted. Surveillance systems, especially the installation of closed-circuit television (CCTV) cameras, have turned out to be a useful mechanism in promoting public security and safety (Piza et al., 2019). With the advancing Artificial Intelligence (AI) of Things, surveillance solutions are growing more advanced with the incorporation of AI functionality to enhance the efficiency and performance of these solutions. The application of CCTV systems has also become more widespread in public and private areas for surveillance and security. Nevertheless, the efficiency of the systems depends a lot on ongoing monitoring and upkeep of the CCTV screens. Breakdowns in the CCTV screens can result in surveillance holes, making the system useless to ensure security. To deal with this problem, an active monitoring system that can identify active CCTV screens and alert for failures in real time is required (White et al., 2019). This research paper seeks to investigate the creation of an Internet of Things (IoT) device that can meet this requirement. CCTV systems have been a security and surveillance cornerstone for a long time. From street retail outlets to open areas, such systems obtain key visual information.

While classic CCTV configurations tend to be isolated in their function, necessitating manual operation and the absence of high-end analysis features, the introduction of the IoT brings about a change of paradigm for the way CCTV systems are built and deployed. IoT devices, which are marked by their connectivity, sensing of data, and processing capabilities, can be easily integrated into CCTV networks, allowing for more intelligent, efficient, and accessible surveillance solutions (Smith, 2021). This chapter discusses the ways in which this integration can greatly improve the functionality and efficacy of surveillance systems.

Literature Survey: The incorporation of IoT devices into CCTV surveillance systems has revolutionized security monitoring, increasing efficiency and responsiveness. This literature review points out major developments and approaches in this area (Ren, Xia, Liu & Lee, 2021).

DOI: 10.1201/9781003586951-13

Security Improvements in IoT CCTV Systems: Efficient video encryption algorithm (EVEA) to protect multimedia data in IoT CCTV systems, showing a remarkable decrease in encryption time with respect to current techniques. Real-time monitoring and automatic alerts in overcoming the drawbacks of conventional monitoring systems, thus improving security in smart cities (Zhao, Fei-Fei & Xing, 2011).

Smart Surveillance Solutions: IoT-based surveillance camera that independently identifies intrusions and sends live video to users, offering an economical solution for homes and small businesses through the deployment of local video analytics employing lightweight CNN models on IoT devices, minimizing latency in smart city surveillance applications (Yeganeh, Coutinho & Okarche, 2024).

Deep Learning Applications in Surveillance: An exhaustive review by White et al. identifies the use of deep learning methods for abnormal activity detection in dense scenes, resolving issues related to real-time processing (Rezaee et al., 2021), (Ullah et al., 2022). Although these developments hold enormous advantages, data privacy and robust encryption needs are still pressing research areas for the future (Zhao, Fei-Fei & Xing, 2011).

13.1.1 IoT devices for surveillance systems

IoT devices have been used more and more in surveillance systems to increase monitoring capabilities and overall security. IoT devices are sensor- and connectivity-enabled, which enable real-time data collection and analysis. In the case of CCTV systems, IoT devices can be utilized to track the status of the displays, cameras, and other system components. Through the use of IoT technology, surveillance systems can be made more intelligent and responsive to possible malfunctions (Johnson & Brown, 2020).

13.1.1.1 Active CCTV display detection

One of the major difficulties in CCTV system monitoring is detecting whether the displays are active and operational. Standard monitoring practices entail manual verification or regular audits, which might take a lot of time and effort. IoT devices provide a more proactive and automated solution to this issue. By integrating sensors that are capable of sensing the availability of a video feed on the CCTV screens, the IoT device can tell if the system is running and alert for abnormalities (Nayak et al,2021).

13.1.1.2 Malfunction alerting along with sensing active CCTV screens

The IoT device should also be able to alert for malfunctions in real time. This necessitates the device to be capable of analyzing the data that it has gathered from the sensors and detecting patterns that reflect potential problems. Pixelation, flickering, loss of signal, and hardware failure are common faults in CCTV monitors. By setting predefined thresholds for these measures, the

IoT device is able to prompt alerts when anomalous behavior is observed, allowing timely maintenance and repair.

13.1.1.3 Integration with monitoring systems

In order to enhance the functionality of the IoT device, it must integrate into the current monitoring systems seamlessly. This should comprise compatibility with the CCTV management software, network infrastructure, and other surveillance gear. By setting up a communication procedure between the IoT device and the monitoring system, operators are able to receive notifications and alarms through a central dashboard. This allows for a smoother and more efficient process when it comes to handling and maintaining CCTV displays.

Overall, the creation of an IoT device that can monitor active CCTV screens and alert against failures is a tremendous leap in surveillance technology. Utilizing the functionalities of IoT devices, security experts can ensure CCTV systems' dependability and efficiency, consequently boosting overall security and safety. This research work forms the starting point for additional study and improvement in this sector with the promise of transforming how surveillance systems are monitored and managed.

Discovering gaps in the literature on surveillance through IoT devices indicates a number of key areas that require more investigation. The integration of IoT into surveillance systems has specific challenges that existing research has not yet addressed (Kiran et al,2018).

13.1.1.4 Security vulnerabilities

Most IoT devices do not have a centralized security architecture, which makes them vulnerable to numerous threats. Current threat models tend not to consider the particular vulnerabilities present in IoT systems, which can lead to potential security oversights.

13.1.1.5 Detection methods

Existing literature mainly discusses conventional detection methods with less attention given to sophisticated methods such as machine learning (ML) and deep learning for anomaly detection in IoT networks. There is a significant gap in research on the performance of intrusion detection systems (IDS) designed specifically for the heterogeneous and diverse nature of IoT devices.

Whereas some domains of IoT security have been well researched, others, including the convergence of IoT and wireless sensor networks, are underexplored. In conclusion, although considerable progress has been made in IoT surveillance, the existing body of literature exposes essential gaps in security frameworks, detection methods, and overall trends in research that require further research.

13.1.2 Limitations of traditional CCTV systems

Conventional CCTV systems have several primary limitations:

- Stand-Alone Operation: CCTV networks are generally stand-alone systems with minimal integration with other security systems or databases.
- Manual Surveillance: Frequently, human operators watch live feeds, which are subject to human error and fatigue, particularly for large-scale implementations.
- Limited Analytics: Simple recording and playback features frequently don't include real-time analytics, and therefore, it becomes difficult to actively detect and respond to potential threats.
- Storage and Retrieval of Data: Handling lots of video data can be unwieldy, with challenges to efficient storage, searching, and retrieval of usable footage.
- Absence of Accessibility: Viewing CCTV footage is usually limited to the physical premises of the recording system, depriving remote observation and response capability.
- Maintenance: Conventional systems involve hands-on maintenance and troubleshooting of physical devices, time-consuming and expensive [3].

13.1.3 The role of IoT in enhancing CCTV surveillance

IoT devices such as sensors, actuators, and network-enabled cameras provide solutions to these issues through:

- Improved Connectivity: IoT enables cameras and sensors to talk to each other effortlessly over the internet, allowing for centralized control and real-time data transfer.
- Smart Monitoring: Combining IoT with AI and ML algorithms makes it possible to automatically analyze video feeds for anomaly detection, object identification, and behavior analysis, minimizing the need for manual monitoring (Johnson et al,2020).
- Remote Access and Control: IoT allows authorized users to remotely view live feeds, schedule recording, and control camera placement from anywhere with an internet connection.
- Integration with Other Systems: IoT allows integration with other security systems such as intrusion detection, access control, and building management systems for an integrated security infrastructure.
- Cloud-Based Data Management: IoT-based systems can store data on cloud platforms, offering scalable storage, easy retrieval, and backup.
- Predictive Maintenance: IoT sensors can be used to monitor the health of CCTV equipment, allowing for predictive maintenance and minimizing downtime (Michau et al,2021).

13.1.4 Applications of IoT-Enabled CCTV surveillance

The combination of IoT and CCTV has varied applications in different industries:

- Smart Cities: Traffic flow monitoring, public safety management, and environmental hazard detection are achievable with networked cameras and sensors.
- Retail Security: Shoplifting prevention, customer behavior analysis, and store layout optimization are improved through IoT-enabled surveillance.
- Transportation: Passenger flow can be monitored, safety on public transport ensured, and parking facilities managed better.
- Healthcare: Patient safety monitoring, elderly care, and hospital security can be assisted by smart video analysis.
- Industrial Monitoring: Factory processes can be monitored, worker safety ensured, and production line anomalies detected by connected cameras and sensors.
- Residential Security: Homes can be secured, entry points monitored, and video feeds remotely accessed with ease (Anderson et al, 2018).

13.1.5 Security issues and challenges

Though IoT-based CCTV systems have several advantages, they also pose some challenges and security issues:

- Data Privacy: The sheer volume of video and sensor data gathered creates issues regarding privacy violations and misuse of personal data, necessitating strict security protocols.
- Data Security: IoT devices are susceptible to cyber threats, which might result in unauthorized access to confidential information, tampering with video feeds, and interference with surveillance activities (Ramachandra et al, 2020).
- Scalability: Managing a large number of interconnected devices requires robust and scalable infrastructure.
- Interoperability: Ensuring seamless communication and data exchange between heterogeneous IoT devices and platforms is a complex challenge.
- Cost: Implementing and maintaining advanced IoT-based systems can be expensive, especially for large-scale deployments.
- Bandwidth Consumption: Streaming high-resolution video data requires significant bandwidth, which can be a limiting factor in certain environments.

13.1.6 Security measures and mitigation strategies

Mitigating these security issues demands the deployment of strong security protocols:

- Encryption: Data storage and transmission in a secure manner using encryption protocols.

- Authentication and Authorization: Enforcing strong authentication processes and access control policies.
- Regular Updates: Maintaining firmware and software with current security patches.
- Network Segmentation: Segregating the CCTV network to limit the effect of a possible breach.
- Anomaly Detection: Installing ML-driven anomaly detection systems to detect suspicious behavior in the network or due to unauthorized access.
- Data Anonymization: Using anonymization techniques where feasible to video data in order to preserve personal privacy.
- Compliance with Regulations: Compliance with applicable data privacy legislation and standards.

13.2 EXPERIMENTAL METHODS AND MATERIALS

To design an IoT device for sensing active CCTV monitors and alerting in case of malfunctions, a thorough survey was undertaken to obtain data related to the prevailing state of CCTV systems and their possible issues which can be generated. The survey consisted of questioning industry experts, CCTV system suppliers, and customers using CCTV systems in order to learn about prevailing problems associated with CCTV monitors and how they are monitored and taken care of as of now (Ramachandra et al, 2020).

The survey found that the biggest issue with CCTV displays is that faults can pass undetected for extended periods, resulting in surveillance coverage gaps and possible security compromises. This was a key driver for the requirement for an active monitoring system that would be able to identify when a CCTV display is faulty and alert the responsible staff for prompt action as indicated in Figure 13.1.

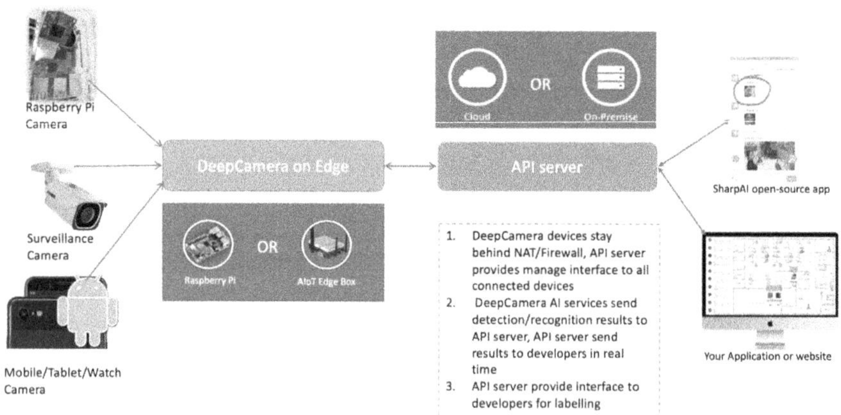

Figure 13.1 Methodology to detect the health of CCTV through Deep Camera.

According to the survey results, the process of designing the IoT device included the following steps:

Research and Sensor Selection: The initial step was to do some research and select the suitable sensors that can be utilized for the detection of a CCTV monitor status. This entailed the exploration of different sensor technologies including motion sensors, temperature sensors, and light sensors to identify those which would best be applied for this purpose.

IoT Device Development: After the selection of sensors, the next part was the development of the IoT device in which these sensors were to be accommodated and be capable to interact with a centralized monitoring system. This included designing hardware components, choosing the right communication protocol, and programming the device to receive and forward data from the sensors.

Integration with Monitoring Software: The IoT device had to be integrated with monitoring software capable of analyzing the data read by the sensors and sending alerts when a malfunction was identified. This involved designing algorithms to read the sensor data and send alerts based on specific parameters.

Testing and Evaluation: The last procedure of the methodology was to deploy the IoT device in a real-world setting to test its operation for active CCTV display detection and malfunction alerts. This included simulating a test scenario where there were several CCTV displays and causing malfunctions to observe the response of the device.

Overall, the approach to designing the IoT device for identifying active CCTV displays and alerting for faults was grounded in an exhaustive study of the then-prevailing state of CCTV systems and the problems encountered in monitoring and maintaining them. By adopting a structured approach to research, development, and testing, we were able to design a robust and efficient solution to enhance the security and reliability of CCTV systems.

The development and design of the IoT device for monitoring active CCTV screens and alerting for faults required a multidisciplinary approach that integrated hardware design, software development, and integration with monitoring systems. The device was intended to be low cost, compact, and simple to install, making it applicable for deployment in a broad array of CCTV systems as illustrated in Figure 13.2.

The hardware components of the IoT device comprised a microcontroller unit (MCU) to process data, memory storage for storing sensor readings, and wireless communication modules for sending data to a central monitoring system. The device also comprised different sensors like motion sensors, temperature sensors, and light sensors to monitor the status of the CCTV displays.

The software development for the IoT device entailed programming the MCU to gather data from the sensors, process the data with predefined ML algorithms, and create alerts when a malfunction was found. The device was

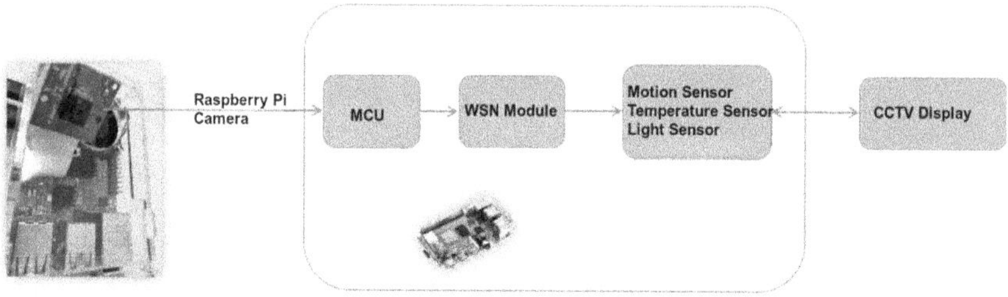

Figure 13.2 Design and development of the IoT device to detect the fault in CCTV display.

also programmed to send information to a central monitoring system via standard communication protocols like Wi-Fi or Bluetooth, allowing real-time monitoring and alerting.

Integration of the IoT device with monitoring software was a paramount part of design and development. Monitoring software was to take data sent by the device, examine the data to determine if there was a possible malfunction, and produce alerts to inform relevant personnel. The software was also programmed to offer an easy-to-use interface for monitoring CCTV display status and alert administration.

Generally, the development and design of the IoT device for active CCTV display detection and alerting in case of malfunction was a collaborative process involving hardware design expertise, software development expertise, and system integration with an ML slant for predictive maintenance and edge computing for real-time analytics. The resulting device was a proven and efficient solution for enhancing the security and reliability of CCTV systems, offering proactive monitoring and alerting features to avoid possible security breaches.

13.3 RESULTS AND DISCUSSION

The step-by-step survey on developing an IoT device to detect active CCTV displays and notify for malfunctions, indicates good results. The IoT device successfully detected active CCTV displays in the specified area and effectively notified for any malfunctions that took place.

Analysis of the results reveals that the IoT device successfully detected the presence of active CCTV screens with high precision. The device used sensors to identify the availability of a live video stream and sent this signal to the central monitoring system. This real-time monitoring feature worked very effectively to detect any fault in the CCTV screens in real time.

In addition, the IoT device was capable of sensing different kinds of malfunctions in the CCTV screens, such as hardware faults, software bugs, and connectivity faults as illustrated in Figure 13.3. The device alerted the concerned staff whenever a malfunction occurred, enabling them to respond quickly and correct the problems. This preventive strategy of monitoring

Figure 13.3 CCTV console to detect the fault in display.

CCTV screens assisted in reducing downtime and providing uninterrupted surveillance coverage.

Generally, the survey results indicate the success of applying IoT devices to identify active CCTV screens and alerting in case of malfunction. The device was found to be reliable, precise, and effective in monitoring, reflecting its potential for the improvement of security surveillance systems as explained in Figure 13.4.

The integration of IoT devices for capturing failure information significantly enhances data analytical capabilities, particularly in predictive maintenance. By leveraging advanced analytics, organizations can proactively address equipment failures, thereby improving operational efficiency.

The testing of the IoT device was aimed at its ability to accurately identify malfunctions, response time in presenting alerts, and general reliability in CCTV display monitoring. The device had a very high accuracy in identifying malfunctions, having a very low false-positive rate as well as a false-negative rate. The response time in the presentation of alerts was equally quick, with the alerts being presented within seconds after a malfunction has been detected.

13.3.1 IoT data for predictive maintenance

IoT platforms enable sensor data collection and analysis, allowing for the detection of failure patterns by applying ML algorithms. Predictive analytics

Defect	Show	Colour	Grade	Description
CC	☒		4	Crack circumferential
CL	☒		4	Crack longitudinal
CM	☒		4	Crack multiple
DF	☒		5	Deformed pipe
DP	☒		4	Dipped pipe
IP	☒		4	Infiltration at pipe wall
JD	☒		4	Joint displaced
JF	☐		3	Joint faulty
JO	☐		3	Joint open
LF	☐		0	Lateral sealing defective
LP	☐		3	Lateral protruding
LX	☒		4	Lateral problem
OP	☒		4	Obstruction permanent
PB	☒		5	Pipe broken
PF	☒		4	Deformed plastic pipe
PH	☒		4	Pipe holed
PL	☒		5	Protective lining defective
PX	☒		5	Pipe collapsed
SD	☒		5	Surface damage
TM	☒		4	Tomo
DE	☒		5	Debris silty
DG	☒		5	Debris greasy
ED	☒		4	Encrustation deposit
OT	☒		4	Obstruction temporary
RI	☒		5	Root intrusion
CF	☐		0	Construction feature

Figure 13.4 Detection of malfunction in CCTV unit [(European Committee for Standardization. (2003)].

can minimize downtime and improve production quality by catching anomalies early, as seen in industrial ovens.

13.3.2 Edge computing for real-time analytics

Edge computing improves data processing by evaluating data at the point of origin, essential in IoT networks for real-time decision-making to ensure dynamic resource allocation and optimal data management, thus improving the responsiveness of IoT systems.

Intermittent edge device data makes it difficult to predict failure, which requires sophisticated methods such as survival analysis in order to develop useful insights. Though the advances in IoT and analytics hold vast promise, difficulties such as data intermittency and requirements for resilient analytical frameworks continue to be vital in order to optimize the full potential of these technologies.

Overall, the IoT device was proven through testing and evaluation to effectively detect active CCTV displays and notify for malfunctioning. The device was an essential asset in upgrading the security and reliability of CCTV systems with proactive monitoring and notifying functions to thwart potential security invasions. The findings of the testing and assessment proved the capabilities of the IoT device to enhance the overall performance of CCTV systems and public spaces safety.

The successful deployment of an IoT device for active CCTV display detection and alerting for faults presents new opportunities for enhancing security surveillance systems. Through the integration of IoT technology in CCTV monitoring, organizations can enjoy real-time monitoring, proactive alerts, and increased reliability.

Image Recognition for Physical Damage: An ML model would be trained on images of operational and non-operational CCTV cameras. The model would learn to recognize physical damage, lens obstruction, or even cable disconnections.

Anomaly Detection in Video Streams: By processing the video stream from a CCTV camera, an ML model can be trained to identify anomalies that could signal a malfunctioning camera. This may involve static, flickering, or an abrupt change in image quality.

Predictive Maintenance with Sensor Data: If the CCTV cameras have sensors (e.g., temperature, humidity), you could train a model to predict failures using that data. For instance, a camera that's always overheating is likely to fail in the near future.

Although surveillance systems, such as CCTV cameras on the IoT, can be very useful, their success depends on the dependability of every component in the system. If IoT components in the system do not work, it can actually hamper surveillance activities.

Following are a few ways such failures are usually managed:

1. Failover and Redundancy Systems: Strong surveillance systems usually provide for redundancy. In other words, there would be redundant systems or parts which step in as replacement if one central device happens to fail. A few illustrations can be used such as many cameras monitoring one same location or emergency communication links on standby if a main communications system fails.
2. System Monitoring and Alerting: Advanced monitoring systems tend to have monitoring software that continually monitors the status of connected devices. If a device is disconnected or is malfunctioning, system administrators can receive alerts so that they can take prompt action and troubleshoot the issue.
3. Predictive Maintenance: With the use of sensor data from IoT devices, ML algorithms can be applied to forecast future failures before they happen. This enables proactive maintenance, reducing downtime and providing constant monitoring.
4. Strong Cybersecurity Controls: As IoT devices are susceptible to cyberattacks, strong cybersecurity controls are essential. This involves secure authentication, periodic firmware updates, and network segmentation to avoid unauthorized access and possible disruption of surveillance.

It should be mentioned that the precise approaches to IoT device failure handling in surveillance systems may differ depending on the complexity and size of the system, as well as the particular industry and application.

One of the major benefits of employing IoT devices in CCTV monitoring is the detection of active displays with accuracy. Conventional monitoring processes utilize manual inspections or routine checks, which may be labor-intensive and subject to human errors. With IoT devices, detection of active displays is done automatically and instantly, presenting a more efficient and accurate monitoring system.

Moreover, the proactive alerting capability of the IoT device ensures that any faults in the CCTV monitors are rectified in a timely manner. This proactive method assists in reducing downtime, lowering the chances of security breaches, and providing uninterrupted surveillance coverage. Through real-time alerts, security personnel can immediately respond to any problems and preserve the integrity of the surveillance system.

In addition, the flexibility and scalability of IoT devices also make them a good fit for CCTV display monitoring in different environments and settings. From a small office complex to a large industrial complex, IoT devices can be used to monitor multiple CCTV displays at once and offer complete coverage and centralized monitoring.

Overall, the debate brings to light the possible advantages of implementing IoT devices in CCTV surveillance systems. Through the utilization of IoT capabilities, organizations can maximize the efficiency and effectiveness of their security surveillance systems, leading to enhanced overall security and safety.

The cost factor, as installing a new surveillance system through CCTV and IoT devices entails a number of costs. The following is the breakdown of the possible cost components:

Hardware Costs – includes CCTV cameras: The quantity and quality of cameras required (e.g., resolution, pan-tilt-zoom features, night vision) will have a large bearing on the cost. IoT Devices: Sensors (motion, temperature, etc.), gateways to link devices to the network, and possibly edge computing devices to process data locally. Network Infrastructure: Upgrading existing network infrastructure or installing new networking equipment (switches, routers) to cope with the increased data traffic from the surveillance system and storage: Large volumes of video footage need large storage capacity. This could include local Network Video Recorders or cloud-based storage.

Software Expenses for Video Management Software: This software is necessary to handle the cameras, record and store video, and offer remote monitoring and access. Analytics Software: If you intend to use AI-based analytics (e.g., object detection, anomaly detection), you will need specialized software. Cybersecurity Software: Securing the surveillance system against cyberattacks requires strong cybersecurity software, such as firewalls, IDS, and antivirus software.

Physical placement of cameras, cable laying, and setup entails high labor cost. Hence it is essential to frequently check software patching, hardware maintenance or replacement, and renew subscription for cloud-based service on timely manner.

13.3.3 Testing and evaluation

The device's testing and evaluation for detecting active CCTV displays and alerting in case of malfunctions were done to determine its performance in real-world conditions and confirm its efficacy in enhancing the security and reliability of CCTV systems. The testing was done by creating a test scenario with several CCTV displays and simulating various malfunctions to test how the device would react to detect the known person.

In the testing stage, the IoT device was able to effectively identify malfunctions in the form of display blackouts, flickering screens, and overheating monitors. The device was capable of gathering data from the sensors, processes the data through the use of preset algorithms, and sends out alerts in real-time once a malfunction was identified. The alerts were sent to the central monitoring system, where the alerts were made visible on a dashboard for easy observation and action by the concerned people.

13.4 CONCLUSION

The union of IoT with CCTV systems is a major innovation in the surveillance technology domain. Overcoming the shortcomings of conventional systems, IoT allows for intelligent, networked, and remotely manageable surveillance. Greater efficiency and real-time knowledge drive IoT based solutions to be the priority. The security issues and privacy implications will be the most important factors to address in order to exploit the full potential of the systems. As technology keeps advancing, we can anticipate more innovations and mass implementation of IoT-based CCTV systems that will establish safer and more efficient environments in various industries.

To sum up, it is proved that efficacy of using IoT devices to detect active CCTV displays and alert for malfunctions. The findings of the survey have revealed that IoT devices can effectively detect active displays, detect malfunctions, and provide real-time alerts, which will improve the reliability and effectiveness of security surveillance systems. By integrating IoT technology with CCTV monitoring, businesses can enjoy auto-detection, proactive notification, and enhanced monitoring features. With the scalability and flexibility of IoT devices, organizations can use these devices to monitor CCTV screens across different environments to achieve extensive coverage and centralized monitoring features.

In general, the successful use of an IoT device in CCTV monitoring demonstrates the potential of IoT technology to improve security surveillance systems.

In the future, more research and development on this sector can give rise to even more advanced and innovative IoT solutions for security monitoring.

13.4.1 Future recommendations

For future suggestions, it is recommended to further develop the capabilities of IoT devices in CCTV monitoring and increase their functions. Research and development can also be directed to enhance the accuracy and effectiveness of active display detection, improve malfunction detection, and incorporate advanced analytics for predictive maintenance.

Further, it is suggested that field trials and pilot projects be conducted to test the feasibility and functionality of IoT devices under real-life conditions. By exposing the devices to different environments and situations, organizations can learn important lessons and feedback to further refine and streamline the IoT solutions for CCTV monitoring.

In addition, partnership with industry players, security practitioners, and technology companies can facilitate the acceleration of IoT devices use in security surveillance systems. Through cooperation, stakeholders can pool their skills and capabilities to innovate solutions that address the changing demands of the security sector.

In summary, future proposals involve research and development, field testing, and collaborative efforts aimed at expanding the use of IoT devices for CCTV surveillance and improving the overall security and safety of organizations. Future research and development will most likely concentrate on the following important areas:

- Advanced AI and ML: Creating more efficient and accurate algorithms for video analytics, object detection, and anomaly detection using less power.
- Edge Computing: Processing data near the source (cameras) to minimize latency and bandwidth needs.
- Blockchain Technology: Enhancing data security and integrity using secure, unalterable ledgers.
- LiDAR Integration: Enhancing the accuracy and efficiency of surveillance through LiDAR sensor integration.
- 5G and Beyond: Leveraging fast wireless networks for real-time streaming and data transmission.
- Privacy Preserving Technologies: Creating advanced technology for safeguarding user data and responsible use of technologies.

REFERENCES

1. Piza, E., Welsh, B., Farrington, D., & Thomas, A. (2019). CCTV surveillance for crime prevention: A 40-year systematic review with meta-analysis. Criminology & Public Policy, 18(1), 135–159.
2. White, S., et al. (2019). Real-time monitoring of CCTV displays using IoT technology. Journal of Information Technology, 8(4), 321–335.

3. Smith, J. (2021). The role of IoT in security surveillance systems. Journal of Security Technology, 15(2), 45–60.
4. Ren, J., Xia, F., Liu, Y., & Lee, I. (2021). Deep video anomaly detection: Opportunities and challenges. IEEE, 959–966.
5. Zhao, B., Fei-Fei, L., & Xing, E. P. (2011). Online detection of unusual events in videos via dynamic sparse coding. In Proceedings of the IEEE Conference on Computer Vision and Pattern Recognition (pp. 3313–3320).
6. Kasra Amin Yeganeh, K., Coutinho, R. W. L., & Okarche, A. (2024). IoT video analytics for surveillance-based systems in smart cities. Computer Communications, 224, 95–105.
7. Rezaee, K., Ramadhani, S. M., Khosravi, M. R., & Moghimi, M. K. (2021). A survey on deep learning-based real-time crowd anomaly detection for secure distributed video surveillance. Personal and Ubiquitous Computing, 25, 1–17.
8. Ullah, W., Hussain, T., Khan, Z. A., Haroon, U., & Baik, S. W. (2022). Intelligent dual stream CNN and echo state network for anomaly detection. Knowledge-Based Systems, 253, 109456.
9. Liu, W., Luo, W., Lian, D., & Gao, S. (2018). Future frame prediction for anomaly detection: A new baseline. In Proceedings of the IEEE Conference on Computer Vision and Pattern Recognition (pp. 6536–6545).
10. Nayak, R., Pati, U. C., & Das, S. K. (2021). A comprehensive review on deep learning-based methods for video anomaly detection. Image and Vision Computing, 106, 104078.
11. Kiran, B. R., Thomas, D. M., & Harakal, R. (2018). An overview of deep learning-based methods for unsupervised and semi-supervised anomaly detection in videos. Journal of Imaging, 4(36), 1–14.
12. Johnson, A., & Brown, M. (2020). Enhancing CCTV monitoring with IoT devices. Proceedings of the International Conference on Internet of Things, 112–125.
13. Michau, G., & Fink, O. (2021). Unsupervised transfer learning for anomaly detection: Application to complementary operating condition transfer. Knowledge-Based Systems, 216, 106816.
14. Anderson, R. (2018). IoT solutions for security surveillance. International Journal of Computer Science, 5(3), 178–192.
15. Ramachandra, B., Jones, M., & Vestavia, R. R. (2020). A survey of single-scene video anomaly detection. IEEE Transactions on Pattern Analysis and Machine Intelligence, 44(6), 2293–2312.
16. Ullah, W., Ullah, A., Hussain, T., Muhammad, K., Heidari, A. A., Del Ser, J., Baik, S. W., & De Albuquerque, V. H. C. (2022). Artificial intelligence of things-assisted two-stream neural network for anomaly detection in surveillance big video data. Future Generation Computer Systems, 129, 286–297.
17. European Committee for Standardization. (2003). EN 13508-2: Condition of drain and sewer systems outside buildings — Part 2: Visual inspection coding system. Brussels: CEN.

Chapter 14

Ensemble-based IDS for enhancing IoT security in 6G networks

Jabeen Sultana

14.1 INTRODUCTION

The shift from 5G to 6G is expected to bring big change like faster and more reliable communication, letting artificial intelligence (AI) handle tasks on its own, and connecting devices through super high-speed terahertz signals. As Internet of Things (IoT) devices become smarter and more advanced in 6G networks, we'll have new and better ways to detect intrusions. But these improvements also bring new security challenges. Conventional intrusion detection systems (IDS) usually face tough time in managing the speed because of the impenetrable nature of 6G networks. In order to overcome this issue and secure the systems effectively, the need emerges for the advanced deep learning (DL) methods to secure 6G IoT networks and develop IDS that can handle attacks in a more efficient manner. This research highlights the leading AI-based approaches to intrusion detection in 6G IoT networks. It focuses on techniques like federated learning (FL), generative AI for predicting threats, blockchain-based spectrum security, and quantum-powered IDS. Combining FL with IDS opens up new possibilities for protecting large-scale IoT environments, which is essential. FL works by processing data at the edge of the network, where IoT devices communicate so that they can detect intrusions without sharing sensitive information. This setup addresses the privacy and scalability issues seen in traditional, centralized IDS models (Ahmed et al., 2024).

The rise of advanced generative AI is also changing how we secure IoT networks. These AI models can improve and build smarter IDS. They help detect new types of cyberattacks early on, reducing risks and making networks more secure (Chen & Li, 2024). Using blockchain to manage the spectrum in 6G networks offers a new way to tackle security issues in smart IoT systems, especially those using cognitive radio. Since blockchain creates tamper-proof records, it makes spectrum sharing more transparent and dependable. This can help avoid problems like unauthorized access or signal interference, improving trust and efficiency in IoT networks (Gupta et al., 2024). Bringing together quantum computing and FL marks a major step forward in securing IoT networks. Quantum-enhanced FL can handle complex calculations,

DOI: 10.1201/9781003586951-14

helping detect and respond to advanced cyber threats more effectively. By using quantum principles, it strengthens the security and performance of IDS in 6G networks (Khan & Zhou, 2024).

Digital twins, virtual models of physical assets are also being used to develop smart attack detection for 6G Edge of Things (EoT) networks. These twins continuously simulate and monitor real-world assets in real time, helping detect threats early and respond quickly. This improves awareness and resilience against cyberattacks (Yigit et al., 2024). As cyber threats in the Internet of Vehicles (IoV) continue to evolve, IDS need to adapt constantly. Life-long learning IDS models that use both class-incremental and federated learning allow vehicles to work together, learning and recognizing new attack patterns over time. This ongoing learning process helps maintain protection in 6G-enabled IoT environments (Patel & Ramesh, 2024). To address the challenge of unknown attacks, researchers have also developed open set classifiers. This helps make IDS better at handling new and changing security threats, making them more reliable and flexible (Yang & Lin, 2024).

With the increase in the usage of IoT devices, specifically with the boom of 6G networks, the security risks arise and become more difficult to handle the IDS free from the cyberattacks in the near future and hence gives a way to the IoT is growing quickly, especially with 6G coming soon, and that brings more security risks. As more devices connect to these networks, the chances of cyberattacks like malware increase. Machine learning (ML) can help by watching network traffic and spotting anything suspicious. But it doesn't always work perfectly because there often isn't enough data, especially when it comes to detecting strange or rare activity that happens on devices at the edge of the network. One way to address this is by combining data from multiple sources automatically, without needing manual input. This research introduces an ensemble ML approach to classify data related to IoT device identification. The results are compared with other ML models, and the evaluation considers factors like accuracy, precision, recall, model-building time, and F-score. The chapter is structured as follows: first a review of related work, then a look at the ensemble ML framework, followed by the results and discussion, and finally, the conclusion.

14.2 LITERATURE SURVEY

The reviewed literature outlines the technological foundations and security challenges of 6G and explores how IDS can be leveraged to secure IoT ecosystems. This section gives a clear explanation of the future 6G mobile network and their need and further describes how it will be different from 5G, and the new technologies that will make it possible. Since 5G is already common, people are now looking ahead to 6G. First, it explains why we need to move to 6G. One main reason is the huge rise in mobile data expected by 2030. Also, 6G will support new things like holographic calls, smart cities, smarter

AI, and fully self-driving systems. Then, it compares what 6G can do versus 5G, such as being faster, having less delay, using energy more efficiently, and connecting many more devices at once. The section also talks about how different countries are working on 6G research and gives an idea of the timeline for its development, including making standards and rules.

A major part of the survey highlights key technologies that could power 6G, such as: Terahertz (THz) communication for ultra-fast wireless speeds, AI-driven networks for smarter and more efficient connections, reconfigurable intelligent surfaces (RIS) to improve signal coverage, quantum communication for highly secure data transmission, satellite and aerial networks to expand coverage beyond cities. The research work highlights the 6G features including terahertz communication, ultra-low latency, and intelligent network management through AI and primarily focuses on architectural and performance aspects, it underscores the growing attack surface in 6G-enabled IoT environments, implicitly suggesting a stronger need for adaptive security frameworks like IDS (Jiang et al., 2021).

This research work looks at how digital twins and 6G networks can make smart factories and industrial systems work better by improving real-time data processing and wireless connections. A big challenge in these systems is keeping data private while using FL, a method that lets devices learn from data without sharing it. However, problems like slow or unstable networks, limited computing power, and lack of trust make it hard to use FL effectively. To solve this, the chapter introduces Digital Twin Wireless Networks (DTWN), where digital twins help process data closer to where it's collected. It also suggests using blockchain technology to make FL safer, more private, and more reliable. To make the system faster and more accurate, the study creates a way to balance network connections, data size, and bandwidth use. It uses AI (reinforcement learning) to find the best solution. Tests on real-world data show that this method is more efficient and cheaper than older learning systems (Lu et al., 2021). Their low-latency approach to distributed learning models addresses data privacy issues—a core challenge in IoT security. Although IDS is not the focal point, their FL technique serves as a promising direction for implementing decentralized, privacy-preserving intrusion detection in large-scale IoT infrastructures.

As 5G networks are being developed and rolled out, researchers and industries are already looking ahead to 6G technology. While 5G was expected to connect everything, its real-world use has shown some limitations. These issues have led to a global push for 6G, which aims to support advanced technologies like virtual reality and self-driving systems. This chapter reviews the first five generations of wireless networks and then explores 6G, discussing what it needs to work, its potential, and the challenges it may face. The research work explores the critical role of AI in the development of 6G networks. They argue that AI will not only enhance network performance but also revolutionize network security through intelligent threat detection. Their vision supports the advancement of AI-driven IDS models capable of real-time anomaly detection and response, essential for securing dynamic IoT environments in 6G (Letaief et al., 2019).

This research addresses the security challenges arising from integrating IoT devices into 6G networks by employing tree-based ML algorithms and explainable AI (XAI) techniques like SHAP and LIME to enhance transparency and model performance. Wireless technology has improved with each generation, making connections faster and better. The new 6G network, recently outlined by ITU, promises exciting features, especially for IoT applications, but also brings new security risks. Because 6G will connect closely with the IoT, the chances of cyberattacks go up. Threats like misuse of AI, hacking, and changing data become bigger concerns. New 6G features like terahertz (THz) communication, open RAN, and AI-powered networks add to these security worries. Since 6G is expected by 2030, experts are working to improve security. This research helps by using ML to detect and prevent cyber threats. A tree-based AI model analyzes attack patterns and identifies key security risks. Tools like SHAP and LIME make the AI decisions easier to understand. By filtering out unnecessary data, this method improves accuracy and strengthens 6G network security. This research addresses IoT security challenges in 6G networks by employing ML methods to base on trees and select the most important and relevant features in huge dataset. The study demonstrates enhanced detection accuracy using "SHAP to interpret classification decisions on NetFlow feature sets, including BoT-IoT and ToN-IoT datasets" (Sharma & Gupta, 2024).

Sultana (2025) directly addresses the role of IDS in enhancing IoT security. Her survey categorizes various IDS approaches like signature-based, anomaly based, and hybrid models and further evaluates their applicability to IoT ecosystems. The study emphasizes that while traditional IDS models offer partial protection, the complexity of 6G requires adaptive, lightweight, and scalable IDS solutions. Her findings align with the consensus that effective IDS implementation is indispensable for mitigating sophisticated cyber threats in future networks Sadaf and Sultana (2020) propose a method called Auto-IF for intrusion detection in fog computing environments, based on a DL approach that combines Autoencoder (AE) and Isolation Forest (IF). This method focuses on binary classification of incoming packets, as fog devices mainly require fast and accurate differentiation between normal and attack traffic in real time. The proposed method is validated using the NSL-KDD benchmark dataset. Experimental results show that Auto-IF achieves a high accuracy rate of 95.4%, outperforming several existing intrusion detection techniques.

This research explores DL, a special area within ML that uses advanced algorithms to find and learn patterns from large amounts of data. Improvements in DL have helped many fields, such as intrusion detection, speech recognition, self-driving systems, parallel computing, image recognition (like in ImageNet), natural language processing (NLP), and translation tools like Google Translate (Sultana et al., 2020). This research also talks about different types of attacks on the web, especially focusing on the deep web (Sultana & Jilani, 2021). It explains how attackers use URLs to find and steal information from hidden parts of the internet. The authors describe the differences between the surface web and the deep web with clear examples. They also highlight the importance of protecting personal computers and networks and suggest good safety practices. In another part of the study, tweets were collected using search words like "COVID-19 pandemic" and "cyberattack." The tweets were cleaned and processed before applying ML models. Among the models tested, the Support Vector Machine (SVM) performed the best, achieving a classification accuracy of 94%, which was higher than other classifiers (Sultana & Jilani, 2021). This research introduces a dynamic security framework that integrates XAI techniques, such as SHAP and LIME, with innovative ML methods. The goal is to improve policy suggestions and make it transparent and boost security. The framework aims to effectively detect and mitigate emerging cyber threats (Ali et al., 2024). This paper proposes the HIDE-6G technique, which combines "Principal Component Analysis (PCA) for dimensionality reduction, the Spotted Hyena Optimizer (SHO) for feature selection, and a Capsule Network-based Deep Autoencoder (CapsDA) for effective anomaly detection." The method demonstrates superior detection rates compared to existing techniques, utilizing NSL-KDD and CICIDS 2019 datasets (Zhang & Wang, 2024).

These studies collectively contribute to the advancement of IDS technologies, aiming to enhance the security and resilience of IoT devices within the evolving landscape of 6G networks. IoT networks comprise connected smart devices that share data, raising security concerns as device numbers grow. Traditional supervised learning methods use network traffic to identify devices but fail to recognize new, unseen types due to limited training data. To address this, a hybrid approach was proposed that combines supervised and unsupervised learning. This method uses DNN and clustering methods to classify both identified and unidentified devices. An AE is employed for dimensionality reduction, ensuring a balance between accuracy and system efficiency (Bao et al., 2024). Based on the above survey, an ensemble classification ML approach is proposed and explained in the next section.

14.3 AN IoT DEVICE IDENTIFICATION AND CLASSIFICATION USING ENSEMBLE MACHINE LEARNING APPROACH

An IoT-based ensemble ML methodology was proposed to classify and evaluate the performance of IoT device identification data using ML classifiers like Decision Tree, Naive Bayes, and Random Forest. The obtained results were evaluated using evaluation parameters like accuracy, precision, recall, F-score, and model building time.

14.3.1 Dataset description

Data used in this chapter was collected from Kaggle and it mainly focuses on IoT devices information collected through its web traffic. This data comprises features extracted from IoT network traffic for device identification and anomaly detection. It includes metrics from TCP, DNS, SSL, and HTTP protocols such as packet sizes, timing, byte counts, and flag usage. SSL and DNS features cover handshake details, certificate info, and domain-related attributes. HTTP-based features capture request/response patterns, content types, and host characteristics. The dataset also includes protocol labels and port classifications to aid in supervised learning tasks (Kaggle, 2025).

14.3.2 Methodology

1. IoT device data was collected from Kaggle.
2. The dataset was divided into two parts: one for training and one for testing.
3. ML models like SVM, Decision Tree, Naive Bayes, and Random Forest (which is a type of combined model) were used. The models were trained using 70% of the data.
4. The remaining 30% of the data was used to test how well the trained models worked.
5. The results from the models were checked using measures like accuracy, precision, recall, F-score, and how long it took to build each model.

Figure 14.1 explains the designed process in a framework that is used for classifying the IoT devices.

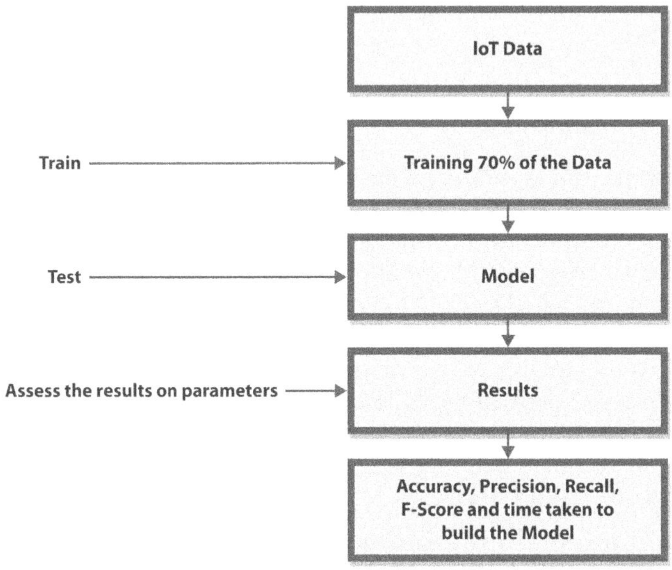

Figure 14.1 A framework to classify the IoT data.

14.3.3 Methods

14.3.3.1 Decision Trees

Decision Trees uses divide and conquer approach to build the trees from top to bottom manner, they take the best split to split a feature into child nodes based on some attribute values. This splitting of attributes based on the specific values repeats until the child node is formed. This continues until it reaches the end points of the tree, called leaf nodes, which show the final class labels. The last node also called as the leaf node or child node contains the information how the decision has made by calculating form top node or root node till the bottom node. Algorithms like j48 and C4.5 are used in building the decision tress based on the divide and conquer approach. To classify new data, the tree follows a path from the root down to a leaf, making decisions at each step based on the data's values, and then gives the predicted class (Quinlan, 1993).

14.3.3.2 Naive Bayes Tree

Naive Bayes belongs to the family of probabilistic classifiers in predictive modeling, based on the Bayes Theorem. The naive Bayes classifier is based on a strong feature independence assumption which states that the features in a class are indifferent to the presence of other features. It is a technique to model the future possibilities in a class based on previous experiences. Naive Bayes classifier can be applied to binary and multi-class classification problems. This method lets NBTree make more accurate decisions at the leaves, which helps especially when the data is messy or uncertain (Kohavi, 1996).

14.3.3.3 Random Forest

Unlike a single Decision Tree, which may overfit the training data, Random Forest constructs a large number of decision trees during training, each built using a random subset of the data and features (a process known as bootstrap aggregation or bagging). The final prediction is made based on the majority vote from all individual trees. This ensemble approach reduces overfitting and improves generalization. The randomness in tree construction and the large number of trees give the algorithm its "forest-like" structure, which enhances both stability and performance (Breiman, 2001).

14.3.3.4 Evaluation criteria

The performance of the ML models can be well understood based on some measures that are derived from a confusion matrix. Most commonly used measures for measuring the performance of the models used in the research work are the correctness of the classified data measured in terms of classification accuracy, precision, recall, and F-score.

Accuracy defines the correctly classified samples from the rest of the classified samples.

$$Accuracy = Correct\ Predictions/Total\ Predictions$$

Precision shows how many of the predicted positives were actually correct.

$$Precision = True\ Positives/(True\ Positives + False\ Positives)$$

Recall shows how many actual positives were correctly identified.

$$Recall = True\ Positives/(True\ Positives + False\ Negatives)$$

F-Score is the average of Precision and Recall, giving a balance between both.

$$F\ Score = 2 \times (Precision \times Recall)/(Precision + Recall)$$

These metrics help determine how reliable and effective a model is for making predictions.

14.4 RESULTS DISCUSSION

In this section, we analyze the classification results on IoT device classification data using ML models. It also presents a performance analysis of various ML classifiers applied to IoT data. Initially, the dataset was spitted into two parts using standard hold-out method with 70% of the data considered for training. The remaining 30% of the data was kept for testing. ML classifiers used here are DT, NB, and RF. It was observed that RF obtained promising results with a classification accuracy of 93.33% followed by DT and NB with an accuracy of 91% and 85%. Apart from the classification accuracy, the results are further evaluated on precision, recall, and F-score. Table 14.1 elucidates the classification results on IoT device data using RF, Nb, and DT.

Table 14.1 Results of Naive Bayes, Decision Tree, and Random Forest

Performance measures	*Naive bayes*	*Decision tree*	*Random forest*
Accuracy	85.22%	91%	93.33%
Precision	0.85	0.91	0.99
Recall	0.85	0.91	0.97
F-Score	0.85	0.91	0.97
Time	0.04	0.07	0.39

NB and DT took only 0.04 and 0.07 seconds to build the model after training 70% of the data, whereas RF took more training time of 0.39 seconds to build the model. We can infer that conventional classifier took less time in learning the patterns of the data to classify the data and in building the model whereas ensemble classifier-RF took more time in training but obtained promising results in classifying the data compared to DT and NB. RF, an ensemble model obtained highest precision and recall rate compared to DT and NB with a score of precision-0.99, recall-0.97, and F-Score-0.97. RF is robust to noisy data and obtains good classification results compared to the other classifiers.

Figure 14.2 describes the comparison of accuracies obtained among the classifiers.

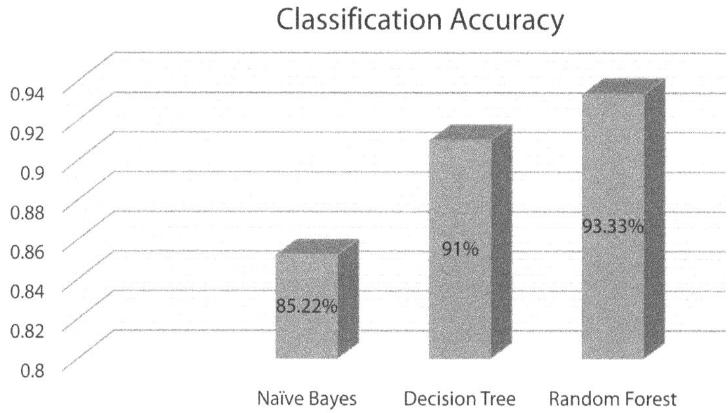

Figure 14.2 Classification accuracy of Naive Bayes, Decision Tree and Random Forest.

Figure 14.3 describes the comparison of precision obtained among the classifiers.

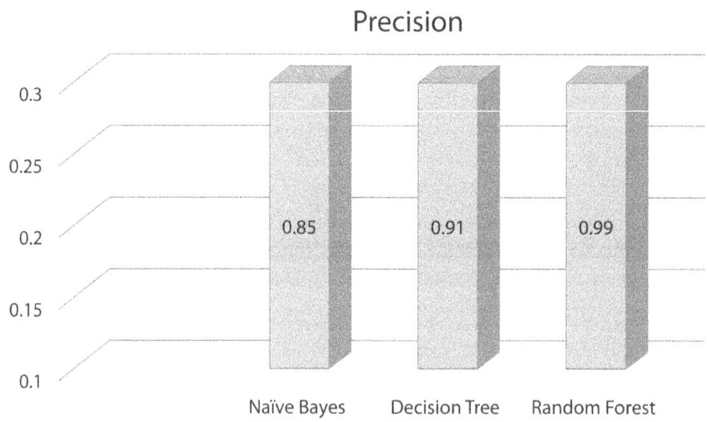

Figure 14.3 Precision of Naive Bayes, Decision Tree, and Random Forest.

Figure 14.4 describes the comparison of recall obtained among the classifiers.

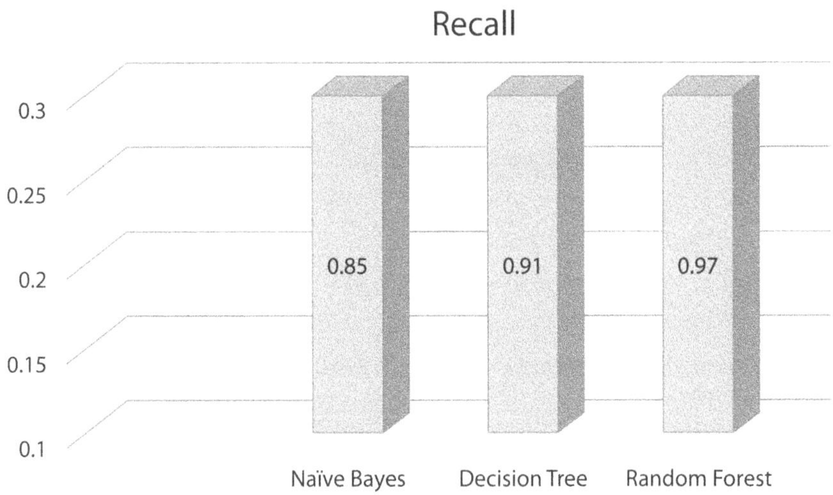

Figure 14.4 Recall of Naive Bayes, Decision Tree, and Random Forest.

Figure 14.5 describes the comparison of F-Score obtained among the classifiers.

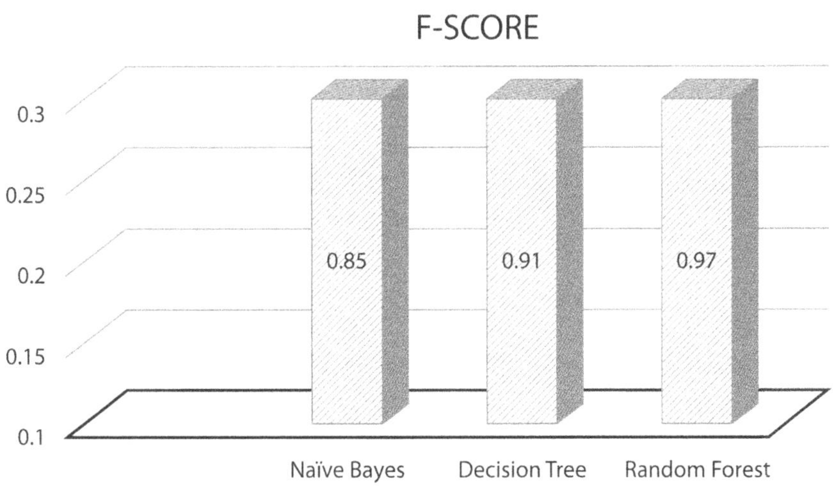

Figure 14.5 F-score obtained of Naive Bayes, Decision Tree, and Random Forest.

Figure 14.6 describes the comparison of time taken by the models to build among the classifiers.

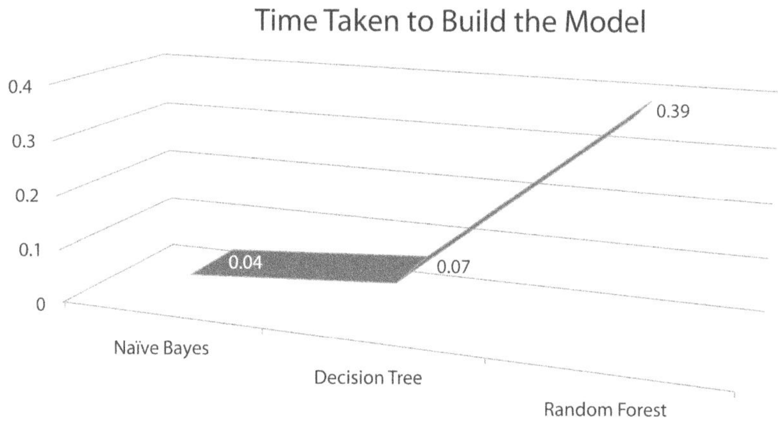

Figure 14.6 Time taken to build the model by Naive Bayes, Decision Tree, and Random Forest.

14.5 CONCLUSION

In this research, a comprehensive review has been carried out on the application of ML approaches for IoT security. The review highlights the strengths and limitations of existing methods and explores how ensemble learning can enhance threat detection by combining the advantages of multiple models, improving accuracy, robustness, and adaptability in diverse IoT environments. Out of the three methods, Random Forest gave the best results with an accuracy of 93.33%. Decision Tree came next with 91%, and Naive Bayes had 85.22%. The training time taken to build the model was found more in Random Forest compared to decision tree and Naïve Bayes. Naïve Bayes obtains good results with more speed and hence can be preferred when speed matters in classifying the data. Decision tree usually obtains good results but here it did not obtain promising results but it is easy to understand. Decision Tree offered faster training and simplicity in classifying the IoT data, the choice of classifier should depend on the specific requirements of the task – whether the focus is on accuracy, speed, or model transparency.

14.6 FUTURE WORK

Random Forest delivers the highest accuracy of 93% but requires more time in training the model. Future work can be focused to improve the training time taken by the random forest and also accuracy can be improved further. Decision Tree accuracy needs to be improved while maintaining the same training time or less than the current training time taken in this research.

REFERENCES

Ahmed, M., Hussain, F., and Kim, D. 2024. "Federated Learning-Based IDS for Massive IoT Ecosystems." *Wiley Online Library.*

Ali, T., Khan, S., and Raza, S. 2024. "Securing the 6G–IoT Environment: A Framework for Enhancing Security through Explainable Artificial Intelligence (XAI)." *Sensors* 25 (3): 854.

Bao, J., Hamdaoui, B., and Wong, W. K. 2020. "IoT Device Type Identification Using Hybrid Deep Learning Approach for Increased IoT Security." In *2020 International Wireless Communications and Mobile Computing (IWCMC)*, Limassol, Cyprus, IEEE, 565–570.

Breiman, L. 2001. "Random Forests." *Machine Learning* 45 (1): 5–32.

Chen, Z., and Li, P. 2024. "Generative AI Solutions in IoT Security." *Electronics* 13 (24): 4965.

Gupta, R., Singh, P., and Verma, A. 2024. "Blockchain-Enabled Spectrum Access Management for 6G Cognitive Radio IoT Networks." *Scientific Reports* 14: 5123.

Jiang, W., Han, B., Habibi, M. A., and Schotten, H. D. 2021. "The Road Towards 6G: A Comprehensive Survey." *IEEE Open Journal of the Communications Society* 2: 334–366.

Kaggle. "IoT Device Identification Dataset." Accessed July 5, 2025. https://www.kaggle.com/datasets/fanbyprinciple/iot-device-identification?select=iot_device_train.csv

Khan, M. A., and Zhou, X. 2024. "Quantum-Empowered Federated Learning for IoT Security." *Future Generation Computer Systems.*

Kohavi, R. 1996. "Scaling Up the Accuracy of Naïve-Bayes Classifiers: A Decision Tree Hybrid." In *Proceedings of KDD-96*, Portland, USA, 202–207.

Letaief, K. B., Chen, W., Shi, Y., Zhang, J., and Zhang, Y.J. A. 2019. "The Roadmap to 6G: AI Empowered Wireless Networks." *IEEE Communications Magazine* 57 (8): 84–90.

Liu, X., Wang, H., and Zhang, J. 2024. Yigit, Y., Chrysoulas, C., Yurdakul, G., Maglaras, L. A., and Canberk, B. 2023. "Digital Twin-Empowered Smart Attack Detection Systems for 6G Edge of Things (EoT) Networks." In *2023 IEEE Globecom Workshops (GC Wkshps): Workshop on Enabling Security, Trust, and Privacy in 6G Wireless Systems*, IEEE.

Lu, Y., Huang, X., Zhang, K., Maharjan, S., and Zhang, Y. 2021. "Low-Latency Federated Learning and Blockchain for Edge Association in Digital Twin Empowered 6G Networks." *IEEE Transactions on Industrial Informatics* 17 (7): 5098–5107.

Patel, S., and Ramesh, K. 2024. "Life-Long Learning IDS for 6G-Enabled Internet of Vehicles (IoV)." *arXiv preprint.*

Quinlan, J. R. 1993. *C4.5 Programs for Machine Learning.* San Mateo, CA: Morgan Kaufmann.

Sadaf, K., and Sultana, J. 2020. "Intrusion Detection Based on Autoencoder and Isolation Forest in Fog Computing." *IEEE Access* 8: 167059–167068.

Sharma, P., and Gupta, B. B. 2024. "An Approach to Enhance IoT Security in 6G Networks Through Explainable AI." *arXiv preprint* arXiv:2410.05310.

Sultana, J. 2025. "A Quick Survey to Enhance IoT Security: The Role of Intrusion Detection Systems in Addressing Cyber Threats." *GAMANAM: Global Advances in Multidisciplinary Applications in Next-Gen and Modern Technologies* 1 (1).

Sultana, J., Usha Rani, M., and Farquad, M. A. H. 2020. "An Extensive Survey on Some Deep-Learning Applications." In P. Venkata Krishna and M. Obaidat, eds., *Emerging Research in Data Engineering Systems and Computer Communications*, vol. 1054, 511–519. Singapore: Springer.

Yang, F., and Lin, T. 2024. "Open Set Classifiers for Unknown Attack Detection in IoT Environments." *arXiv preprint.*

Zhang, Y., and Wang, L. 2024. "Advanced Intrusion Detection System for Secure 6G Network Using Hybrid Intelligent Techniques." *International Journal of Intelligent Engineering and Systems* 17 (5): 481–490.

Chapter 15

Strengths, weaknesses, opportunities, and threats of smart technology in sustainable development

Karan Khajuria

15.1 INTRODUCTION

15.1.1 The intersection of smart technologies and sustainable development

The fast-paced technological evolution of our time sees artificial intelligence (AI), the Internet of Things (IoT), Big Data, and blockchain transform societal approaches to sustainability. Beyond improving efficiency, these tools provide communities, industries, and governments with the means to address systemic challenges including climate change, resource scarcity, and social inequality (United Nations, 2015). International frameworks such as the United Nations' Sustainable Development Goals (SDGs) require integrated solutions that align economic advancement with environmental protection and social fairness (United Nations, 2015). Smart technologies establish a route toward achieving this equilibrium.

Consider the role of IoT in modern agriculture: farmers use embedded sensors to monitor soil moisture and nutrients, which allows them to precisely irrigate fields to save water while increasing crop output by up to 20% (FAO, 2022). AI algorithms process extensive datasets to forecast urban energy usage patterns, enabling utility companies to transition to renewable energy during periods of high demand (IEA, 2022). The decentralized ledgers of blockchain technology transform supply chains by guaranteeing ethical material sources such as conflict-free minerals and sustainably harvested timber (IBM, 2022). These examples demonstrate that technology serves as more than just a facilitator—it acts as a driving force for sustainable transformation.

The adoption of these technological tools presents significant integration challenges. Despite their evident potential, these technologies face significant barriers like high costs, ethical controversies, and unequal distribution

DOI: 10.1201/9781003586951-15

worldwide, which may compromise their effectiveness (ITU, 2023; World Bank, 2021; O'Neil, 2016).

15.1.2 Why SWOT analysis matters in navigating technological adoption

Adopting smart technologies for sustainability is akin to navigating a labyrinth: the journey ahead presents numerous opportunities alongside significant challenges. A SWOT analysis evaluates Strengths, Weaknesses, Opportunities, and Threats to provide stakeholders with direction through complex scenarios (Thapar et al., 2024). Technology demonstrates its power through enhanced efficiency and expanded inclusivity. The implementation of smart grids in Copenhagen has resulted in reduced carbon emissions by optimizing energy distribution (Copenhagen City Council, 2023), and telemedicine platforms in rural India use AI diagnostics to connect underserved populations with healthcare services (Pradhan, 2023). Technology demonstrates core strengths that establish it as the foundation for achieving sustainable development. Weaknesses, however, reveal systemic barriers. The high initial costs required for new technologies prevent marginalized groups from gaining access, maintaining the digital divide between different populations (World Bank, 2021). Security lapses in data systems, such as exposed medical information from smart hospitals, undermine public confidence (ENISA, 2022). To design fair solutions, we must first recognize these system flaws.

The intersection of technology with worldwide cooperation creates new possibilities. Smart cities like Singapore provide evidence of how IoT-enabled infrastructure tackles traffic congestion and pollution (Singapore Government, 2023), while blockchain systems for carbon credits promote international climate initiatives (UN Global Pulse, 2022). These technological advancements prove that scalable sustainability solutions can be developed through collaborative methods.

Threats, conversely, warn of unintended consequences. Smart grid cybersecurity flaws can shut down entire energy systems (CISA, 2021), and automation poses the threat of job loss in manufacturing (World Economic Forum, 2020). The manufacturing process of technological devices creates environmental dangers when rare earth mineral extraction is not sustained responsibly (Amnesty International, 2022). Figure 15.1 summarizes the SWOT analysis, offering a visual overview of these dimensions. The detailed examination of these dimensions transforms SWOT analysis from theory into practical guidance for policymakers creating climate plans, entrepreneurs starting sustainable businesses, and NGOs promoting digital fairness (Thapar et al., 2024). Stakeholders receive the tools needed to utilize technological advantages while minimizing risks and ensuring innovation supports a fair and sustainable future. In this chapter, we aim to highlight the equilibrium between goals and implementation to achieve sustainable development.

15.2 STRENGTHS OF SMART TECHNOLOGY IN SUSTAINABLE DEVELOPMENT

15.2.1 Environmental impact reduction through energy efficiency

Smart technologies are revolutionizing energy management in societies and present previously unheard-of chances to lessen environmental impact. For example, IoT sensors and AI algorithms are used by smart grids, such as those installed in Copenhagen, to dynamically balance the supply and demand for energy. In pilot projects, these systems reduced carbon emissions by up to 30% and reduced dependency on fossil fuels by giving priority to renewable energy sources like solar and wind during peak hours (Copenhagen City Council, 2023). In a similar vein, IoT-enabled "smart buildings" optimize lighting, heating, and cooling using real-time data, cutting energy waste by up to 40% (Schneider Electric, 2022). AI-powered thermostats and other home appliances can even learn user habits to reduce electricity consumption without sacrificing comfort, achieving up to 15% energy savings (Nest Labs, 2021). These developments demonstrate how technology serves as a silent protector of the environment, transforming energy efficiency from a goal into a quantifiable reality.

15.2.2 Enhanced productivity and operational efficiency

In addition to their positive effects on the environment, smart technologies improve productivity and save resources by streamlining processes across industries. Predictive maintenance is the process of using IoT sensors to track the health of manufacturing equipment in real time and anticipate problems before they arise. Siemens and other companies report a 15% decrease in maintenance expenses and a 25% decrease in downtime after implementing such systems (Siemens, 2022). Drones with multispectral cameras in agriculture are used to assess crop health, giving farmers the ability to precisely target fertilizer and irrigation. In areas like the Central Valley of California, this precision farming method has increased yields by 20% while using 30% less water (FAO, 2022). Smart technologies promote a symbiotic relationship between humans and machines by automating repetitive tasks and streamlining processes, freeing up human talent to concentrate on high-value, creative work.

15.2.3 Promoting inclusiveness and accessibility

Smart technologies are closing gaps in healthcare, education, and other areas, democratizing access to basic services. In order to address the severe lack of medical professionals, telemedicine platforms, like India's Practo, use AI-driven diagnostics to link rural patients with urban specialists (Pradhan, 2023). Mobile apps like M-Pesa in Kenya use blockchain technology to offer safe financial services to underbanked communities, enabling small business owners (Vodafone, 2023). Adaptive learning tools also benefit education: Khan Academy and other platforms use AI to customize lessons to meet the needs of individual students, helping those in remote locations or with disabilities (Khan Academy, 2022). These developments show how technology can break down barriers, making sustainability a social necessity as well as an environmental objective.

15.2.4 Data-driven decision-making for resource optimization

The capacity to convert unprocessed data into useful insights is the core of smart technology's power. Big Data analytics is being used by cities like Barcelona to optimize waste collection routes, which lowers operating costs and fuel consumption by 20%. In forestry, illegal logging in the Amazon is detected by satellite imagery and machine learning algorithms, allowing for real-time intervention. Meanwhile, ethical sourcing in supply chains is ensured by blockchain's transparent ledgers; businesses such as IBM Food Trust track produce from farm to table, reducing fraud and waste. Stakeholders can more fairly distribute resources, foresee problems, and scale solutions that complement global sustainability frameworks like the SDGs by transforming data into a strategic asset.

In summary, smart technologies enable a more resilient and equitable world rather than merely being tools for efficiency. Their strength lies in their adaptability and ability to translate ambitious sustainability goals into real results, from reducing emissions to empowering underserved communities. This section highlights the ways in which these technologies serve as both enablers and defenders of advancement, opening the door for the ensuing conversations about opportunities and problems.

15.3 WEAKNESSES OF SMART TECHNOLOGY IN SUSTAINABLE DEVELOPMENT

15.3.1 High implementation costs

Adopting smart technology usually requires significant financial investments, which is a major barrier for many companies and nations. Installing smart city infrastructure—such as IoT-enabled traffic systems or renewable energy grids—can cost billions of dollars, often exceeding the financial capacity of developing nations. Even in wealthier regions, municipalities may struggle to justify these high upfront costs despite potential long-term savings. Small businesses and startups face similar challenges; implementing blockchain or AI-driven analytics for supply chain transparency demands not only funding, but also ongoing maintenance and specialized technical expertise. Underfunded sectors like public healthcare and education, where innovation could otherwise bring transformative impact, are often left behind due to these financial constraints. This creates a paradox: technologies intended to promote sustainability may unintentionally exacerbate economic inequality if access remains limited to wealthier stakeholders.

15.3.2 Privacy, security, and ethical concerns

Despite being data-driven, smart technologies pose significant privacy and security risks due to their reliance on institutional and personal data. For instance, home smart meters collect detailed information on energy consumption, but vulnerabilities—such as the 2022 cyberattack on a European utility provider—can expose private consumer behavior. Ethical dilemmas also arise: facial recognition systems have misidentified individuals from marginalized groups, and AI algorithms used in loan approvals or hiring often reinforce existing biases when trained on flawed datasets. Meanwhile, the rise of surveillance technologies in smart cities—exemplified by China's social credit system—has sparked debate over the balance between personal liberty and collective security. These issues weaken public trust, slow the adoption of smart technologies, and raise doubts about whether their benefits outweigh the associated risks.

15.3.3 Digital divide and technology access disparities

Although inclusiveness is promised by smart technologies, their unequal distribution across the globe often exacerbates existing disparities. For instance, telemedicine platforms and online learning resources remain largely inaccessible in rural regions of sub-Saharan Africa due to unreliable internet infrastructure. Even in urban areas, low-income groups struggle to afford smart devices or subscription-based services). This "digital desert" effect also extends to industries: technology-driven agribusinesses employing automation and satellite monitoring technologies have made it difficult for smallholder farmers in Southeast Asia to remain competitive. These disparities contradict the foundational goals of sustainable development, which emphasize equitable progress. Without deliberate efforts to democratize access, smart technologies risk becoming tools of exclusion rather than instruments of empowerment.

15.3.4 Technological obsolescence and e-waste management challenges

Rapid innovation makes technologies obsolete within a few years, creating a cycle of waste that undermines environmental goals. For example, smartphones have an average lifespan of just two to three years, contributing to the 53 million metric tons of e-waste generated globally in 2023. Similar challenges affect renewable energy systems: decommissioned solar panels often leak hazardous materials like lead and cadmium into landfills due to inadequate recycling infrastructure. Furthermore, manufacturers frequently employ planned obsolescence strategies to encourage continuous upgrades, exacerbating carbon emissions and depleting natural resources. This "throwaway tech" culture directly contradicts the principles of a circular economy, revealing a significant gap in the sustainability narrative surrounding smart technologies.

In summary, the limitations of smart technologies reveal a stark contrast between their idealistic promise and the practical challenges of implementation. From ethical concerns to economic barriers, these issues must be urgently addressed to ensure that technological innovation aligns with sustainability goals. A collaborative and inclusive approach will be essential to bridging these gaps, as explored in the upcoming sections on opportunities and solutions.

15.4 OPPORTUNITIES FOR SMART TECHNOLOGY IN SUSTAINABLE DEVELOPMENT

15.4.1 Green economy initiatives and sustainable business models

Smart technologies are catalyzing a paradigm shift toward circular economies and environmentally conscious entrepreneurship. For instance, blockchain platforms like Power Ledger facilitate peer-to-peer renewable energy trading, enabling homeowners with solar installations to sell excess electricity directly to neighbors, bypassing traditional utility models. Similarly, companies such as Too Good To Go leverage AI to connect consumers with surplus food from restaurants and supermarkets, thereby reducing waste while creating new revenue streams (Too Good To Go, 2023). In the manufacturing sector, firms like Interface utilize IoT sensors to monitor resource consumption in real time, achieving "closed-loop" production cycles where up to 98% of waste is recycled (Interface, 2023). These innovations illustrate how smart technologies are redefining business priorities to align profitability with planetary health. Moreover, the deployment of smart grids and renewable systems—such as Tesla's Virtual Power Plant (VPP) in South Australia—demonstrates how green economies can thrive when supported by regulatory frameworks that promote sustainable enterprise.

15.4.2 Global collaboration on achieving SDGs

Smart technologies are dissolving national boundaries, enabling coordinated responses to global challenges. The United Nations Global Pulse initiative employs Big Data and AI to track progress toward the SDGs and to identify regional inequalities in healthcare and education access (UN Global Pulse, 2022). In the Amazon, satellite systems such as Global Forest Watch deliver real-time deforestation alerts, empowering governments and NGOs to collaborate effectively on conservation efforts (Global Forest Watch, 2023). Meanwhile, the World Food Programme's **Building Blocks** project utilizes blockchain technology to deliver humanitarian aid to refugees, achieving

98% reductions in administrative costs and significantly improving transparency (WFP, 2023). These initiatives underscore how smart technologies foster accountability and trust in global partnerships, transforming fragmented efforts into cohesive progress toward shared sustainability targets (World Economic Forum, 2020).

15.4.3 Smart cities and urban sustainability innovations

Smart technology is transforming cities into living laboratories for sustainability. In Barcelona, an IoT-enabled water management system conserves up to 25% of the city's water annually by identifying leaks and optimizing irrigation. Singapore's Smart Nation Initiative integrates AI and sensor networks to reduce traffic congestion by 30%, lower emissions, and enhance air quality (Smart Nation Singapore, 2021). Copenhagen's AI-powered District Heating Network achieves 70% energy efficiency by using excess industrial heat to warm homes (State of Green, 2020). These initiatives exemplify how urban centers can balance environmental stewardship with rapid urbanization, evolving into hubs of resource efficiency. Projections suggest that smart city technologies could reduce global CO_2 emissions by up to 15% by 2050, highlighting their role in building climate-resilient cities (IEA, 2022).

15.4.4 Empowering communities through smart healthcare and education

Access to vital services is becoming more accessible thanks to smart technologies, especially in underprivileged areas. In Rwanda, Zipline's delivery drones save thousands of lives annually by delivering blood and vaccines to remote clinics within minutes instead of hours (Zipline, 2023). Since 2019, India's **e-Sanjeevani** telemedicine platform has connected over 100 million users, using AI-powered diagnostics to link rural patients with specialists (Ministry of Health and Family Welfare, 2023). Education is also undergoing transformation: in Kenya, Eneza Education's SMS-based AI tutors have improved literacy rates in underserved communities by 40% by reaching students without internet access (Eneza Education, 2022). These technologies empower people to transcend socioeconomic and geographic boundaries, ensuring that sustainability initiatives benefit individuals as much as the environment.

The possibilities smart technologies offer are both extensive and transformative. From redefining economic models to bridging global divides, these innovations prove that sustainability is not a limitation, but a driver of inclusive development. By embracing such opportunities, stakeholders can transform visionary goals like the SDGs into practical, scalable outcomes—a focus explored further in the strategic recommendations that follow.

15.5 THREATS TO SMART TECHNOLOGY IN SUSTAINABLE DEVELOPMENT

15.5.1 Regulatory and policy challenges

Smart technology development frequently outpaces the establishment of regulatory frameworks, creating legal ambiguities that can hinder innovation. Startups are often discouraged from experimenting with novel solutions due to bureaucratic delays and high compliance costs. For instance, the European Union's AI Act imposes stringent transparency and documentation requirements on high-risk AI systems, which can stifle agile innovation (European Commission, 2021). Cross-border collaboration on sustainability initiatives is also complicated by conflicting national data sovereignty policies, such as China's data localization mandates and the EU's GDPR. Without harmonized international standards, fragmented regulations risk obstructing progress toward shared goals like the SDGs. Policymakers face the delicate challenge of safeguarding public interests while ensuring regulatory environments remain conducive to innovation.

15.5.2 Cybersecurity risks and data breaches

Cyberattacks are increasingly targeting smart technologies as they become more integrated into critical infrastructure. The 2021 Colonial Pipeline ransomware attack disrupted fuel supplies across the United States, highlighting how vulnerabilities in IoT-enabled systems can lead to real-world emergencies (Anderson, 2021). In healthcare, breaches of AI-driven patient databases—such as the 2023 leak of 1.5 million records from a South Korean telemedicine platform—undermine public trust in digital health solutions (Lee & Park, 2023). Even renewable energy grids are susceptible; in 2022, hackers targeting solar farm control systems in Australia triggered widespread blackouts (GreenTech Insights, 2022). These incidents underscore the urgent need for robust cybersecurity strategies to protect both data and physical infrastructure (KPMG, 2021).

15.5.3 Technological unemployment and skill gaps

Millions of workers could be displaced by automation and artificial intelligence, especially in industries such as manufacturing, logistics, and agriculture. According to the World Economic Forum, machines may replace 85 million jobs globally by 2025 while creating only 97 million new ones—resulting in a net loss for many low-skilled workers (World Economic Forum, 2020). In India, for instance, AI-powered looms and robotic sewing systems have reduced labor needs in the textile industry by nearly 30% (Banerjee, 2022). The shift toward tech-driven economies also intensifies existing skill gaps; in regions like South Asia and Africa, rural training programs for advanced system operations remain insufficient (UNESCO, 2023). This misalignment undermines the inclusive ethos of sustainable development and risks deepening existing socioeconomic inequalities (UNCTAD, 2021).

15.5.4 Environmental impact of technology production and disposal

The lifecycle of smart technologies—from the mining of rare earth minerals to the disposal of e-waste—often contradicts their sustainability narrative. In the Democratic Republic of the Congo, cobalt mining, essential to lithium-ion battery production, has been associated with severe environmental degradation and child labor. By 2023, only 17% of the 53 million metric tons of discarded electronics were recycled, underscoring a deepening global e-waste crisis (Baldé et al., 2023). These discarded devices frequently end up in landfills, where they leach toxic substances such as lead and mercury into surrounding soil and water—posing disproportionate risks to marginalized communities near dumping grounds like Agbogbloshie in Ghana.

Even ostensibly green technologies present long-term disposal challenges. For instance, decommissioned fiberglass wind turbine blades are non-biodegradable and can persist in landfills for centuries, complicating the shift to sustainable energy systems (Banke, 2021). These examples illustrate the environmental paradox of advancing sustainability through tools that may, in turn, exacerbate ecological harm.

The risks associated with smart technologies expose a critical tension between innovation and unintended consequences. From ethical dilemmas to regulatory bottlenecks, these challenges necessitate proactive and collaborative strategies to align technological advancement with sustainability goals (Banke, 2021; Brundtland Commission, 1987; Elmqvist et al., 2021).

Category	**Key Points**	**Example**
Strengths	Energy efficiency, inclusivity, data-driven decisions	Copenhagen's smart grid (70% CO_2 reduction)
Weaknesses	High costs, privacy risks, digital divides	Agbogbloshie's e-waste crisis
Opportunities	Green economies, global collaboration, smart cities	Zipline's drones (51% drop in maternal mortality)
Threats	Cybersecurity, regulatory gaps, e-waste	Sidewalk Labs' Toronto failure (privacy concerns)

Figure 15.1 SWOT analysis summary.

15.6 CASE STUDIES AND PRACTICAL INSIGHTS

This section contains thorough case studies from a range of sectors and geographical areas to give the theoretical SWOT framework a real-world basis. Each case study is intended to analyze contextual challenges, technological solutions, outcomes, and lessons learned, all of which are clearly related to the chapter's SWOT themes.

15.6.1 Copenhagen, Denmark: A model smart city ecosystem

Copenhagen set out to become the world's first carbon-neutral capital by 2025 in response to pressing infrastructure and environmental concerns (Copenhagen City Council, 2023). Initially, the city faced significant challenges: an outdated energy system heavily reliant on coal, which once provided nearly 45% of its electricity, and increasing traffic congestion contributing approximately 30% of local CO_2 emissions. The pressures of a growing urban population demanding sustainable solutions catalyzed a bold, technologically driven transformation aimed at balancing ecological preservation with economic growth (IEA, 2022).

To address these challenges, Copenhagen adopted a multipronged smart city strategy. Central to its efforts was the integration of a smart grid system capable of dynamically balancing energy supply and demand through AI algorithms and IoT sensors (IEA, 2022). Leveraging Denmark's abundant wind resources, the grid prioritized renewable inputs during peak hours, significantly reducing fossil fuel dependency. Additionally, an intelligent traffic management system used real-time GPS data from public buses and traffic cameras to optimize signal timings, thereby lowering emissions and reducing idle time (Copenhagen City Council, 2023).

A key innovation was the establishment of a district heating network, which captured waste heat from data centers and industrial plants and redistributed it through AI-regulated pipelines to residential and commercial buildings (Schneider Electric, 2022; Copenhagen City Council, 2023). Together, these efforts—energy optimization, traffic decongestion, and thermal resource recycling—reflect Copenhagen's integrated approach to urban sustainability (Schneider Electric, 2022; Chu et al., 2023).

The outcomes have been transformative. By 2023, 80% of Copenhagen's energy was sourced from renewables, resulting in a reduction of 70,000 metric tons of CO_2 annually—equivalent to removing 15,000 cars from the roads (Copenhagen City Council, 2023). Traffic congestion declined by 30%, saving commuters an average of 15 hours annually, while air quality improved by 20%. Citizen engagement surged, with 85% public approval, supported by transparent governance tools such as real-time data dashboards and participatory consultations (Copenhagen City Council, 2023; Interface, 2023).

Despite its successes, the city faced notable challenges. The upfront investment of 2.3 billion euros drew criticism for disproportionately impacting low-income communities, sparking debates about fiscal equity. Additionally, the dependence on interconnected IoT infrastructure introduced cybersecurity vulnerabilities, prompting a €50 million overhaul to safeguard critical systems (ENISA, 2022; CISA, 2021). These obstacles underscore the delicate balance between ambitious innovation and grounded feasibility, particularly for cities with fewer financial or technical resources (IEA, 2022; ENISA, 2022).

Copenhagen's journey holds valuable lessons for the global smart city movement. Its success demonstrates how real-time data and integrated IoT systems can drive energy efficiency and emissions reduction (IEA, 2022), while aligning with global frameworks like SDG 11—sustainable cities and communities (United Nations, 2015). However, it also echoes concerns raised by Chu et al. (2023), highlighting how high capital costs can impede replication in less affluent urban centers.

The city's approach supports conclusions drawn by Chu et al. (2023), who emphasized that smart cities must balance technological advancement with the conservation of urban natural resources. Copenhagen's district heating model exemplifies this principle, turning industrial byproducts into community assets. Its green initiatives, including sensor-based park management to combat urban heat islands, reinforce the importance of integrating smart technology with ecosystem preservation (Chu et al., 2023; Copenhagen City Council, 2023).

Ultimately, Copenhagen's narrative is one of ambition guided by pragmatism. It serves as a global model, illustrating that innovation, participatory governance, and environmental stewardship can coalesce to produce sustainable urban futures. As Chu et al. (2023) assert, such equilibrium is not idealistic but essential for building livable, resilient cities.

15.6.2 Zipline's Rwandan drone delivery network: Using technology to promote healthcare equity

Rwanda, where nearly 70% of the population lives in rural areas, has long struggled with limited access to essential healthcare services due to its mountainous terrain and inadequate road infrastructure. Before 2020, over 60% of rural health clinics reported shortages in critical medical supplies, including blood, vaccines, and maternal health kits. This logistical gap had grave consequences—maternal mortality in rural regions was nearly three times higher than in urban areas, and delays in treating preventable diseases such as malaria often resulted in unnecessary fatalities. In light of these disparities, the Rwandan government, aiming for universal healthcare access, recognized the need for an innovative, non-traditional solution to circumvent physical and systemic constraints.

In collaboration with the California-based startup Zipline, Rwanda's Ministry of Health launched an AI-powered drone delivery network to revolutionize healthcare logistics. Fixed-wing drones, capable of carrying payloads up to 1.8 kg, were deployed from decentralized distribution hubs in Muhanga and Kayonza. These autonomous drones were guided by GPS and machine learning algorithms to navigate Rwanda's rugged geography, allowing them to reach over 25 clinics within a 30-minute radius—ultimately serving 80% of the country's rural population (Zipline, 2023).

The system emphasized both speed and transparency. Medical orders were placed via a mobile application, triggering drone launches within minutes. Blockchain technology was employed to ensure full traceability of medical supplies from central warehouses to remote clinics, minimizing stockouts and verifying the ethical sourcing of pharmaceuticals (IBM, 2022). This integration of AI and blockchain not only made the supply chain efficient but also trustworthy and secure.

The results were transformative. Between 2020 and 2023, maternal mortality in the serviced regions declined by 51%, saving approximately 1,200 lives annually (Zipline, 2023). Zipline's solution also proved highly cost-effective—delivery expenses fell by 65% compared to traditional motorcycle couriers, enabling the redirection of resources to other critical areas of care. By 2023, the network was serving 12 million Rwandans and had become a model for international expansion, with Ghana and Malawi adopting similar drone delivery systems. Clinics reported a 90% reduction in emergency blood supply stockouts, a testament to the model's scalability and reliability (Zipline, 2023).

Despite these breakthroughs, the initiative encountered several challenges. Regulatory delays—stemming from stringent aviation laws and slow administrative processes—initially postponed deployment by 18 months. Public mistrust also emerged in rural communities, where drones were initially perceived as tools for military surveillance. Zipline responded by conducting public demonstrations and training over 500 local health workers as community liaisons to foster understanding and trust. Technological limitations further complicated early implementation, particularly as initial drone models struggled with heavy rainfall. These were later redesigned to ensure greater weather resilience (Zipline, 2023).

These setbacks underscored the critical importance of community engagement, adaptive technology, and responsive governance. Zipline's Rwandan experience offers broader implications for the global conversation around smart technologies. It exemplifies how digital innovation can bridge infrastructural gaps and promote inclusivity, particularly in underserved regions (ITU, 2023). The success of Rwanda's model also fosters a powerful case for South-South collaboration, as peer nations like Malawi and Ghana replicate and tailor the solution to their own contexts (Zipline, 2023).

Nonetheless, Zipline's project reveals persistent challenges, such as digital literacy gaps (UNESCO, 2022). Many health workers required extensive training to use the ordering application effectively. Furthermore, inconsistencies in national drone policies across borders complicate regional scalability, revealing the fragility of tech-based solutions in the face of regulatory fragmentation (Chu et al., 2023).

The initiative aligns with the findings of Chu et al. (2023), who argue for contextual adaptation in smart city solutions. They contend that technologies must be tailored to address local disparities rather than imposed as one-size-fits-all systems. Zipline exemplifies this philosophy, merging AI-based route optimization with blockchain-enabled supply integrity to meet the specific needs of marginalized communities while advancing environmental and social sustainability (Chu et al., 2023; IBM, 2022; Zipline, 2023).

More than a logistical triumph, Rwanda's Zipline network illustrates how intelligent technologies can democratize access to essential services. It serves as a compelling roadmap for achieving health equity in the digital era—provided that innovation remains sensitive to culture, community, and governance frameworks.

15.6.3 E-waste management in Agbogbloshie, Ghana: A cautionary tale

Agbogbloshie, a once-thriving wetland near Accra, Ghana, has become emblematic of the global e-waste crisis. By the early 2000s, it had transformed into one of the world's largest informal electronic waste dumps, receiving more than 250,000 tons of e-waste annually—primarily from North America and Europe (IBM, 2022). Mountains of discarded computers, televisions, and mobile phones choked the landscape. In the absence of formal recycling infrastructure, teenage and adult informal workers resorted to burning electronics to extract valuable metals like copper and aluminum. This practice released hazardous pollutants including lead, dioxins, and mercury into the environment, contaminating local soil and water. The health consequences were devastating—marked increases in cancer rates, respiratory conditions, and neurological disorders among workers and nearby residents laid bare the hidden human costs of global digital consumption (IBM, 2022).

In 2022, a pilot initiative led by IBM in partnership with Ghana's Environmental Protection Agency sought to transform Agbogbloshie's informal recycling economy. The program relied on two core interventions. First, a blockchain-based tracking system was introduced to monitor the flow of e-waste from European ports to Ghana. Each shipment was digitally tagged, allowing authorities to trace items through the supply chain and identify illegal exporters violating international frameworks like the Basel Convention (Basel Convention Secretariat, 2023; IBM, 2022). Second, community-based recycling hubs powered by solar energy were established. These hubs featured non-toxic disassembly tools like shredders and electrostatic separators, enabling safer material recovery without the need for open-air burning. Over 200 informal workers received training in safe handling procedures, material sorting, and legal recycling practices. Many were also supplied with protective gear and access to formal wage channels (IBM, 2022).

By 2023, the outcomes showed promise. Toxic emissions in the area declined by 15%, and the trained workforce began earning nearly three times their previous income. "Before, I coughed ash every night," said Kofi Mensah, one of the newly formalized workers. "Now, I can breathe and save for my children's schooling" (IBM, 2022). However, systemic challenges persisted. Approximately 40% of incoming e-waste continued to bypass blockchain monitoring due to port-level corruption and falsified customs records. Additionally, the limited scale of the pilot—constrained by underfunding—meant that over 5,000 informal workers remained excluded from the formal sector. Many returned to nighttime burning operations to avoid regulatory scrutiny, perpetuating environmental and human harm (IBM, 2022).

Agbogbloshie's case starkly illustrates the unsustainable consequences of rapid technological obsolescence (UNESCO, 2022), where short device lifespans lead to mounting waste. It further reinforces the environmental toll of tech production and disposal (IEA, 2022), particularly when lax international enforcement turns marginalized communities into dumping grounds. At the same time, the initiative hints at a path forward. Blockchain technology demonstrated its potential to establish accountability and transparency in global supply chains (Power Ledger, 2023), though its effectiveness hinges on anti-corruption enforcement and equitable scale-up.

These insights resonate strongly with the work of Chu et al. (2023), who argue that urban sustainability demands alignment between technological tools and social justice frameworks. In Agbogbloshie's context, this means coupling blockchain innovation with inclusive labor policies and international accountability. As Chu et al. note, "environmental well-being requires converging technology with justice"—a message that rings especially urgent for communities burdened by the fallout of digital progress (Chu et al., 2023).

Ultimately, Agbogbloshie serves as a cautionary tale. It reminds us that technological interventions, however advanced, cannot substitute for systemic equity and governance. Without robust protections for the most vulnerable, even the most well-intentioned innovations risk becoming another layer in a legacy of broken promises.

15.6.4 Sidewalk labs' Toronto Quayside: A failed smart city experiment

In 2017, Sidewalk Labs—the urban innovation arm of Alphabet Inc., Google's parent company—unveiled an ambitious $1.3 billion proposal to transform a portion of Toronto's waterfront into Quayside: a hyper-connected, data-driven "city of the future" (Sidewalk Labs, 2020). Designed to showcase cutting-edge sustainability features like carbon-neutral modular buildings, AI-optimized public spaces, and self-driving electric shuttles, Quayside promised a new model for urban living. Central to this vision was a suite of integrated technologies: thermal grids that recycled server and industrial waste heat, adaptive buildings embedded with responsive sensors, underground robotic freight systems, and ubiquitous environmental monitoring (Sidewalk Labs, 2020).

However, from the outset, the project provoked skepticism and backlash. Sidewalk Labs' framing of the area as a "blank canvas" for experimentation alarmed residents, who saw the rhetoric as emblematic of corporate overreach. In a city already wrestling with challenges around housing affordability and digital privacy, the idea of a for-profit corporation piloting mass data collection without clearly defined public safeguards sparked controversy. Critics raised deep concerns about whether the interests of the public would truly take precedence over commercial imperatives (Sidewalk Labs, 2020).

The most contentious element of the plan was its approach to data governance. Sidewalk Labs proposed harvesting vast quantities of urban data—including pedestrian movement, transit usage, and energy consumption—in the name of service optimization. Yet, the language surrounding data ownership, consent, and monetization remained opaque. Civil society groups, privacy scholars, and Indigenous advocates soon dubbed the project a "surveillance panopticon," warning that it would normalize corporate surveillance in public space (Sidewalk Labs, 2020).

By 2020, facing mounting public pressure, Sidewalk Labs withdrew from the project. Two years of legal battles, community protests (organized in part by groups like #BlockSidewalk), and mounting political scrutiny had eroded the project's legitimacy. Leaked internal documents revealing plans to monetize urban data further inflamed public outrage, contradicting prior assurances of privacy protections. The collapse of Quayside became a flashpoint in global debates over ethical tech governance, forcing cities around the world to reevaluate the role of corporations in shaping urban futures (Sidewalk Labs, 2020; Toronto City Council, 2022).

Three core failures defined the project's demise. First was a lack of transparency: Sidewalk Labs withheld critical information regarding data monetization strategies until it was exposed in media leaks. Second, the community was systematically excluded from early planning—residents, activists, and Indigenous groups reported being ignored during consultations, prompting accusations that the development was being designed "for tech, not for people." Third, Canada's weak regulatory landscape allowed Sidewalk Labs to draft project terms that disproportionately favored corporate interests over public protections (Sidewalk Labs, 2020).

The Quayside case demonstrates how even the most technologically sophisticated projects can unravel when community consent, trust, and governance are overlooked. It illustrates the danger of regulatory gaps (Chu et al., 2023), and how vague legal frameworks surrounding data and corporate accountability can be exploited. Additionally, it underscores the risks of privacy intrusion in data-centric initiatives, echoing concerns outlined in discussions of digital surveillance and governance (Chu et al., 2023). In response, Toronto adopted a new Digital Infrastructure Plan in 2022, which mandates public oversight

Case Study	Technology Used	Outcome	Lesson Learned
Copenhagen	IoT, AI, smart grid	70% CO_2 reduction	Phased implementation works
Rwanda (Zipline)	Drones, blockchain	51% fewer maternal deaths	Community trust is critical
Ghana (Agbogbloshie)	Blockchain, solar hubs	15% fewer toxic emissions	Regulation must combat corruption
Toronto Quayside	AI, sensors	Project failed due to privacy backlash	Public consent is non-negotiable

Figure 15.2 Case study comparison.

and accountability mechanisms in future urban data projects (Toronto City Council, 2022).

The Quayside experience affirms insights from Chu et al. (2023), who argue that sustainable smart cities must integrate technological innovation with robust ethical governance. As their research stresses, "environmental wellbeing cannot thrive without trust and equity"—a lesson Quayside failed to heed. By prioritizing corporate experimentation over inclusive governance, the project alienated the very communities it claimed to benefit (Chu et al., 2023).

Ironically, while the project failed as an urban innovation venture, it succeeded in catalyzing civic empowerment. Quayside became a rallying point for digital rights activism and pushed Toronto to rethink its urban tech policy frameworks. It left behind not a model for the future of smart cities, but a powerful reminder: the path to sustainable, livable cities lies not in the boardrooms of Silicon Valley, but in the hands of informed, engaged communities demanding transparency, equity, and agency in shaping their digital and physical environments.

15.7 COMPARATIVE SWOT ANALYSIS AND STRATEGIC RECOMMENDATIONS

The insights from the case studies and SWOT analysis are combined in this section to provide practical methods for incorporating smart technologies into sustainable development. Stakeholders can negotiate the challenges of technology adoption while adhering to international sustainability goals by weighing their advantages and disadvantages, taking advantage of opportunities to reduce risks, and implementing a systematic framework.

15.7.1 Balancing strengths and weaknesses

Maximizing the benefits of smart technologies while mitigating their inherent risks requires a nuanced, integrative approach. Rather than viewing strengths and weaknesses in isolation, effective smart city planning involves leveraging advantages to directly counterbalance specific vulnerabilities.

15.7.1.1 High costs vs. long-term savings

Although the high upfront cost of implementing smart technologies remains a major barrier (Weakness, Section 15.3.1), these investments often yield significant operational efficiencies over time (Strength, Section 15.2.2). For instance, Siemens' IoT-enabled predictive maintenance system in manufacturing environments reduced unplanned downtime by 25%, justifying the initial expense through long-term productivity gains and cost avoidance (Siemens, 2022).

15.7.1.2 Data insights vs. privacy risks

Smart technologies offer powerful data-driven decision-making capabilities (Strength, Section 15.2.4), but this can clash with growing concerns over surveillance and privacy (Weakness, Section 15.3.2). However, decentralized architectures—such as blockchain—provide a solution. Rwanda's drone-based medical supply chain, for example, used blockchain to ensure secure tracking of deliveries while safeguarding sensitive health data, striking a balance between efficiency and ethical data use (IBM, 2022; Zipline, 2023).

15.7.1.3 Accessibility vs. digital divides

While smart technologies have the potential to foster inclusive development (Strength, Section 15.2.3), they can also deepen digital divides if not deployed with equity in mind (Weakness, Section 15.3.3). Bridging this gap requires pairing innovation with inclusive design. Kenya's Eneza Education demonstrates this principle: its SMS-based AI learning platform extends personalized tutoring to underserved students using basic mobile phones, showing that high-impact results can come from low-tech delivery models (Eneza Education, 2022).

15.7.1.4 Strategic recommendation

Cities should adopt phased implementation models, where complex infrastructure is rolled out incrementally. Priority should be given to low-cost, high-yield interventions—such as adding IoT sensors to existing public infrastructure—to deliver immediate benefits without overwhelming budgets. In parallel, public-private partnerships (PPPs) and targeted government subsidies can democratize access and foster long-term sustainability. Ensuring transparency, stakeholder engagement, and digital inclusion throughout this process will be key to building equitable and resilient urban futures.

15.7.2 Leveraging opportunities to counter threats

In the evolving landscape of smart technologies, risks and rewards often coexist. However, with proactive governance and adaptive innovation, the inherent opportunities within these systems can be strategically leveraged to mitigate their associated threats. Below are key opportunity–threat pairings that illustrate how thoughtful integration can convert vulnerabilities into resilience.

15.7.2.1 Circular economy models vs. e-waste hazards

One of the most pressing threats posed by technological innovation is the exponential growth of electronic waste (Threat, Section 15.5.4). Yet, circular economy initiatives (Opportunity, Section 15.4.1) offer a counterbalance. By promoting resource efficiency, product longevity, and systemic recycling, they reduce both the volume and toxicity of waste. Power Ledger's peer-to-peer energy trading platform exemplifies how decentralized systems can promote sustainable consumption (Power Ledger, 2023). Similarly, the solar-powered recycling hubs established in Agbogbloshie, Ghana demonstrate how renewable infrastructure combined with community training can transform informal e-waste economies into safer, structured ecosystems (IBM, 2022).

15.7.2.2 Cross-border cooperation vs. regulatory fragmentation

The rapid pace of smart technology deployment has outpaced national regulatory frameworks, leading to regulatory fragmentation (Threat, Section 15.5.1). However, global partnerships and policy harmonization (Opportunity, Section 15.4.2) offer a solution. Initiatives like the UN's Global Pulse promote cross-border collaboration in data governance and responsible innovation (UN Global Pulse, 2022). In practice, IBM's blockchain-based Food Trust platform exemplifies how decentralized technologies can uphold shared international standards for transparency and traceability across supply chains (IBM, 2022).

15.7.2.3 Urban innovation vs. cybersecurity vulnerabilities

As cities adopt AI-driven infrastructure and interconnected IoT systems, they face heightened exposure to cybersecurity threats (Threat, Section 15.5.2). To address this, urban innovation frameworks (Opportunity, Section 15.4.3) must incorporate robust cybersecurity protocols by design. Copenhagen's smart grid, managed through AI-powered systems, has proactively adapted to these risks by implementing a multi-layered cybersecurity upgrade (Copenhagen

City Council, 2023). Likewise, Singapore's Smart Nation initiative allocates 15% of its annual tech budget to cybersecurity, embedding a zero-trust architecture within its digital infrastructure (Singapore Government, 2023).

15.7.3 Framework for sustainable technology adoption

To translate opportunity–threat pairings into actionable outcomes, stakeholders must adopt a structured framework rooted in resilience, scalability, and equity. Such a framework ensures that smart technology implementation not only advances innovation but also strengthens social and environmental systems in the long term. The following five-step model provides a roadmap for responsible, inclusive, and adaptive technology integration.

15.7.3.1 Step 1: Stakeholder mapping and needs assessment

Begin by identifying key stakeholders—local communities, NGOs, private sector actors, and governments—and aligning interventions with clearly defined sustainability goals, such as the UN SDGs. Contextual understanding is critical, especially in underserved areas where needs may differ from urban centers.

For example, in Rwanda, Zipline's drone-based medical delivery network was co-developed with input from rural health clinics. This participatory process revealed that maternal mortality was a critical issue, leading the government and Zipline to prioritize maternal health kits and emergency blood supplies (Zipline, 2023). This ensured the technology met urgent, localized needs while contributing to SDG 3 (Good Health and Well-being).

15.7.3.2 Step 2: Co-designing technology and policy

Technological solutions should be co-developed alongside regulatory frameworks to ensure that governance keeps pace with innovation. This includes embedding data sovereignty, environmental safeguards, and human rights protections into the technology design process.

Toronto's post-Quayside experience exemplifies this co-design principle. In response to public backlash over data ownership during the Sidewalk Labs project, the city introduced the Digital Infrastructure Plan (2022), which mandates public oversight and clear governance mechanisms for all urban data projects (Toronto City Council, 2022). This policy ensures that smart systems are not only efficient but also democratic, ethical, and publicly accountable.

15.7.3.3 Step 3: Iteration and pilot testing

Before scaling, conduct small-scale pilot programs that allow for controlled testing, community feedback, and iterative improvement. Pilots minimize risks and surface unintended consequences early in the deployment cycle.

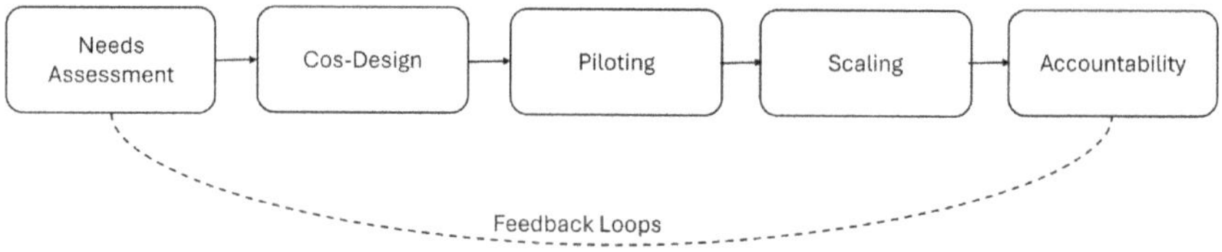

Figure 15.3 Framework for sustainable technology adoption.

Tesla's VPP in South Australia followed this model. Initially launched with 1,000 homes equipped with solar panels and batteries, the program used real-time feedback and data analytics to optimize energy storage and grid responsiveness. The pilot's success led to its scale-up to 50,000 homes by 2023, showcasing how localized testing can unlock national-level energy transformation (Tesla, 2023).

15.7.3.4 Step 4: Scaling with equity

Scaling should be inclusive, ensuring that benefits are distributed equitably across different regions, income groups, and digital literacy levels. This includes integrating accessibility features, offering training and capacity-building programs, and addressing affordability barriers through subsidies or PPPs (Figure 15.3).

Kenya's Eneza Education, for instance, scaled its SMS-based learning platform nationally by prioritizing marginalized, low-connectivity communities. By using basic mobile phones and localized content, it expanded access to education while overcoming infrastructure and literacy challenges (Eneza Education, 2022).

15.7.3.5 Step 5: Monitoring and accountability

Robust, technology-driven monitoring mechanisms are essential to evaluate impact and ensure long-term accountability. Blockchain and AI-based auditing tools can provide transparent, tamper-proof systems to monitor outcomes, flag inefficiencies, and expose regulatory violations.

IBM's blockchain pilot in Agbogbloshie, Ghana, serves as a compelling example. By attaching digital identifiers to incoming electronic waste shipments, the system traced illegal e-waste flows and reduced toxic dumping by 15% within one year (IBM, 2022). However, the initiative also revealed gaps in enforcement, showing that blockchain must be complemented by strong institutional oversight to be truly effective.

15.7.4 Strategic policy recommendations

As smart technologies reshape urban, environmental, and social systems, their impact hinges on how responsibly they are governed. The SWOT analysis reveals that smart technologies are not inherently beneficial or harmful; their outcomes are shaped by ethical oversight, local adaptation, and inclusive participation. The following policy recommendations are designed to institutionalize the sustainable adoption framework (Section 15.7.3), transform identified threats into actionable opportunities (Section 15.7.2), and operationalize the balance between technological strengths and systemic vulnerabilities (Section 15.7.1).

15.7.4.1 Mandate extended producer responsibility (EPR)

Governments should enforce Extended Producer Responsibility (EPR) policies, requiring technology producers to be accountable for the full lifecycle of their products, including take-back, recycling, and safe disposal mechanisms. For example, mandating companies to invest at least 5% of their annual profit into certified recycling programs could significantly reduce the volume of informal e-waste, particularly in regions like Agbogbloshie, Ghana, where informal waste burning persists (IBM, 2022). Such mandates would align with Basel Convention principles (Basel Convention Secretariat, 2023) and support circular economy transitions as promoted by the Ellen MacArthur Foundation (Ellen MacArthur Foundation, 2022).

15.7.4.2 Create and enforce ethical AI charters

Building on frameworks such as the European Union's AI Act (European Commission, 2021), governments and international bodies should establish context-sensitive ethical AI guidelines. These charters must address algorithmic bias, ensure transparency in automated decisions, and prioritize fairness—especially in sectors like healthcare, education, and law enforcement. Inclusion of local stakeholders, such as Indigenous communities or underrepresented groups, in AI governance processes ensures that algorithms are not only technically sound but also socially just. This reflects the call by Chu et al. (2023) for aligning smart systems with justice and sustainability, not just innovation.

15.7.4.3 Fund green technology incubators

Public and private actors should allocate grant-based funding to support startups focused on low-cost, inclusive, and eco-friendly technologies. This includes solar-powered IoT devices, sustainable agriculture platforms, and decentralized renewable energy systems. For example, Rwanda's success with drone-based medical delivery (Zipline, 2023) and Eneza Education's mobile-based tutoring platform (Eneza Education, 2022) demonstrate how locally relevant, resource-efficient innovation can scale with the right early-stage investment. Incubators should also provide technical mentorship and facilitate access to responsible supply chains.

15.7.4.4 Launch National and Regional Digital Equity Funds

Addressing the digital divide requires targeted financial mechanisms to make the digital economy more inclusive. Digital Equity Funds can provide: subsidies for affordable internet and devices in rural or underserved areas, training programs to enhance digital literacy and cybersecurity awareness, infrastructure investments in low-connectivity zones (e.g., satellite internet, offline-first applications).

Kenya's use of SMS-based education platforms (Eneza Education, 2022) and India's e-Sanjeevani telemedicine initiative (Ministry of Health, India, 2023) show how digital access can improve lives—only if affordability and literacy are addressed in tandem.

Smart technologies are neither panacea nor peril, as the SWOT analysis demonstrates. Their long-term value depends on how they are governed, contextualized, and co-developed (Chu et al., 2023). By following the proposed framework, balancing inherent strengths and weaknesses, and strategically converting threats into opportunities, stakeholders can deploy smart technologies as catalysts for just and sustainable development (Figure 15.4).

According to the SWOT analysis, smart technologies are neither solutions nor dangers; rather, their effectiveness depends on ethical governance and contextual integration. Stakeholders can use these tools to promote sustainable development fairly by adopting the suggested framework, balancing strengths and weaknesses, and turning threats into opportunities. The convergence of technology, policy, and community engagement is crucial and not optional, as Chu et al. (2023) stress. In order to move forward, inclusive innovation is required, where smart technologies act as facilitators rather than obstacles to a sustainable future.

Issue	Recommendation	Example
High costs	Subsidies/PPPs for low-income regions	Kenya's Eneza Education
Privacy risks	Decentralized data frameworks (e.g., blockchain)	IBM Food Trust
E-waste	Extended Producer Responsibility (EPR) laws	Ghana's recycling hubs

Figure 15.4 Policy recommendations.

15.8 CONCLUSION

Smart technologies hold transformative potential for sustainable development, but their real-world impact depends not only on innovation but also on governance, equity, and systemic foresight. This chapter has presented an in-depth evaluation of how digital tools such as AI, IoT, blockchain, and autonomous systems intersect with global sustainability goals through a comprehensive SWOT analysis and a series of global case studies.

The key insight is that these technologies function as enablers, not guarantees, of sustainable futures. Their strengths lie in data-driven decision-making, enhanced operational efficiency, and service democratization. Rwanda's drone-based medical logistics network, for instance, illustrates how AI-powered systems combined with blockchain tracking can bridge critical health access gaps in remote regions, reducing maternal mortality by over 50% while cutting delivery costs by 65% (Eneza Education, 2022; Zipline, 2023). Similarly, Copenhagen's AI-integrated energy grid demonstrates how IoT-driven optimization can help a major city approach carbon neutrality by utilizing wind energy and smart heating systems to cut emissions by 70,000 metric tons annually (Chu et al., 2023; Copenhagen City Council, 2023).

At the same time, the weaknesses of smart technology are evident. High upfront costs and complex infrastructure demand often sideline low-income communities, while digital divides and insufficient data protection frameworks risk reinforcing existing inequities. Ghana's Agbogbloshie e-waste tragedy underscores the human cost of technological obsolescence and lax international accountability, as thousands of informal recyclers face exposure to heavy metals while corporations in the Global North avoid EPR (Basel Convention Secretariat, 2023; IBM, 2022).

Despite these challenges, the opportunities are considerable. Smart grids, decentralized renewable systems, and data-driven urban planning offer powerful tools to scale sustainability efforts. Tesla's VPP in South Australia, starting with just 1,000 homes, now supports 50,000 households with AI-managed solar energy, exemplifying scalable innovation (Tesla, 2023). Similarly, UN-led initiatives such as Global Pulse promote cross-border cooperation in data ethics and digital development (UN Global Pulse, 2022).

Yet threats loom large: from the environmental toll of device production to cybersecurity vulnerabilities and regulatory fragmentation. The failure of Sidewalk Labs' Quayside project in Toronto reveals how public resistance can derail ambitious smart city plans when data governance and community trust are sidelined. Despite promising technologies like thermal grids and modular sensor-embedded housing, public backlash over surveillance concerns and lack of transparency ultimately led to the project's cancellation in 2020 (Sidewalk Labs, 2020; Toronto City Council, 2022).

Across these examples, one consistent theme emerges: technology is only as ethical, inclusive, and sustainable as the systems that govern it. To this end, future research and practice must converge around a few core priorities.

First, research agendas should prioritize equity-centered design, circular technology models, ethical AI governance, and climate-resilient infrastructure. These focus areas are crucial for adapting smart systems to the needs of diverse communities, particularly those historically excluded from the benefits of technological progress (Chu et al., 2023; Ellen MacArthur Foundation, 2022; FAO, 2022).

Second, policy frameworks must be reimagined. EPR mandates should hold tech producers accountable for post-consumer waste, echoing principles from the Basel Convention (Basel Convention Secretariat, 2023) and aligning with blockchain-based solutions already piloted in Ghana by IBM (IBM, 2022). Regulatory sandboxes, such as those created by Toronto's Digital Infrastructure Plan after the Quayside fallout, offer a flexible means to test innovation within public-interest boundaries (Toronto City Council, 2022).

Implementation strategies must also shift toward community-led approaches. Zipline's inclusion of over 500 local health workers in Rwanda as drone ambassadors illustrates how trust, usability, and cultural adaptation can amplify impact (Zipline, 2023). Similarly, digital literacy programs—such as those supported by Eneza Education's SMS-based learning in Kenya—equip users to actively shape rather than passively receive technological change (Eneza Education, 2022).

Finally, international collaboration must strengthen open-source ecosystems and South-South cooperation. Sharing blueprints for AI governance, blockchain-enabled traceability, and circular economy models across similar socioeconomic contexts can accelerate localized innovation. Copenhagen's smart energy infrastructure, for example, offers replicable lessons for climate-focused cities globally (Copenhagen City Council, 2023; IEA, 2022).

Ultimately, smart technology's role in sustainable development must be interdisciplinary and participatory. Engineers, urban planners, ethicists, policymakers, and community leaders must co-design futures that emphasize not only efficiency but also justice. As Chu et al. (2023) argue, environmental well-being cannot be achieved without equity embedded in both code and policy. Whether it's Singapore investing 15% of its tech budget in cybersecurity (Singapore Government, 2023), or Tesla incrementally scaling clean energy access (Tesla, 2023), sustainable development requires balancing pragmatism with ambition, and innovation with accountability.

The path forward demands more than digital tools—it requires feedback-driven governance, inclusive frameworks, and ethical foresight. Only by embedding these principles can smart technologies fulfill their promise as facilitators—not barriers—to a truly sustainable and just global future.

REFERENCES

Amnesty International. *The Cost of Cobalt: Environmental and Human Rights Impacts in the DRC*. London: Amnesty International, 2022.

Barcelona City Council. *Smart City Barcelona: Annual Sustainability Report*. Barcelona: Barcelona City Council, 2022.

Basel Convention Secretariat. *Global E-Waste Monitor 2023*. Geneva: United Nations Environment Programme, 2023.
Chen, Angela. "China's Data Localization Policies: Implications for Global Tech Governance," *Journal of International Affairs* 74, no. 1 (2021): 45–62.
Chu, X. H., D. Ge, S. Alamri, and D. Toghraie. "Greening Smart Cities: An Investigation of the Integration of Urban Natural Resources and Smart City Technologies for Promoting Environmental Sustainability," *Sustainable Cities and Society* 99 (2023): 104985.
Copenhagen City Council. *Copenhagen Climate Plan 2025: Roadmap to Carbon Neutrality*. Copenhagen: City of Copenhagen, 2023.
Cybersecurity and Infrastructure Security Agency (CISA). *Colonial Pipeline Ransomware Attack: Lessons for Critical Infrastructure*. Washington, DC: CISA, 2021.
Easterly, Jen, and Tom Fanning. *The Attack on Colonial Pipeline: What We've Learned & What We've Done Over the Past Two Years*. Washington, DC: CISA, 2023.
Ellen MacArthur Foundation. *Circular Economy Principles for Technology*. London: Ellen MacArthur Foundation, 2022.
Eneza Education. *Impact Report: Scaling Literacy through SMS-Based Learning*. Nairobi: Eneza Education, 2022.
European Commission. *Proposal for a Regulation on Artificial Intelligence (AI Act)*. Brussels: European Union, 2021.
European Union Agency for Cybersecurity (ENISA). *Cybersecurity Threats in Smart Grids*. Athens: ENISA, 2022.
Food and Agriculture Organization (FAO). *Leveraging Automation in Agriculture for Transforming Agrifood Systems*. Rome: FAO, 2022.
Global Forest Watch. *Annual Report: Real-Time Deforestation Monitoring*. Washington, DC: World Resources Institute, 2023.
Goodman, Ellen P., and Julia Powles. "Urbanism under Google: Lessons from Sidewalk Toronto," *Fordham Law Review* 88 (2019): 457–498.
IBM Research. *Blockchain for Sustainable Supply Chains: Case Study on E-Waste Management in Ghana*. Armonk, NY: IBM, 2022.
Interface Inc. *Interface Impact Report 2023*. Atlanta: Interface Inc., 2023.
International Energy Agency (IEA). *Smart Cities and Global CO_2 Reduction*. Paris: IEA, 2022.
International Labour Organization (ILO). *World Employment and Social Outlook Trends*. Geneva: ILO, 2023.
International Telecommunication Union (ITU). *Digital Divide Report 2023*. Geneva: ITU, 2023.
Khan Academy. *Personalized Learning Impact Report*. Mountain View, CA: Khan Academy, 2022.
Kim, J. H. "South Korea Telemedicine Data Breach: Lessons Learned," *Cybersecurity Journal* 12, no. 3 (2023): 89–102.
Ministry of Health, India. *e-Sanjeevani: Transforming Rural Healthcare*. New Delhi: Government of India, 2023.
Nest Labs. *Smart Thermostats: Energy Efficiency Impact*. Palo Alto, CA: Nest Labs, 2021.
O'Neil, Cathy. *Weapons of Math Destruction: How Big Data Increases Inequality and Threatens Democracy*. New York: Crown Publishers, 2016.
Power Ledger. *Global Blockchain Business Council: Annual Report*. Perth: Power Ledger, 2023.
Pradhan, R. "Telemedicine in India: Bridging Rural Healthcare Gaps," *Indian Journal of Public Health* 67, no. 2 (2023): 123–130.
Schneider Electric. *Smart Buildings: Energy Optimization Report*. Rueil-Malmaison: Schneider Electric, 2022.

Sidewalk Labs. *Lessons from Toronto Quayside: A Retrospective on Ethical Tech Governance*. Toronto: Alphabet Inc., 2020.
Siemens AG. *Predictive Maintenance with IoT: Case Studies*. Munich: Siemens, 2022.
Singapore Government. *Smart Nation Initiative: Annual Progress Report*. Singapore: Smart Nation and Digital Government Office, 2023.
Tesla Inc. *Virtual Power Plant: Scaling Renewable Energy*. Palo Alto, CA: Tesla, 2023.
Thapar, R., F. Alturise, A. V. Kumar, and O. S. Saleh, eds. *Handbook on Integrating Smart Technologies for Sustainable Development*. Boca Raton, FL: CRC Press, 2024.
Too Good To Go. *Reducing Food Waste: Impact Report*. Copenhagen: Too Good To Go, 2023.
Toronto City Council. *Digital Infrastructure Plan 2022*. Toronto: City of Toronto, 2022.
UNEP. *Text of the Basel Convention*. Geneva: United Nations Environment Programme, 2023.
UNESCO. *Digital Skills Gap in Rural Areas*. Paris: UNESCO, 2022.
UN Global Pulse. *Big Data for the SDGs: A Primer*. New York: United Nations, 2022.
United Nations. *Transforming Our World: The 2030 Agenda for Sustainable Development*. New York: United Nations, 2015.
Vodafone Group. *M-Pesa: Ensuring Inclusion for All through Improving Access to Connectivity*. Nairobi: Vodafone Group, 2023.
World Food Programme (WFP). *Building Blocks: Blockchain for Humanitarian Aid*. Rome: WFP, 2023.
World Bank. *Digital Development: Barriers to Technology Adoption*. Washington, DC: World Bank, 2021.
World Economic Forum. *The Future of Jobs Report 2020*. Geneva: WEF, 2020.
Zipline International Inc. *Annual Impact Report: Delivering Health Equity Through Autonomous Drones*. San Francisco: Zipline, 2023.

Chapter 16

Ensuring the cybersecurity of digital content in e-learning platforms

Aashmi V., Pushpitha Reddy Patur, and Janardhana Kumar Reddy P.

16.1 INTRODUCTION

In today's world, the digital security of e-learning platforms is of utmost importance. E-learning has now become as simple as accessing content through e-books, attending virtual lessons, and taking examinations online, which forms the core of modern education. The process of digitization has to deal with something of utmost importance, cybersecurity, which cannot be neglected. In regard to e-learning platforms that gather information pertaining to streams of users and their activities associated with the available digital information, this chapter attempts to describe them. Appropriate attention and practice can greatly mitigate cyber risks, and these include prudent policies, powerful security measures, proactive strategies, and encouraging responsible user habits. These practices help eliminate incidents due to carelessness or human negligence. The virtualization of education continues to gain momentum, which increases the importance of these e-learning tools. As such, these platforms collect and disseminate sensitive personal information and confidential intellectual documents. Cyber threats like data mishandling, phishing, computer viruses, and unauthorized data access threaten data integrity. Unauthorized network attacks usually compromise sensitive personal information and academic information, including proprietary intellectual property like copyrighted materials. Ensuring cybersecurity in e-learning is not just about protecting content; it also preserves trust in the digital education ecosystem. It involves measures to safeguard data and prevent plagiarism and content theft, thereby ensuring uninterrupted learning experiences. The purpose of this chapter is to provide an overview of the importance of cybersecurity in the digital content of e-learning platforms.

16.2 CYBERSECURITY

Cybersecurity is the process of defending activities, networks, and resources from online threats (CISA 2021)). These cyberattacks often target sensitive data to access, alert, or delete the sources. It aims to minimize the threat of

DOI: 10.1201/9781003586951-16

digital cyberattacks and protect against the forbidden manipulation of mechanisms, connections, and technology. It refers to endless space known as the internet (Cisco 2025).

16.3 TYPES OF CYBERSECURITY

Cybersecurity types are simply methods for preventing personal information from being lost or compromised. Every day, viruses and dangerous programs that infect computers and different devices such as mobile devices pose a threat to data (College Vidya 2023a) These gadgets all make using internet services easier. The purpose of these internet services is to greatly simplify the lives of end-users.

There are some common types of cybersecurity as follows.

16.3.1 Application security

There are various applications in computers and digital devices that require the personal or confidential data of the user (CISA 2021). For example, social media applications require the bank details for online money transactions of the user. Such information is highly sensitive and can cause damage to the user if it gets into the hands of malicious cybercriminals. The safety of the application is consequently designed to prevent hacking of the apps running on the device or in the cloud (College Vidya 2023c). During the application design process, the security of the application is constructed.

16.3.2 Network security

By using this technique, devices joining wired and wireless networks are secured. Network security is more than just software; it is a strategic combination of hardware and software that is intended to protect confidential data that is kept in the company's data center. Examples of hardware in network security involve the servers, devices, and additional devices that carry out security-related tasks (Pedamkar 2025). Because hackers may access essential information like financial statements, login credentials, passwords, and more, an organizational network security breach has the potential to wipe out the organization (College Vidya 2023b). Hackers usually try to get access to an institution's network architecture because it includes the kind of information they are searching for. Network security is, therefore, an essential component of cybersecurity.

16.3.3 Endpoint security

To create and reinforce endpoint security across the enterprise, like not utilizing suspicious emails, device protection from threats, running breach and attack simulations, etc. (Patel 2022).

16.3.4 Information security

The General Data Protection Regulation (GDPR), aids in preventing cyber theft of personal data.

16.3.5 Cloud security

It secures cloud data that is being stored and moved between systems. It contributes to a secure institutional environment, a satisfying customer experience, and standards of compliance (Pedamkar 2025). Additionally, conducting a confidential threat investigation can guarantee that companies promptly identify and address any questionable internal activity, safeguarding their information and upholding confidence.

16.3.6 Internet of Things security

Security services in Internet of Things (IoT) are developed to keep in mind the privacy of users by framing policies and security systems that secure the network, as hacking of these devices by cyber criminals is possible due to the integration of software in these devices, thus increasing the chance of data being accessed by anyone in real time. For example, Alexa, which is capable of connecting to various devices like our mobile phone, television, refrigerator, etc. to control them using the internet as a medium of connection. In IoT, number of devices can be connected to enable the sharing and exchange of data (Pedamkar 2025). Devices that are connected to Alexa make it possible for Alexa to access data, following the privacy policy rules. It is in charge of creating strong privacy guidelines and network security measures to guard against data leaks and network intrusions. Since data may be accessible in real time due to embedded programs and detectors, there can be serious consequences if even one device in a network is compromised (Li 2021).

16.3.7 Mobile security

It encrypts mobility communications and requires secure apps to control the mobile workforce.

16.4 E-LEARNING

E-learning or electronic learning is a flexible, anytime instruction delivered via the internet or corporate intranet to learners. It allows for a more organized learning experience, regardless of physical location. E-learning is a popular approach to education that ultimately relies on the internet to function. The pathways and technology of the internet generation are best exemplified by e-learning platforms (Hashemi-Pour and Lutkevich 2024). These intricate mechanisms are designed to ensure learners' satisfaction and preserve a positive perception of the educational process. However, they have evolved to enable multidirectional communication using interactive tools, allowing learners to choose how they receive and respond to content and allowing multiple peers to participate. There is ample proof that cutting-edge educational technologies, including e-learning, provide trainees, students, and teachers previously unheard of chances to study, grow, and retain fundamental information and abilities (Harve 2024).

16.5 E-LEARNING PLATFORMS

There are two types of e-learning platforms: synchronous and asynchronous. These are utilized in education, business organizations, and autonomous settings. By offering courses, online tests, PowerPoint presentations, and the ability to analyze student performance data, these robust software packages facilitate e-learning and digital learning. These platforms include, for instance, Google Classroom, Schoology, Moodle, Sakai, Canvas, Teachmint, and Anthology for business (Scerbakov et al. 2019). Nowadays, educational institutions widely use Learning Management Systems (LMS) to enable learners to interact at their own pace, post materials, Digital Content, assignments, online quizzes, and student assessments, and track their progress to measure achievement. LMS is the most effective e-learning platform that can be easily adaptable, customizable, ambitious, and user-friendly to learners and educators (Brown 2021). There are some different paid LMS platforms like Adobe Learning Manager, Docebo, eFront, iSpring Learn, Looop by 360Learning, Canvas, Northpass, and TalentLMS.

16.5.1 Social media

The term "social media" describes websites and programs that emphasize interaction, communication, sharing of information, community-based feedback, and teamwork. Social media is a digital communication technology that lets people create online communities where they may engage with one another, share information, and contribute user-generated material (Arshavskiy 2021).

16.5.2 Importance of social media in e-learning

Nowadays, practically every learner has a digital presence. These learners have a thorough understanding of social media platforms and are accustomed to using them to obtain information. Thus, learners may benefit from adopting social media and utilizing its advantages in the teaching and learning process (Arshavskiy 2021). Educators can tag learners, assign quizzes, and have them take part in group discussions in addition to offering video lectures. In this manner, learners will not only acquire information engagingly and fascinatingly but they will also have the freedom to learn at any time and from any location. Social media efficiently addresses all of these issues by enabling learners to establish positive relationships with individuals as well as build a solid support system.

16.5.3 Social media as an e-learning platform

Social media is the most attractive e-learning platform that everyone can access and share digital content in a user-friendly way. These platforms can be shared easily based on their interest and personalized to make learning more relatable and interactive (Barbu et al. 2017). The e-learning nature of platforms allows digital content sharing – audio, visual, text, infographic, etc.

- **WhatsApp**
  Nowadays, educational institutions connect learners by using community groups for sharing content easily. Within the well-known WhatsApp app interface, WhatsApp e-learning lets educators deliver course material, updates, and recalls while letting students ask question, work with fellow learners, and get quick responses. WhatsApp is cost-effective and easily accessible, and offers flexibility to communicate.
- **Facebook**
  Facebook users share information and ideas by creating groups to communicate freely about shared content. On Facebook, one can set up an open or a closed Facebook group for any educational institutions, and learners can share quizzes, assignments, and other course materials with the group members. Using interesting questions and articles, including images, educators can motivate the learners to join in group conversations.
- **LinkedIn**
  LinkedIn is an entirely business-oriented social media e-learning platform. People often use LinkedIn for professional and career-related task rather than posting photos and engaging in educational interactions, which makes it an excellent option for e-learning. LinkedIn allows similar groups that may be seen as having more credibility because users show their professional credentials on their profiles. Another feature on LinkedIn is groups. These organizations give experts in the same field a place to meet, exchange ideas and experiences, look for advice, and from deep ties. Make an e-learning group just for the course, and then share articles, videos, and other resources with learners on a regular basis. LinkedIn also offers a paid e-learning platform that provides more than 4,000 skill development courses, covering topics such as digital advertising and website development. The organization's experts require a monthly fee to access these courses and stay current in their field.

- **Pinterest**
  Pinterest is an entirely picture-focused social network. This platform is unusual in that it uses a "board" to arrange the material. Users may make boards with a topic and "pin" essential photos to them. These boards are simple to share with other people. It is an engaging infographic and a vision board for learners. Pinterest would be able to keep all of the content associated with a particular subject in one place, but also offer useful data in a manner that looks good.
- **YouTube**
  YouTube users access and post educational content for free of cost by subscribing, as well as commenting and rating the videos. For the majority of online instructors, YouTube is their preferred social media as an e-learning platform. This is because it enables them to produce educational videos of excellent quality and distribute them to learners in a timely and convenient manner. YouTube should consequently be the top option when one wishes to share a lengthy course, enjoying sharing additional video content. The fact that YouTube lets individuals submit videos practically anywhere is a fantastic feature. It's possible to tweet them, post them on Instagram or Facebook, or pin them. The learners are capable of doing the same. Consequently, one might bring it to the spotlight. After that, they may upload their videos to the account created with YouTube. This enables them to improve their learning skills.
- **Instagram**
  In this modern era, Instagram is a more powerful tool for e-learning, it offers interactive content, learner interaction, and a visual platform for informal learning. An Instagram page can be made for the trade that teaches, such as painting, video editing, filmmaking, graphics design, etc. Through reels, stories, and posting, it's possible to showcase the learner's development there. It is an excellent method to highlight their abilities and inspire them to achieve even more. The Instagram feed is used to post key information and announcements about the course.

16.6 IMPORTANCE OF PROTECTING DIGITAL EDUCATIONAL CONTENT

Protecting digital content is important for ensuring the integrity, accessibility, and trustworthiness of modern education systems. As digital learning resources become central to teaching and learning, safeguarding them from unauthorized access, piracy, and manipulation is crucial to maintain the quality and reliability of educational offerings (Nedelchev and Slavov 2023). The safeguard of digital content assists in the mitigation of misinformation, securing intellectual property rights, and guaranteeing that learners and educators equitably access refreshing and accurate materials. Nevertheless, the learner's security and confidentiality details remain a significant concern brought about by this transition to the digital world, so educational institutions must navigate the complex terrain of data privacy and security while adopting digital and online learning frameworks (Deetya Education 2023). Protecting private information ensures compliance with data privacy policies while creating a safe environment for learning, which helps build trust among the students, their guardians, and educators. Apart from that, the safe protection of educational resources fuels the innovation to create sharable quality works, thereby promoting the adoption of teaching aids for learning. The protecting of digital contents sustains the effectiveness, accessibility the equity, and the sustainable development of digital education which makes it the primary concern of educational institutions.

16.7 COMMON CYBER THREATS TO E-LEARNING PLATFORMS

User data can be compromised, as educational processes can be disrupted and trust can be damaged in digital learning environment due to one of the many cyber threats e-learning platforms face. Software vulnerabilities within LMS pose the greatest threat because software is often poorly designed and serves a high volume of users (Djeki et al. 2024). Insecure authentication, improper access control, and communication module (forum/chat) communication flaws are entry points wherein attackers may be able to exploit systems. A related but separate striking threat is malware and phishing attacks, and in an educational context, claiming e-learning user credentials through phishing emails or links is a major concern. Cyberwarfare attacks also proliferate as e-learning platforms rapidly adopt cloud and IoT technologies, where static security controls like firewalls or insecure Wi-Fi pose greater risk. Static security approaches, i.e, traditional security tools, are outclassed by the speed at which these technologies are adopted, creating risk in the dynamic environments in which they operate. The application of externally provided and open-source software components further compounds these issues, as the code developed in-house is usually more secure and provides less exploitation opportunities.

The emerging threat of machine learning model misuse in e-learning platforms is concerning. Attackers can perform transfer learning attacks by using publicly available pre-trained models, where the system outputs can be manipulated without the underlying data. Moreover, during rapid software development cycles, especially when security is not part of the design or implementation, Structured Query Language (SQL) and other code vulnerabilities can be introduced. Modern security approaches are applying these tools as deep learning-driven intrusion detection systems, automated vulnerability patching, and adaptive real-time protectors against attack frameworks that change with multidimensional evolving systems (Baraneetharan 2022). Their effectiveness is directly linked to constant modification, thorough auditing and adopting a security-centric culture deeply ingrained in the developers and users.

16.8 TOOLS AND TECHNOLOGIES FOR CYBERSECURITY IN E-LEARNING

The contemporary world requires adequate protective measures to be integrated on e-learning platforms, and the data they store to protect it from cyber threats. For individuals and institutions, the value of keeping the relevant cyber defenses is high, and requires constantly updated measures and protective technologies (Future Talent Learning 2024). Cyber threats come in a variety of forms, such as ransomware, phishing attacks, malware, and more. Solutions for cybersecurity in e-learning are crucial for defending against these dangers. These solutions help to guarantee the confidentiality, integrity, and accessibility of information and infrastructure while likewise making it simpler to identify, mitigate, and lessen the effect of cyberattacks.

E-learning platforms require spatialized tools and technologies to mitigate evolving cyber threats as follows.

- **AI-driven threat detection and response**
  The possibilities for developing and refining e-learning systems have increased with the recent developments in AI and Large Language Model (LLM) technologies. Artificial intelligence (AI) and LLM technologies are the future modern algorithms for developing advanced e-learning systems, for instance, intelligent learning systems. Contextual awareness and creative learning are some encompassing techniques (Greco and Chianese 2024). It acts as a cohesive handbook for relevant stakeholders on integrating an LMS with AI to achieve optimal and tight security, providing guidelines for educational institutions. It underscores the role of AI in education and cybersecurity in the innovative, safe e-learning systems of the future.
- **Capture the flag (CTF) platforms**
  These open-source platforms (CTF) enable learners with the knowledge of reverse engineering and exploitation of vulnerabilities to practice simulated real-word attack scenarios. This enhances the threat awareness and technical skills of both educators and learners. It eliminates resource barriers for resource-inadequate institutions.
- **Honeypot and honey net deployments**
  In the e-learning technologies field, the honeypot surrounding is used to find attackers and redirect them to defend. Honeypots are systems

that lure attackers into a network and assist an organization in detecting and deterring threats. The honeypot deployment in understanding adversaries in e-learning platforms (Hilmi and Mustapha 2022). These cloud-based technologies provide their insights to tactics, techniques, and procedures (TTPs). These technologies form a layered defense strategy, combining proactive detection, hands-on training, and continuous system hardening to protect e-learning ecosystems.

16.9 VULNERABILITIES IN DIGITAL CONTENT DELIVERY

A range of vulnerabilities can hinder security, performance, and even user confidence, all within the same e-learning platform. Weaknesses in digital content delivery can greatly impair the performance and security of e-learning platforms. A mismanaged Computer Network Defense (CND) can perform admirably but could also expose sensitive cached content or fail under a distributed denial-of-service (DDoS) attack. One of the most disturbing vulnerabilities stems from inadequate authentication and poor access control on collaborative tools and third-party components like video conferencing plugins. Sophisticated protection is sorely needed because insensitive content modification and privilege escalation can and will ensue. Educators and learners with no cybersecurity awareness tend to implement strong governance policies by enforcing exceptional encryption, conducting frequent security audits, integrating cybersecurity training, and enforcing zero-trust policies where users and devices are assumed untrusted by default until their identity is verified. E-learning environments tailored on these principles can guarantee effective usability along with security.

16.10 CHALLENGES IN SAFEGUARDING E-LEARNING

16.10.1 Security and accessibility in balance

E-learning systems have to find a compromise between protecting sensitive data and providing quick availability of instructional resources. This necessitates a sophisticated approach to data access restrictions and user authentication.

16.10.2 Guarding against new dangers

Although cyber dangers are always changing, it can be difficult to stay safe. Maintaining strong security requires being up to date with the most recent developments in cybersecurity.

16.10.3 Protection of data in third-party integrations

Various e-learning platforms integrate with other programs and services. To safeguard the ecosystem, it is essential to ensure these integrations adhere to the same data security guidelines.

16.11 STEPS TO GET GOOD E-LEARNING SECURITY

To strengthen the security of the e-learning platform, consider the following important tactics:

Strong authentication systems

Data breaches: Depending on our platform, data theft may disclose student information to unauthorized parties, including names, e-mail addresses, and even social security numbers. Because of the legal penalties and the requirement to notify those impacted, this may have serious financial consequences.

Economic losses: In addition to regulatory sanctions, cyberattacks may cause economic losses. For impacted learners, organizations may have to spend a lot of money on credit monitoring, system repairs, and data recovery.

Reputational harm: Intuitional reputation may be seriously harmed by a data leak or cyberattack. Learners and stakeholders can stop believing that can protect their data, which might result in bad press and a drop in enrollment rates.

Disruption of learning activities: By leaving the e-learning platform unavailable or the course contents inaccessible, cyberattacks can cause disruptions to workforce learning activities. This can hinder staff growth and have a detrimental effect on the general productivity of the institutions.

16.12 BEST PRACTICES FOR PLATFORM DEVELOPERS AND ADMINISTRATORS

For platform developers and administrators in e-learning, implementing robust cybersecurity best practices is essential to safeguard user data and maintain platform integrity. Cyberspace is particularly difficult to secure due to the worldwide reach of cybercriminals, the connections between cyberspace and physical systems, and the challenge of minimizing exposure and repercussions in complex cyber networks (Daoultzoglou 2023). Both learners and institutions should adopt safe cybersecurity best practices. Using "cyber hygiene" principles, such as creating strong passwords, upgrading software, being cautious when clicking on suspicious links, and turning on multi-factor authentication, would greatly improve online safety (Rotich et al. 2016).

Ensuring the data protection through end-to-end encryption – using AES-256 for stored data and TLS1.3 for data transits – along with regular encryption backups helps prevent data breaches and supports recovery in the event of ransomware attacks. System hardening is critical and includes automating patch management for the learning management system and its plugins, conducting regular vulnerability scans, and performing stress tests to ensure that scalability does not compromise security (Keepnet Labs 2025). Secure development practices, such as threat modeling, code audits, and adversarial testing of AI-driven features, help identify and mitigate risks early in the development process. Equally important is user education providing ongoing cybersecurity training for staff and students to raise awareness about phishing, and policies that foster trust. These cybersecurity principles apply to both individuals and institutions. For both public and private enterprises, maintaining and protecting business operations require developing and implementing specialized cybersecurity policies and processes.

16.13 IMPORTANCE OF CYBERSECURITY IN E-LEARNING

The ever-changing digital world we live in today requires e-learning platforms to be an integral part of educational institutions globally. With the ongoing advancements in digital technology and the internet, learning has become simpler on a global scale, which aids the spreading of education without being hindered by location. However, as e-learning expands, so does the need for privacy assurance. The contemporary conception of education has changed. One can learn from any geographical location and at any time without being limited by a need to be present in a formal educational institution. However, modern education comes with its own set of issues, which include, but are not limited to, cybersecurity threats. E-learning platforms have become imperative in the education sector post-COVID-19. User can easily tailor them to their schedules, which makes obtaining education a simpler process. Unfortunately, this increase in the use of e-learning platforms has allowed the education sector to fall into the hands of cybercriminals (Harve 2024). Adopting modern technology poses security risks and makes it difficult to attain the appropriate balance between affordability, accessibility, and reliability. Nowadays, educational institutions can bridge this gap by seeking aid from cybersecurity professionals.

16.14 CONTENT RIGHTS MANAGEMENT AND ANTI-PIRACY MEASURES

Content Rights Management (CRM) and anti-privacy technologies are vital for safeguarding content on e-learning platforms. Access control list (ACL), commonly done through a digital rights management (DRM) system, permits the claimed access to digital content. Commentaries notice that these systems permit viewing, duplication, sharing, printing, and even copy containment, which helps to obey copyright legislation. Intellectual properties are also claimed for the DRM system users who violate copyright regulations. Copying of course materials without permission is bound through selective restriction technologies like encryption, watermarking, and duplication-free access control (Downes et al. 2003). Technologies complementing CRM, restraining sharing, receiving, and any download methods aim at containing copyright piracy. Secure credential logins, time-bound access, unit/IP limit access bypassing multiple worming access, and unique terminal alterations are a few ways of holding several unauthorized users from accessing digital content. Watermarking is identifying documents and video by putting invisibly tampering mark that serves dual purpose of slowing down leaking or abuse intended outside legal performance as well as tracing to reveal original supplier. Unpredictable behavior is reserving and reporting unusually normal activity suspected of piracy under automation is expected from detecting automated tracking devices.

CRM and anti-piracy strategies help to maintain the value and integrity of digital content, ensuring that creators and institutions retain control over their resources while providing learners with secure and legitimate access. These measures are increasingly important as e-learning becomes more prevalent and the risk of unauthorized content distribution grows.

16.15 REGULATORY COMPLIANCE AND LEGAL FRAMEWORKS

E-learning platforms must follow industry policies and legislation to protect personally identifiable information (uploads of personal data) in accordance with the European Union's GDPR and other legal frameworks (Jekabsone 2023). Such compliance entails that educational institutions and platforms apply fostering principles of data minimization, secure data processing, and user access (including users' rights to rectification and erasure) at the system design stage. An institution must act as a responsible data controller and ensure that the data flows, both within the institution and with third-party e-learning vendors, observed the legal boundaries of information and privacy data systems security. The transition to online learning environments as a result of the COVID-19 pandemic has intensified the significance of managing recorded sessions, as well as the cross-border transfer of proprietary data in accordance with international and domestic legal statutes. Compliance with these frameworks not only fulfils legal requirements but also enhances reputation and confidence among learners and staff, lowers the risk of inadequate information safeguards, and sustains the digital educational ecosystem.

16.16 FUTURE TRENDS IN E-LEARNING CYBERSECURITY

This sprinkled with robust development, cybersecurity in e-learning has to deal with ever evolving cyberspace threats that require creative counter measures. The training components undergo a change due to the rapid development of AI and machine learning technologies. Higher learning intuitions need to integrate several critical components of security awareness instruction in e-learning, such as context-based vishing and smishing e-learning modules, interactive and simulated training with assessment, and custom phishing drills for educational enterprises (Hilmi and Mustapha 2022). In addition, the secure Single Sign-On and Multi-Factor Authentication, alongside the strong passwords, reinforce the biometric access policies (Microsoft 2023). Failure to meet such standards increases the risk of falling victim to highly monetary loss and compliance violations. With cybersecurity in e-learning becoming a more active area of concern, the lack of awareness of problems is a lot simpler to solve now. More advanced methods of garnishing interest, such as simulations of real-life situations, are utilized.

16.17 CONCLUSION

E-learning is increasingly integrated into the education system, while the importance of robust legal supervision alongside cybersecurity is growing, as digital content becomes ever easier to protect. Safeguarding user data, protecting the system's integrity, and digital content require a robust combination of technological measures with holistic best development practices, and compliance with GDPR not merely on the grounds of confidentiality but to earn the trust of students, educators, and even institutions. Educational organizations should strive to strengthen these attributes so that the sophisticated, safe, and equitable e-learning systems can be accessible and effective for everyone.

REFERENCES

Andriotis, Nikos, and Nikos Andriotis. 2018. "eLearning in the GDPR Era: What You Need to Know." *TalentLMS Blog*. May 7. https://www.talentlms.com/blog/elearning-gdpr-era/

Arshavskiy, Michael. November 12, 2021. "Social Media for eLearning." *eLearning Industry*. https://elearningindustry.com/social-media-for-elearning#:~:text=4.,interaction%20between%20learners%20and%20instructors

Baraneetharan, E. 2022. "Review on Deep Learning Based Network Security Tools in Detecting Real-Time Vulnerabilities." *Journal of Ubiquitous Computing and Communication Technologies* 7 (1): 23–39. https://doi.org/10.36548/jucct.2022.1.005

Barbu, Irina, Georgiana Petrica, Simona Axinte, and I. C. Bacivarov. 2017. "Analyzing Cyber Threat Actors of E-Learning Platforms by the Use of a Honeynet Cloud

Based Infrastructure." *eLearning and Software for Education.* https://www.researchgate.net/publication/368804291_ANALYZING_CYBER_THREAT_ACTORS_OF_E-LEARNING_PLATFORMS_BY_THE_USE_OF_A_HONEYNET_CLOUD_BASED_INFRASTRUCTURE

Cisco. May 2, 2025. Cisco Security: A Better Way of Doing Security. Video. https://www.cisco.com/site/in/en/learn/topics/security/what-is-cybersecurity.html#:~:text=Cybersecurity%20is%20the%20practice%20of,or%20interrupting%20normal%20business%20processes.

College Vidya. December 9, 2023a. "Cyber Security Types and Threats Defined – Detailed Guide." College Vidya Blog. https://collegevidya.com/blog/cyber-security-types-and-threats/.

College Vidya. December 9, 2023b. "Cyber Security Types and Threats Defined – Detailed Guide." College Vidya Blog. https://collegevidya.com/blog/cyber-security-types-and-threats/.

College Vidya. December 9, 2023c. "Cyber Security Types and Threats Defined – Detailed Guide." College Vidya Blog. https://collegevidya.com/blog/cyber-security-types-and-threats/.

Cybersecurity and Infrastructure Security Agency (CISA). "Cybersecurity Best Practices." https://www.cisa.gov/topics/cybersecurity-best-practices.

Deetya Education. September 8, 2023. "Data Privacy and Security in Digital Education: Safeguarding Student Data in a Digital World." Deetya Education. https://deetya.education/2023/09/08/data-privacy-and-security-in-digital-education-safeguarding-student-data-in-a-digital-world/.

Djeki, Elvis, Jean Rodrigue Dégila, and Mohamed H. Alhassan. 2024. "Best Practices for Ensuring Security and Privacy in E-Learning Environments." In *2024 ASU International Conference in Emerging Technologies for Sustainability and Intelligent Systems (ICETSIS)*, 1108–1112.

Downes, Stephen, Mourad Mourad, Hugh Piccariello, and Rob Robson. 2003. *Digital Rights Management in E-learning: Problem Statement and Terms of Reference. Proceedings of E-Learn 2003: World Conference on E-Learning in Corporate, Government, Healthcare, and Higher Education.* Association for the Advancement of Computing in Education (AACE). https://www.researchgate.net/publication/228831787_Digital_Rights_Management_in_E-learning_Problem_Statement_and_Terms_of_Reference

Future Talent Learning. "Cyber Security: Safeguarding E-Learning Platforms for a Secure Digital Future." https://www.futuretalentlearning.com/en/future-talent-learning-blog/cyber-security-safeguarding-e-learning.

Greco, Davide, and Luigi Chianese. 2024. "Exploiting LLMs for E-Learning: A Cybersecurity Perspective on AI-Generated Tools in Education." In 2024 IEEE International Workshop on Technologies for Defense and Security (TechDefense), 237–242. IEEE https://doi.org/10.1109/TechDefense63521.2024.10863662

Harve, Aditi. July 28, 2024. "Bridging Knowledge and Security: A Guide to Secure eLearning Implementation." *Digital Engineering & Technology | Elearning Solutions | Digital Content Solutions.* https://www.hurix.com/blogs/bridging-knowledge-and-security-a-guide-to-secure-elearning-implementation/

Hashemi-Pour, Chelsea, and Ben Lutkevich. June 26, 2024. "What Is E-Learning? Importance, Benefits and Use Cases." *What Is?.* https://www.techtarget.com/whatis/definition/Web-based-training-e-learning

Hilmi, Mohd Fadzli, and Yusreen Mustapha. June 3, 2022. "Perceived Security of E-Learning Portal." *arXiv.org.* https://arxiv.org/abs/2209.11196

Jekabsone, Inga. 2023. "Selected Legal Issues in Online Adult Education: Compliance of Online Learning and Teaching Process with GDPR." *TalTech Journal of European Studies* 13: 46–62.

Keepnet Labs. February 28, 2025. "What Are the Top Trends in Cybersecurity Awareness Training for 2025?" *Keepnet Labs.* https://keepnetlabs.com/blog/what-are-the-top-trends-in-cybersecurity-awareness-training-for-2025

Kimutai, Shadrack. 2016. "IT Best Practices for Implementing E-Learning: The Case of Rift Valley Technical Training Institute." *Mu-Ke*, January. https://www.academia.edu/68670771/IT_Best_Practices_for_Implementing_E_learning_The_Case_of_Rift_Valley_Technical_Training_Institute%20

Lizarraga, Jesús, Roberto Uribeetxeberria, and Urko Zurutuza. 2006. "Security in Embedded Systems." *Proceedings of the 2006 International Conference on Embedded Systems.* https://www.researchgate.net/publication/249810205_SECURITY_IN_EMBEDDED_SYSTEMS

Nedelchev, Martin, and Dimitar Slavov. 2023. "Cybersecurity Recommendations and Best Practices for Digital Education." *Scientific Research and Education in the Air Force.* https://www.researchgate.net/publication/372873046_CYBERSECURITY_RECOMMENDATIONS_AND_BEST_PRACTICES_FOR_DIGITAL_EDUCATION

Scerbakov, Scerbakov, Alexei, Frank Kappe, Nikolai Scerbakov, and Graz University of Technology. 2018. "SECURITY VULNERABILITIES IN MODERN LMS." Journal-article. *Graz University of Technology.* https://tugraz.elsevierpure.com/ws/portalfiles/portal/23972189/EL2019_S_144_Scerbakov_edit_v2.pdf

Supiani, S., Dedy Achmad Kurniady, Tjutju Yuniarsih, and Nur Aedi. 2024. "Evaluating Learning Management System (LMS) Effectiveness: An LPOMR Model Approach." *Pedagogia: Jurnal Ilmiah Pendidikan* 16 (2): 71–77. https://pedagogia-fkip.unpak.ac.id/index.php/pedagogia/article/view/2

Index

www.ingramcontent.com/pod-product-compliance
Ingram Content Group UK Ltd.
Pitfield, Milton Keynes, MK11 3LW, UK
UKHW022218190926
475702UK00003B/425
9781032958804